17eme CONFERENCE INTERNATIONALE SUR LE RAYONNEMENT COSMIQUE
17th INTERNATIONAL COSMIC RAY CONFERENCE · PARIS · FRANCE JULY 13·25 1981

# CONFERENCE PAPERS

# INVITED PAPERS
# VOL. 12

17eme CONFERENCE INTERNATIONALE SUR LE RAYONNEMENT COSMIQUE
17th INTERNATIONAL COSMIC RAY CONFERENCE · PARIS · FRANCE JULY 13·25 1981

# COMMISSARIAT A L'ENERGIE ATOMIQUE
# INTERNATIONAL UNION OF PURE AND APPLIED PHYSICS

scientific secretariat :
section d'astrophysique
centre d'études nucléaires de saclay

postal address :
CEN Saclay DPh/EP/Ap
91191 Gif-sur-Yvette Cedex. France

ISBN 978-2-7272-0070-3        ISBN 978-3-662-25523-0 (eBook)
DOI 10.1007/978-3-662-25523-0

# FOREWORD

The Proceedings of the 17th International Cosmic Ray Conference held in Paris, July 15 to 25, 1981, appear in two sets. The Regular Volumes, 1 to 8, contain contributed papers received at the Secretariat by April 1st, 1981. They were issued at the opening of the Conference. The Late Volumes, 9 to 14, contain contributed papers received after that date, Invited and Rapporteur Talks, and the General Index.

The assiduous reader will notice several changes with respect to the well-established traditions of the Conference.

1/ Following a recommendation of the Commission on Cosmic Rays of IUPAP, and although an increase in the total number of papers submitted was noticed as compared to the 16th ICRC (Kyoto, 1979), the total number of pages has been significantly reduced, thanks to introduction of three new rules for publication. (i) None of the first "Preliminary" Abstracts was published. These abstracts had to be confirmed, either by a new "Confirming Abstract" or by a Full Paper. The Confirming Abstracts are included in the Proceedings. (ii) The sum of the "fractional" contributions of each author should not exceed 3 papers, and each author should not appear in more than 10 papers. (iii) The maximum number of pages per paper was reduced from 6 to 4.

The Organizing Committee thanks all authors who have, in their vast majority, very efficiently cooperated by kindly complying with these new rules.

The papers we selected on the basis of the Preliminary Abstracts. As usual, however, there was no refereeing, and the Full Papers and the Confirming Abstracts were reproduced as they were received from the authors. The Organizing Committee expresses its thanks to O.C. Allkofer, A. Raviart, T. Thambyahpillai and A.A. Watson for their collaboration in sorting the Preliminary Abstracts.

2/ After reviewing the Preliminary Abstracts, and with the approval of the Commission of Cosmic Rays of IUPAP and the help of outside experts, the Organizing Committee felt it appropriate to rearrange the sessions and subsessions to take into account a rapid development in various fields. The former MG and SP sessions were merged into a new SH ("Sun and Heliosphere") session, and the former OG-1 to OG-5 sessions were taken out of the OG session to form a new XG ("X and Gamma-Rays, non-solar") session. Within the other sessions, the changes were introduced to follow more closely the details of their content.

The headings of the sessions were defined and arranged as follows in the Regular Volumes :

| XG | X-and Gamma Rays (non-solar) | Vol. | 1 |
| OG | Origin and Galactic Phenomena | Vol. | 2 |
| SH | Sun and Heliosphere | Vol. | 3,4 |
| HE | High-Energy Interactions | Vol. | 5 |
| EA | Extensive Air Showers | Vol. | 6 |
| MN | Muons and Neutrinos | Vol. | 7 |
| T | Techniques | Vol. | 8 |

The Late Volumes are arranged as follows :

| XG, OG, T | Vol. | 9 |
| SH, MN | Vol. | 10 |
| HE, EA, Errata | Vol. | 11 |
| Invited Papers | Vol. | 12 |
| Rapporteur Papers | Vol. | 13 |
| General Index | Vol. | 14 |

## FINANCIAL SUPPORT AND SPONSORSHIPS

The organization of the Conference would have been impossible without the major financial support of the *Commissariat à l'Energie Atomique* (CEA, the French Atomic Energy Commission), to which the *Section d'Astrophysique*, local organizer, belongs. Support was also provided by the *Centre d'Etudes Spatiales des Rayonnements* (CESR, Toulouse).

Invited speakers and rapporteurs were supported in part by the *Centre National d'Etudes Spatiales* (CNES, the French Space Agency) and the *Centre National de la Recherche Scientifique* (CNRS, the French Research Council). Support came also from *Air France*, Official carrier of the Conference, and from the Companies which were advertised in the Conference program.

The financial help of the International Union for Pure and Applied Physics, (IUPAP), which patronizes the Conference, and the sponsorship of the European Physical Society are also acknowledged.

December 1st, 1981        Charles Ryter
Chairman, National Organizing Committee

TABLE OF CONTENTS

VOLUME 12

INVITED PAPERS

```
*********************
*                   *
* OPENING CEREMONY  *
*                   *
*********************
```

```
*********************
*                   *
* HEAO-3 SYMPOSIUM  *
* (INVITED PAPERS)  *
*                   *
*********************
```

```
******************
*                *
*  INVITED PAPERS  *
*                *
******************
```

```
**********************
*                    *
*  OPENING CEREMONY  *
*                    *
**********************
```

WELCOME ADDRESS TO THE

17th INTERNATIONAL COSMIC RAY CONFERENCE


by Louis Leprince-Ringuet


Both Pierre Auger and I are reckoned as being among the last representatives, the last witnesses, the last vestiges of the remarkable days of cosmic-ray research in the thirties. One half-century ago, thanks to the work of Hess, a very weak radiation was already known, whose intensity increased with altitude, but only very little information had been obtained on its nature and its effects. At that time, the detection technique of fast moving charged particles was just beginning to be useful for the study of this radiation. Coincidence arrangements of Geiger counters were developed first by Bothe and Kolhörster, and then by Rossi; the use of Wilson chambers, which were previously triggered at random, was widely extended thanks to the triggering counters conceived by Blackett and Occhialini. A few years later, the practice of "thick" photographic emulsions was completed by Powell, Lattes and Occhialini.

But fifty years ago, only global cosmic-ray measurements with ionization chambers were achieved, and the first indication on the latitude effect was obtained by Clay with such a device. At that time, Pierre Auger and I decided to join our efforts in developing primitive wire counters in coincidence. We crossed the ocean between Hamburg and Buenos Aires with these counters, in order to detect cosmic particles and to estimate their penetration power and the reduction of their flux in the equatorial zone.

Before the second world war, by means of sea-level and mountain based experiments using counters or Wilson chambers, it was possible to identify the main characteristics of the observed radiation: penetration power, time variation, secondary showers, energy. I will always remember waking up old Professor Aimé Cotton, on a Sunday evening, to show him the first picture obtained with our 75 cm high Wilson chamber mounted inside his 13000 Gauss magnet at Bellevue, a picture featuring a perfectly rectilinear cosmic-ray track, which gave the evidence of a particle with an energy in excess of 10 billion electron-Volts, the world record at that time.

This pioneering era expanded just after the war with balloon-borne emulsions, Wilson chambers at mountain altitudes and large arrays of counters. The nature of the primaries was recognized, in particular by Bradt and Peters, as well as the secondary effects in the atmosphere. The discovery of new particles, the mesons and the first hyperons, began at that same time. The Bagnères-de-Bigorre Conference held in 1953, was the swan song of this period. There the word "hyperon" was pronounced for the first time, and this event is commemorated by the "Place de l'Hypéron" in the city of Bagnères. But it became obvious that particle accelerators would supersede the cosmic rays and Powell exclaimed during the closing session: "we are in danger, we are pursued". Cosmic rays had been used mainly as very high energy projectiles to study interactions with nuclei in the atmosphere or in the detectors. But particle accelerators started to produce a much more abundant source of high-energy projectiles. All my group at Ecole Polytechnique moved towards CERN. Nevertheless, some astrophysical problems had already been evoked and they remain the prerogative of the cosmic-ray community: the composition of the radiation, the maximum energy, the production processes and the mean age; and at the time of the Bagnères Conference, some ideas regarding these subjects were already proposed.

Since then, how many breakthroughs in astrophysics with such powerful devices! Cosmic rays require more and more sophisticated experimental apparatus, either on the ground or aboard space vehicles. They have an important message to deliver. This is the purpose of this Conference. I give my kind regards to all my young colleagues and I wish that this Conference will bring forward many new discoveries to increase our knowledge of astrophysics.

(Translated by the Organizing Committee)

# HIGH ENERGY COSMIC RAYS

by Pierre Auger

Some eighty years ago, C.T.R.Wilson, observing the so-called "residual ionization", concluded to the existence of an unknown radiation; and ten years later Victor Hess demonstrated the strong increase of the ionization with altitude, and so was the cosmic radiation discovered under the name of "Höhenstrahlung" as a radiation issuing from outside the atmosphere. The only characteristics of this radiation was its slow absorption by the atmosphere and its ionizing power. Its exact nature as a corpuscular radiation coming from the space outside the terrestrial magnetosphere was demonstrated by Clay, when he observed the strong decrease of the ionization it produced when the apparatus was transported from high to low latitudes. Finally, the corpuscular nature was definitely proved when Geiger counters in vertical coincidence were used for the measurement, showing the decrease of the number of coincidences in equatorial regions, far from the magnetic poles, by Louis Leprince-Ringuet and myself in 1934.

A few years earlier, in 1929, D.Skobelzyne, working with a Wilson cloud chamber in a magnetic field, had observed corpuscular tracks which were not noticeably curved by the field, and I confirmed his attribution of these tracks to cosmic ray particles. The very high energy of these cosmic rays and their extraterrestrial origin was so demonstrated, a justification of their name as cosmic ray particles. During the following year, the dual nature of these rays, with a soft and a hard group characterized by their penetrating powers, was ascertained by Bruno Rossi and myself, working with coincidence counters and cloud chambers at various altitudes and underground. The new phenomenon of the shower production in solid screens by the cosmic particles was made visible in the experiments of Blackett and Occhialini, with a counter-controlled cloud chamber. The energy of the incoming cosmic ray particles was then estimated at a large number of MeV (million electron volts), somewhat above the range of energies of the alpha and beta radioactivity particles, and of the accelerated ion beams produced in the cyclotrons of Lawrence. And the action of the earth's magnetic field pointed to the existence of a fraction of the primary particles with energies of the order of the GeV, billion electron volts.

Wanting to study the showers produced by these most energetic particles by increasing the horizontal distance of the coincidence counters under a shower producing screen, I encountered the difficulty arising from the so called "spurious" or chance coincidences, and asked my collaborator Roland Maze to reduce that background by increasing the selection power of the coincidence system from milliseconds to microseconds. We could then observe coincidences between counters separated by distances of several meters: the simultaneously ionizing particles responsible for these coincidences could not originate from a local source, and I suspected an atmospheric phenomenon. To test this hypothesis, I displaced one of the counters into another building, at a distance of more than 150 meters, and could attribute the coincidences, several per hour, to the existence of large, extensive showers taking their origin high in the atmosphere, and due to the action of cosmic ray particles of very high energy. Together with Roland Maze, we could increase the extension to 300 meters, working in the Jungfraujoch Laboratory, Switzerland, at an altitude of more than 3000 meters. The same results were obtained with even greater distances in the Laboratory of the Pic du Midi, in the Pyrénées.

The measurement of the density of shower particles per square meter, and the evaluation of their energy allowed me to announce in 1938 the existence in the primary cosmic radiation of a component with energies of at least $10^{15}$ eV, one million of giga electron volts, one million times more than what was so far attributed to "high energy" cosmic ray particles. I could study the penetrating power of the extensive shower particles, and found that these showers included a small but measurable proportion of so called "hard" particles, presently called muons, thus proving the secondary origin of this cosmic ray component.

As you will hear during this Conference, the extensive atmospheric showers, E.A.S., have been studied since their discovery by numerous scientific teams, using much more powerful detectors than my Geiger counters and increasing the surface covered by the shower from hectares to square kilometers, elevating the energy of the primary cosmic particles up to $10^{18}$ eV, and in some recent measurements to $10^{20}$ eV. The problem of the origin of such particles is still unsolved, after more than forty years: perhaps the best theory is still the one proposed by Fermi, involving the action of very extensive and variable magnetic fields. I had pointed out, at the time of my first evaluation, that such high energies could not be transferred to a particle by one single operation, but only by the action of extensive fields, for instance electric. Many of the great astrophysical phenomena have been suggested for the source of these particles, such as supernova shock waves, explosive radiogalaxies or high speed jets from quasars, and I can only hope that some conclusive results will be announced during this Conference.

```
*********************
*                   *
* HEAO-3 SYMPOSIUM  *
* (INVITED PAPERS)  *
*                   *
*********************
```

THE HEAO-3 HIGH RESOLUTION GAMMA-RAY SPECTROMETER

Allan S. Jacobson

Jet Propulsion Laboratory
California Institute of Technology
Pasadena, CA 91109, U. S. A.

## ABSTRACT

The gamma ray astronomy experiment launched aboard HEAO 3 performed an all-sky exploratory survey for sources of gamma-ray line emission in the energy range of .05 to 10 MeV. The detectors were four high purity germanium crystals, each of about 100 cm$^3$ active volume and energy resolution line width of about 3 keV FWHM at 1.33 MeV. Observations are reported on 22 solar flares, one of which emitted a narrow deuterium formation line at 2.225 ± .002 keV; galactic plane line upper limits; a variable source of .511 MeV radiation in the vicinity of the galactic center and long term variations in the source Cygnus X-1. The experiment performed successfully and when analysis of the entire data set is completed, the experiment will achieve the narrow line sensitivity of $10^{-4}$ photons/cm$^2$-sec over most of the sky.

## 1. Introduction

The gamma-ray spectroscopy experiment flown aboard the HEAO-3 spacecraft was designed to perform the first all-sky survey with high spectral resolution for gamma-ray emissions in the energy range of .05 to 10 MeV. While all sources radiating in this energy range are of interest, the primary emphasis of the experiment is the study of sources of gamma-ray lines. Such lines can originate in the processes of nuclear de-excitation, radiative capture, positron annihilation and quantized cyclotron emission. The scientific objectives are to detect the products of these processes as they are related to the structure and evolution of the galaxies; nucleosynthesis; novae and supernovae; accretion of matter onto compact objects such as neutron stars and black holes; cosmic ray interactions in the interstellar medium and extragalactic sources.

As distinct from previous measurements made by scintillators in this spectral region, the high resolution germanium detectors used in this experiment allow more precise measurement of the energy of spectral features, the detection of possible redshifts and line broadening and the resolution of complex line structure.

## 2. The Instrument

Figure 1 is an exploded view of the instrument. The primary detectors were four high purity germanium crystals, each of which was approximately 55 mm in diameter by 45 mm long. They were surrounded by a cesium iodide anticoincidence shield composed of four quadrants of well-crystal and a collimator crystal which defined the instrument aperture by positioning a drilled hole above each of the germanium crystals. The FWHM aperture response was energy depen-

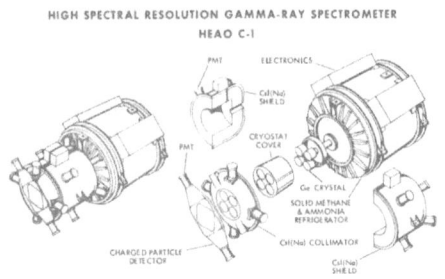

HIGH SPECTRAL RESOLUTION GAMMA-RAY SPECTROMETER
HEAO C-1

Figure 1

dent with a geometric value of about 30°. A thin plastic scintillator was placed over the aperture to complete the $4\pi$ shielding against charged particles. The minimum shield thickness was 6.62 cm, and the shield isotropic geometry factor was 907 cm$^2$. It was segmented in such a way that relative count rates in the various pieces would provide some directional information for strong sources outside the aperture.

The total effective area of the germanium crystals was about 70 cm$^2$ at 100 keV and 26.4 cm$^2$ at .511 MeV. The spectral resolution at launch was about .23% at 1.33 MeV. Under the bombardment of cosmic rays and trapped protons there was some gradual degradation of the resolution so that six months after launch, the resolution at 1.332 MeV was about 1%. The effect was energy dependent and the degradation below several hundreds of keV was negligible. Radiation damage during the mission is thoroughly discussed in Mahoney et al., 1981.

Cooling to a temperature of about 100°K was provided for the germanium detectors by solid methane and ammonia in a sublimation refrigerator. The instrument is described fully in Mahoney et al., 1980.

## 3. The Mission

The HEAO-3 spacecraft was launched on 20 September 1979 into a circular orbit of 500 km altitude and an inclination of 43.6 degrees. With the instrument viewing axis perpendicular to the spin axis, which nominally pointed at the sun, a great circle scan was made every twenty minutes. In this manner, a full sky survey was completed in six months. There also existed the capability to move the spin axis within limited angular and temporal values and twice during the mission the spin axis was aligned to coincide with the galactic poles, resulting in a total of four weeks of galactic plane scans, half in the Fall of 1979 and half in the Spring of 1980. Depletion of the cryogens limited the

lifetime of the germanium crystals to 8.5 months of operation.
The shield crystals functioned as isotropic detectors for the
remainder of the mission, monitoring solar flares and gamma-ray bursts.
Mission termination was on May 30, 1981 due to the depletion of the
attitude thruster fuels.

## 4. Performance

Cosmic ray interactions in the instrument, spacecraft and atmosphere result in the generation of a very complex background spectrum
with numerous lines.  A spectrum measured in orbit is shown in Figure 2.

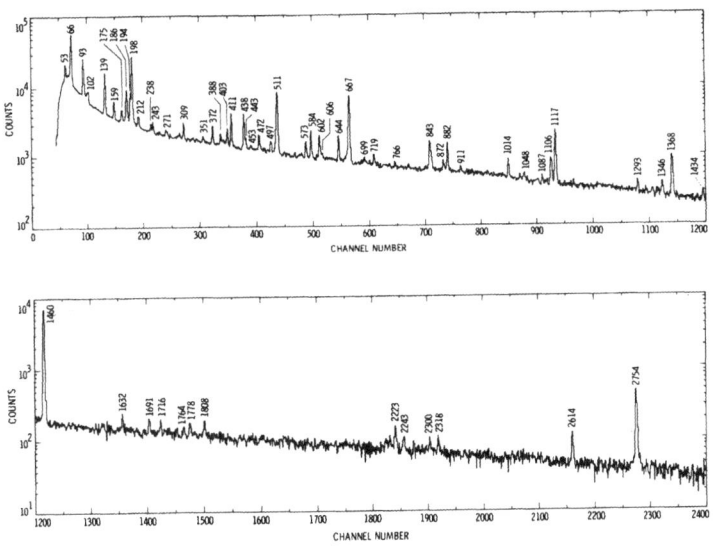

Figure 2

This represents a thirteen day accumulation of background data in which
more than 120 lines are detected.  Of these about 60 percent originate
from interactions in the germanium and cesium iodide, about 15% from
interactions in other materials in the instrument and spacecraft, and
some 10% originate in the decays of natural radioactive contamination in
these materials.  Virtually every line of astrophysical importance has
a component in the background so that their detection requires careful
study of the systematic background effects.

An estimate of the experiment sensitivity to line fluxes is shown
in Figure 3.  It is based upon the background spectra accumulated in
orbit, and performance parameters measured before launch.  Assumed are
one 6 month celestial scan with $10^5$ secs observing time per point source.
In the presence of strong background lines the sensitivity is somewhat
less than shown.  For instance, at .511 MeV, it is approximately $3 \times 10^{-4}$

photons/cm²-sec. This sensitivity is roughly an order of magnitude improvement over the typical balloon measurements made in the past.

## 5. Results

The data reduction and analysis of steady or quasi-steady point sources is of special interest but pose significant problems of interpretation because observations are typically made over a 10 percent signal-to background ratio. Thus to study such sources, long accumulations of data are required making a thorough understanding of the numerous background modulations and time varying features critical. Because

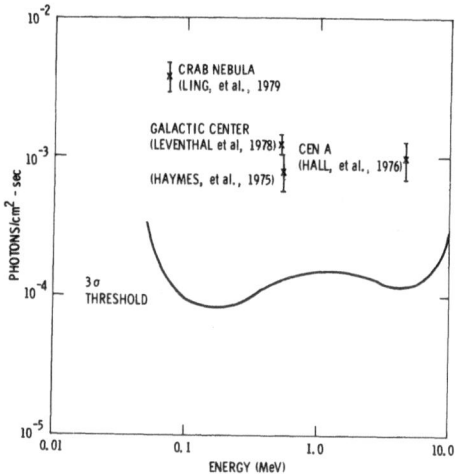

Figure 3

of this, the earliest results obtained dealt with transient phenomena of relatively short duration such as solar flares and gamma-ray bursts.

The experiment was not designed to observe the sun and the instrument never looks directly at it, nominally pointing 90° from the solar direction. Solar flare measurements are therefore made primarily in the scintillation shield crystals with directionality furnished by relative count rates in the various shield crystal components. Integral count rates are measured in each shield piece for events greater than .08 MeV, and read out every 1.28 seconds; and between .41 and .58 MeV, and above 3.8 MeV, sampled every 10.24 seconds. In addition, a 128-channel pulse height spectrum is read out for the collimator every 40.96 seconds with a 50% duty cycle. Because it has the most regular geometry, the collimator crystal rates are used for the absolute measurements quoted.

At least 37 solar flare-like events were observed in the period from 1 October 1979 through 1 July 1980, of which 22 had emissions at or above 0.5 MeV and 13 had flux at energies greater than 3.8 MeV. The criterion for selecting each event was that it was simultaneously observed by at least one other instrument monitoring the sun (Riegler et al., 1981). It was found that most flares of x-ray class M5 or stronger were detectable by the experiment. The strong gamma-rays tended to occur with flares of class X1 or stronger. There is very little correlation between the gamma-ray intensity and radio intensities in the 2.7 to 8.8 GHz band. In general, the flares observed had a variety of spectral shapes and time structures, and the gamma-ray emissions lasted anywhere from 10 seconds to several minutes.

All of the flares, observed to have flux at energies greater .4 MeV were fit with power law spectra. For the 13 flares with flux of energy greater the 3.8 MeV, the data are best fit by power law distributions

indicating probable origin in non-
thermal processes. Single or double
power laws were fit to the data and
Figure 4 shows the distribution of
spectral indices. The shaded events
represent those with greater than 3.8
MeV flux. In some two-thirds of the
cases, it is found that there is a
spectral hardening above .5 MeV. The
high energy indices cluster between -2
and -2.5, while lower energy indices
have a much wider distribution.
Gamma-rays and x-rays seem to be produced simultaneously.

FREQUENCY OF OCCURRENCE
OF SPECTRAL INDICES

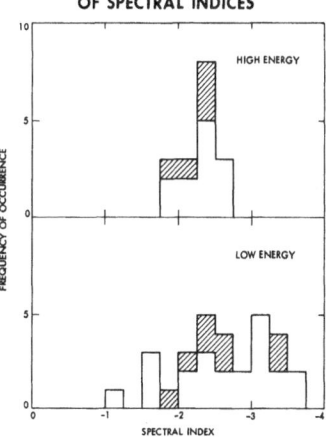

Figure 4

Two of the flares, those on
4 June 1980 and 20 June 1980 are
different in that they show
strong evidence for time separation
of up to 30 seconds between the peak
hard x-ray emission and the peak
gamma-ray emission. They also show
significant gradual spectral hardening.
The 4 June 1980 flare is shown in
Figure 5. Quasi-periodic impulsive
bursts are seen on top of a slowly
rising and relatively quick decaying
high energy component. The maximum
of the high energy flux occurs some
30 seconds after the peak of the lower
energy activity. There is very little
change in the high energy spectral index,
over the course of the flare, while the
low energy spectrum softens significantly.
This type behavior has been observed in
about 10 percent of the flares and seems
to indicate the presence of distinct two
stage acceleration separated in time.

COMPOSITE TIME PROFILE FOR
FLARE OF 4 JUNE 80

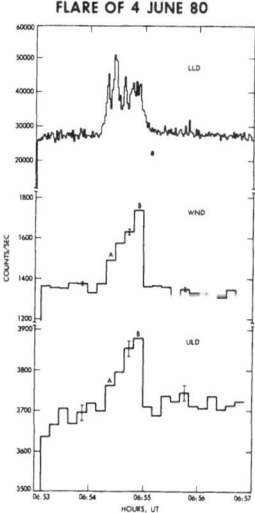

Figure 5

If the flare is sufficiently
intense, the flux can penetrate the
shield and be detected in the germanium
crystals. This occurred in the flare of
0305 UT, 9 November 1979, furnishing the
first opportunity to observe a gamma ray
line from a solar flare with high spectral
resolution. Lines in solar flares have
been observed before with scintillation
detectors (Chupp et al., 1973; Hudson et al.,
1979). The gamma-ray experiment on the
Solar Maximum Mission has added more instances
and in general, the SMM and the HEAO-3 missions, have increased the number of observed solar flare gamma-ray events dramatically.

On 9 November 1979, a very narrow line at energy 2.225 ± .002 MeV, clearly the line resulting from neutron capture on hydrogen forming deuterium, was observed. Theoretical studies have indicated that this line will be the most intense gamma-ray line from flares and indeed, it has been observed in several flares to be so. As predicted, the observed line width is narrow, giving a measure of the nucleon temperature in the capture region. The event lasted for 40 seconds and was concurrent with a 1B optical and an x-ray class M9 flare from active region 2099 at latitude S12 and longitude W02. This same region produced a flare the previous day (Riegler, et al. 1981).

The time history of the 2.225 MeV line and the shield count rate greater than .08 MeV are shown in Figure 6. The gamma-ray line builds up rapidly some 20 seconds from the flare onset. Such a delay and subsequent line build up is consistent with the time of flight between the site of production and the neutron capture region in the photosphere. The line flux should then decay in a quasi-exponential fashion characteristic of neutron capture and decay lifetimes. The full time history was not observed because of earth occultation of the spacecraft.

There is no evidence of two stage acceleration separated in time. In spite of the short duration of this event compared to line emitting flares observed previously, the integrated flux is comparable. This particular flare represented fast energy release although there were no large increases in interplanetary particle fluxes following it.

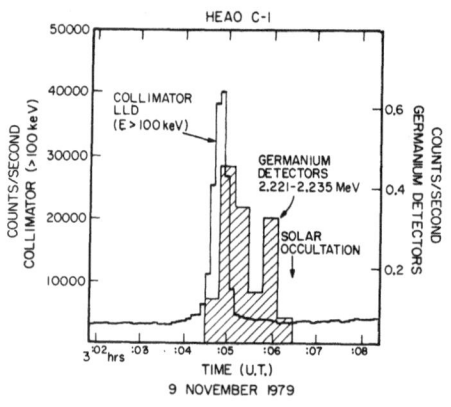

Figure 6

Figure 7 shows the line spectrum. Doppler broadening is expected to be small since capture takes place in a relatively cool photosphere. The observed line width is less than 2 keV, implying a temperature less than $\sim 10^{6}$ K, and the average flux over 80 seconds was 0.25 ± .06 photons/cm$^2$-sec. This compares to the 2.2 MeV peak line flux of 0.28 photons/ cm$^2$-sec in the flare of 4 August 1972.

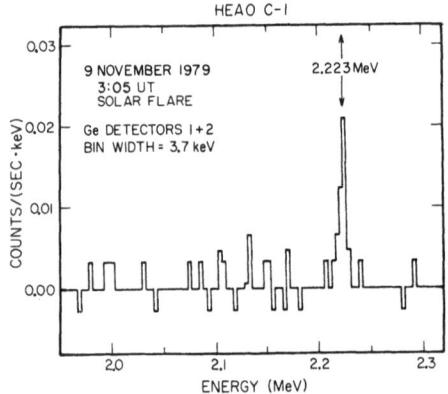

Figure 7

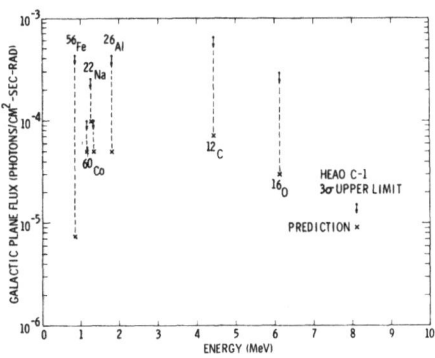

**DIFFUSE GALACTIC PLANE GAMMA-RAY LINE EMISSION-PREDICTIONS AND HEAO C-1 LIMITS**

Figure 8

Two gamma ray bursts observed during the mission have been studied. The results are presented elsewhere in this publication (Wheaton et al., 1981).

The 2 week galactic scans performed twice in the mission afforded an opportunity to study possible line distributions related to galactic structure. Ramaty and Lingenfelter (1981) have pointed out numerous lines, many of which are narrow, expected from the galactic plane. Three of the strongest originate in the decays of $^{60}$Fe and $^{26}$Al synthesized in novae and supernovae. This emission should be concentrated in the galactic plane with intensity enhancement toward the galactic center. A preliminary search for these lines has been completed and upper limits for plane averages are shown in Figure 8. The data used to obtain these results were limited to one of the two week galactic scan periods. The total data which ultimately will be available is about four times more than was used in this sample, allowing a factor of two improvement in the upper limits, and possible detection of some of the lines. In the data set used, there was no evidence of any net cosmic line emission with the possible exception of the $^{26}$Al line at 1808 keV. The 1808 keV result is not statistically significant at this time and will be studied further when more data are added to the sample. All limits shown in the figure apply to average galactic plane emission and not to point sources or to sources of limited spatial extent. In no case do the upper limits fall below the present predictions although those for $^{60}$Co, $^{22}$Na and $^{26}$Al are quite close. In general these limits are considerably better than those derivable from previous experiments.

The galactic center region is of great interest and .511 MeV gamma-ray line emission from there has been the subject of intense study. It has been found to be time varying, having decreased in intensity significantly between the Fall 1979 and Spring 1980 scans. Observations of a .511 MeV line from the vicinity of the galactic center have spanned about a decade (Johnson and Haymes, 1973; Haymes et al., 1975; Albernhe, et al., 1981). The first measurement with a high resolution spectrometer was by the Ball/Sandia Laboratories consortium (Leventhal et al., 1978; Leventhal et al., 1980). The reported flux values have ranged from $0.8 \times 10^{-3}$ to $4 \times 10^{-3}$ photons/cm$^2$-sec. The HEAO-3 measurements were the first to give good information on spatial extent of the source, and reveal apparent time variations. Details of these observations are reported in Riegler, et al., 1981.

Figure 9 shows data from the two galactic scan periods in the light of .511 MeV radiation. It can be clearly seen that there is a significant difference in these two data sets. Superimposed on the scan data are curves representing the sum of the aperture response at .511 MeV and a constant background. The best fitting curve for the Fall 1979 measurement corresponds to a flux level of $(1.85 \pm 0.21) \times 10^{-3}$ photons/cm$^2$-sec, with the source centered at $\ell^{II} = 3.9° \pm 4.0°$ if a point source in the galactic plane is assumed. If an extended source is assumed, the position of the source center is $\ell^{II}$ 3.5° $\pm$ 4.0° with a source extent of 19° $\pm$ 8°.

Performing the same fit for the Spring 1980 data set yields a flux of $(0.65 \pm 0.27) \times 10^{-3}$ photons/cm$^2$-sec, a decrease of $(1.20 \pm 0.35) \times 10^{-3}$ photons/cm$^2$-sec. The probability that such a change in observed flux is due to chance is $\sim 5.0 \times 10^{-4}$ for a normal distribution.

A net spectrum of the .511 MeV line for the fall 1979 observation of the galactic center region is shown in Figure 10. This was obtained by subtracting data gathered in the ranges of $97 < \ell^{II} < 139$ and $217 < \ell^{II} < 265$ from that in the range $331 < \ell^{II} < 19$. The net line is centered at 510.90 $\pm$ 0.25 keV and has an observed width of 3.3 $\pm$ 0.57 keV FWHM. Considering the instrument resolution, this gives an intrinsic line width of $1.6 \begin{smallmatrix} +0.9 \\ -1.6 \end{smallmatrix}$ keV FWHM.

The variation in this source over the six months between observations suggests that the size of the annihilation region is no greater than of order $10^{18}$ cm. The narrow line width implies that the region is partially ionized with a temperature $< 10^{5}$°K. The annihilation rate is $10^{43}$/ sec. These observations put some rather stringent constraints on source models, some of which, along with a discussion of hard x-ray emissions from the galactic center region are presented elsewhere in this publication (Riegler et al., 1981).

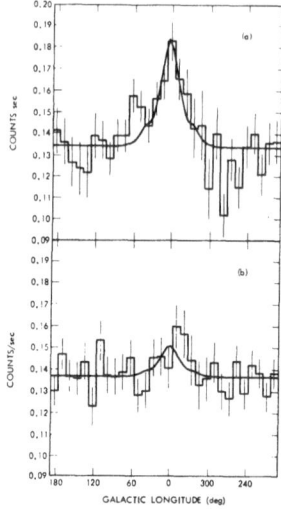

Figure 9

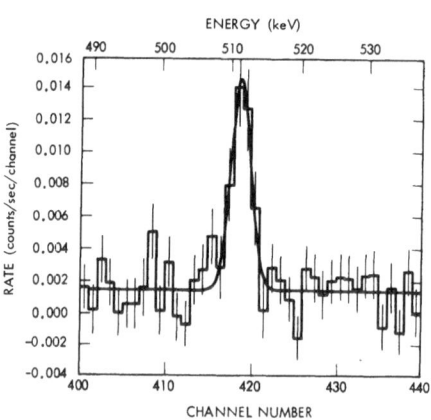

Figure 10

Because our understanding of background effects now seems well in hand, our analysis of point sources has begun and one of the first to be studied in detail is Cygnus X-1 (Ling, et al. 1981). Because of its high ecliptic latitude, and the galactic plane scan maneuvers, it was subject to the extraordinarily long observation time of 148 days between 27 September 1979 and 2 June 1980. These observations reveal temporal behavior patterns which have not been previously observed. These observations are discussed elsewhere in this publication (Ling, et al., 1981).

## 6. Conclusion

In this paper, I have presented an overview of the results from the HEAO-3 gamma-ray spectrometer which have been developed to date. We anticipate that further important results will be forthcoming as known point and localized sources are studied and unknown sources explored with all-sky maps generated in the light of the more prominently expected lines such as the .511 MeV positron-annihilation line.

## 7. Acknowledgements

I wish to acknowledge my colleagues who have been responsible for various aspects of the material presented here. These are J. C. Ling, W. A. Mahoney, T. A. Prince, G. R. Riegler, W. A. Wheaton, and J. B. Willett.

The research described in this paper was carried out at the Jet Propulsion Laboratory, California Institute of Technology, under NASA Contract NAS7-100.

## References

Albernhe, F., et al., 1981, Astron. Astrophys., 94, 214.

Chupp, E. L., et al., 1973, Nature, 241, 333.

Haymes, R. C., et al., 1975, Ap. J., 201, 593.

Hudson, H. S., et al., 1980, Ap. J., 236, L91.

Johnson, W. N. and Haymes, R. C., 1973, Ap. J., 184, 103.

Leventhal, M., et al., 1980, Ap. J., 240, 338.

Leventhal, M., et al., 1978, Ap. J., 225, L11.

Ling, J. C., et al., 1981, 17th ICRC Proceedings.

Mahoney, W. A., et al., 1981, Nucl. Instr. and Meth. (in press).

Mahoney, W. A., et al., 1980, Nucl. Instr. and Meth., 178, 363.

Ramaty, R., and Lingenfelter, R., 1981, Phil. Trans. Royal Soc. (in press).

Riegler, G. R., et al., 1981, Ap. J. Letters, (in press).

Riegler, G. R., et al., 1981, Ap. J., (in press).

Riegler, G. R., et al., 1981, 17th ICRC Proceedings.

Wheaton, W. A., et al., 1981, 17th ICRC Proceedings.

# SOURCE ABUNDANCES AND PROPAGATION OF RELATIVISTIC COSMIC RAYS UP TO Z = 30 : HEAO 3 RESULTS

L. Koch-Miramond [1]

Section d'Astrophysique, Centre d'Etudes Nucléaires de Saclay, France

## 1. Introduction

The end of the 1970 decade has provided exciting new results in cosmic ray astrophysics both in theory and observations. To mention only two of the many results : (i) the discovery of large isotopic anomalies in cosmic ray neon and possibly Mg and Si compositions as compared to the solar system matter (Fisher et al., 1976 ; Garcia-Munoz et al., 1979 ; Mewalt et al., 1980 ; Wiedenbeck et al., 1981) has provided for the first time a direct proof that a specific component of cosmic rays is accelerated near active sites of nucleosynthesis ; (ii) the great advances made in understanding shock wave acceleration have given a great impetus to the problem of the origin of cosmic rays in allowing a direct comparison between theory and observations in a wide variety of physical conditions (Axford 1981). Yet as exciting and astrophysically rewarding as these results may be, they fall short of achieving the full potential of cosmic ray astrophysics in the study of sources and propagation of cosmic rays in the Galaxy. This information can best be obtained at relativistic energies, where composition changes due to solar modulation and ionization losses are negligible and where secondary cosmic ray production in the interstellar medium is minimized, with nuclear interaction cross sections being constant or varying only slowly with energy. However at relativistic energies, where the results are simpler to interpret, the mass measurements are more difficult to perform. In order to achieve these aims a radical improvement in satellite instrumentation was necessary, namely the transition from the small solid state telescope working beautifully in the lower energy range ($\sim$ 100 MeV/n) to the very large Cerenkov telescope providing high resolution and high statistics in the relativistic energy range. The mass measurement is achieved through the filtering effect of the geomagnetic field on the cosmic ray incoming particles (Peters 1974).

Thus the launch of the HEAO 3 satellite in September 1979 just after the Kyoto Cosmic Ray Conference, has coincided with a very exciting period in cosmic ray astrophysics. I will attempt to describe the results obtained so far with the cosmic ray isotope experiment on HEAO 3 and their impact on several domains of high energy astrophysics.

[1] Representing the CEN Saclay - DSRI Copenhagen Collaboration.

## 2. The HEAO 3 cosmic ray isotope experiment

The experiment results from a collaboration started in 1968 between the Danish Space Research Institute and the Centre d'Etudes Nucléaires de Saclay in France. This is not the place to give a description of the developments which led to the design and realization of this experiment, nor to give a detailed description of the instrument, which can be found in Bouffard et al. (1981). I will only briefly summarize the historical background. The method of using a combination of Cerenkov counters with different refractive indices to determine the charge and momentum of relativistic particles traversing the telescope was first tested in a series of balloon flights in 1968-70 (Corydon-Petersen et al. 1970). Since 1973 a series of balloon flights has been carried out with a full HEAO 3 size instrument (Lund et al. 1975) together with extensive accelerator testing at the Bevalac (Cantin et al. 1975). During the same period, triggered by an idea by Linney and Peters (1972) and a second one by Cassé, a successful development of silica aerogel Cerenkov radiators was achieved (Cantin et al. 1974, Engelmann and Cantin 1978). These new Cerenkov detectors with adjustable refractive indices in the interval not covered by ordinary transparent material allowed us to match the momentum range covered by geomagnetic cut-offs along the HEAO 3 orbit and to realize a very compact telescope with large aperture.

The telescope consists of 5 Cerenkov detectors and a flash tube hodoscope with 4 trays, as shown in figure 1. The hodoscope serves the dual purpose of allowing corrections for geometrical variations in the Cerenkov counter response and to determine the particle arrival directions in the geomagnetic field (Rotenberg et al., 1981 ; Lund et al.,1981a). Both ends of the telescope are open to space and particles from both front and back are analysed, their direction through the instrument being determined by a time of flight measurement. The telescope has a useful geometry of $0.07 \ m^2$. ster for each direction of propagation and has collected 7 million useful events during the 1.6 yr of HEAO 3 lifetime.

The energy range covered is 0.7 to 20 GeV/n for elements between Be and Sn. The method used for momentum determination is presented by Lund et al.(1981c). The momentum resolution was calculated using in flight measured parameters (Cantin et al. 1981). It lies between 0.5 and 10 per cent in the 1 to 10 GeV/n range and varies with charge and energy as shown in figure 2. One sees that there exists three ranges of momentum, near the thresholds of the three momentum sensitive Cerenkov counters where the calculated accuracy in momentum determination is better than 3 per cent for all nuclei and reaches 1 per cent for iron nuclei. The

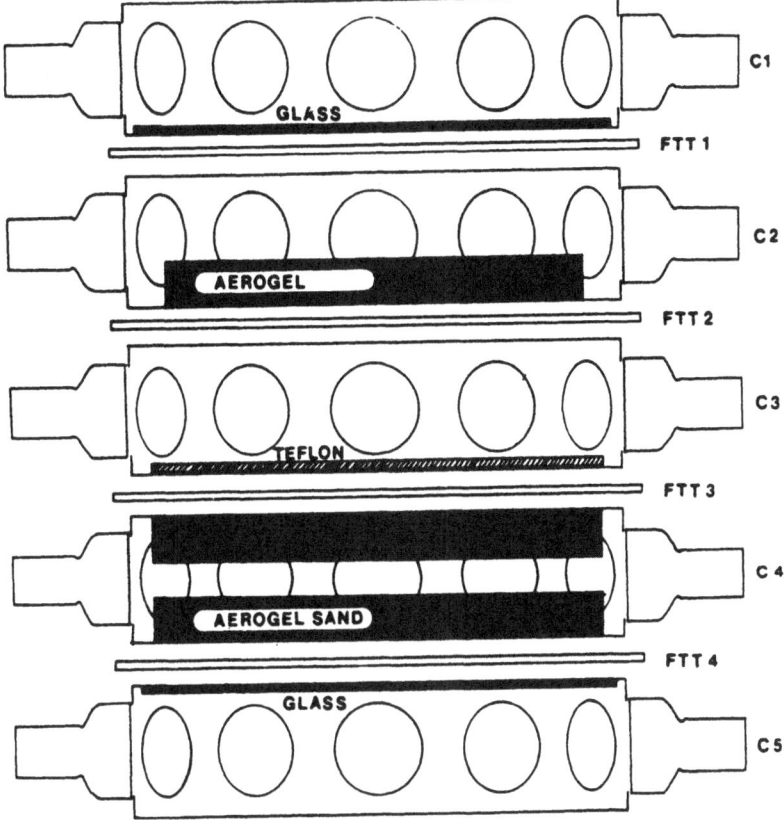

**Fig.1.** The HEAO-3 C2 detector configuration. C1 to C5 Cerenkov counters with refractive indices : $n_1 = n_5 = 1.64$ ; $n_2 = 1.053$ ; $n_3 = 1.33$ ; $n_4 = 1.012$ and diameters around 60 cm. FTT 1 to 4 flash tube hodoscope with four trays, each of which contains two layers of 128 flash tubes. Time of flight system between C1 and C5 counters.

method used to determine the isotopic composition of a given species relies on the filtering effect of the geomagnetic field on the momentum spectrum of this element. It was first described by Peters (1974) and will be discussed below. The mass resolution resulting from the momentum uncertainties only is better than 0.5 mass unit near 1.5 and 3 GeV/n for all nuclei. At 7 GeV/n it is ~ 0.8 mass unit for C to Si nuclei, and ~ 1.5 mass unit for Fe. The charge resolution of the instrument is 0.2 charge unit over the entire energy range. The good separation of even and odd neighbouring nuclei is shown in figure 3a, 3b, 3c ; it is more clearly visible in figure 4 where the very difficult region from Fe to Zn is displayed. This is the first

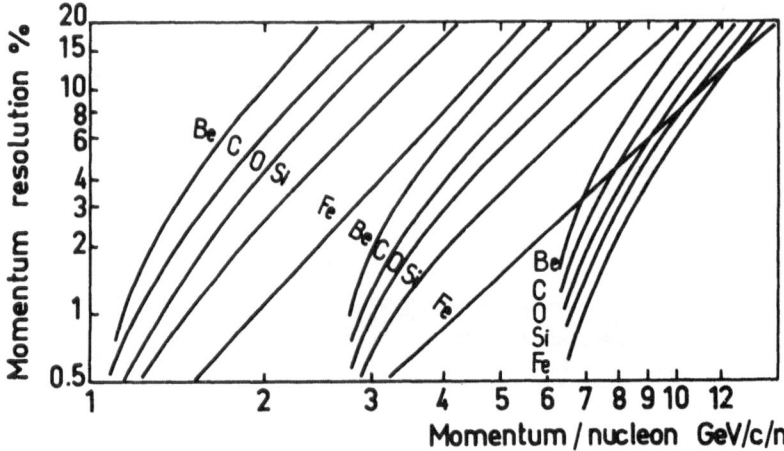

**Fig.2.** Momentum resolution in per cent versus momentum/nucleon in GeV/c/n in the three velocity counters for Be, C, O, Si and Fe cosmic ray nuclei. The teflon counter threshold is at 1.06 GeV/n, the aerogel block counter threshold at 2.84 GeV/c/n and the aerogel sand threshold at 6.1 GeV/c/n for $^{16}$O nuclei.

time that a Cu peak has been resolved, the last element with $Z < 30$ to be identified in the cosmic ray radiation.

## 3. The elemental composition of galactic cosmic rays at Earth

Abundance values for all elements between Be and Zn have been obtained after a careful selection of the registered events. The event selection criteria and the coefficients of correction for nuclear interaction losses in the instrument, for hodoscope charge dependent efficiency and for residual overlap are described in Engelmann et al. (1981). The resulting relative abundances versus energy per nucleon are shown in figures 5 to 8 together with the abundances calculated in the framework of an homogeneous diffusion model described below. The results clearly show that the ratios of mainly secondary to mainly primary species decrease steeply with increasing energy, as was observed before e.g. Juliusson (1974), Lezniak and Webber (1978). Figures 5 and 6 show a progressive steepening of the energy dependence of abundance ratios from Al/Si and K/Fe. This variation can be interpreted as a progressive decrease of the amount of primary component from Al to Na and K. On the contrary, the ratios of two mainly primary nuclei are nearly constant with energy as shown in figure 7, the slight decrease observed in the

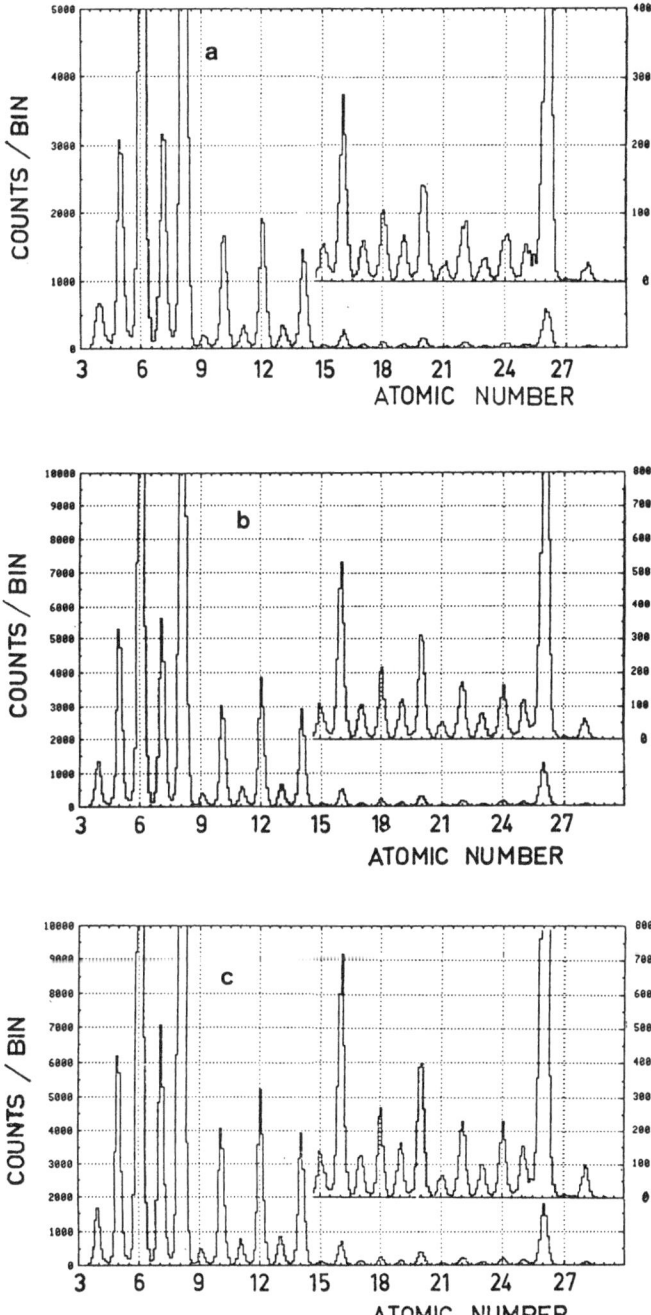

Fig. 3a, b, c. Charge histograms from Be to Zn in 3 energy ranges : a) from .8 to 2.5 GeV/n. b) from 2.5 to 6 GeV/n. c) above 6 GeV/n. In the inset, the vertical scale has been multiplied by a factor of 10.

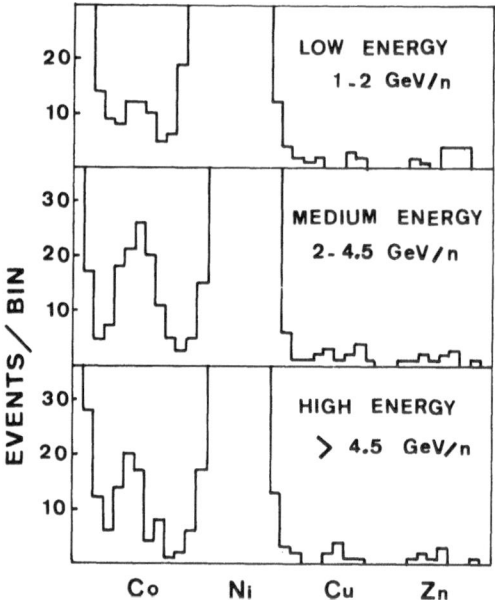

**Fig. 4.** Charge histograms from Co to Zn in 3 energy ranges of the HEAO 3-C2 instrument. Note resolved peaks at Co, Cu and Zn.

abundance ratio of a lighter to a heavier primary nucleus reflecting the combined effects of interaction length dependence on atomic mass and escape length variation with energy. Figure 8 shows the observed abundances relative to iron of the elements between Scandium and Manganese. The general decrease of abundance of Iron secondaries relative to Iron in the GeV energy range has been observed before (Juliusson 1974, Caldwell 1977, Lezniak and Webber 1978, Simon et al. 1980) but can now be studied in detail on individual elements in the HEAO 3 data (Koch-Miramond et al., 1981a). A progressive steepening of the relative abundance versus energy is observed from Cr to Sc which reflects the increasing amount of tertiary component (Perron, 1975). The special behaviour of the Mn/Fe abundance ratio will be discussed later.

4. Propagation of relativistic cosmic rays in the framework of the energy dependent leaky-box model

An uniform distribution of cosmic ray sources is assumed in a confinement volume with rapid diffusion of particles and relatively slow leakage across the boundary, this is the so-called leaky-box model (Davis 1959, Cowsik et al. 1967),

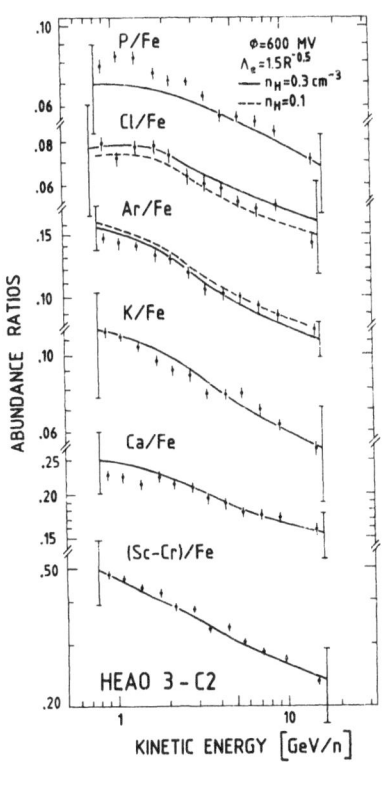

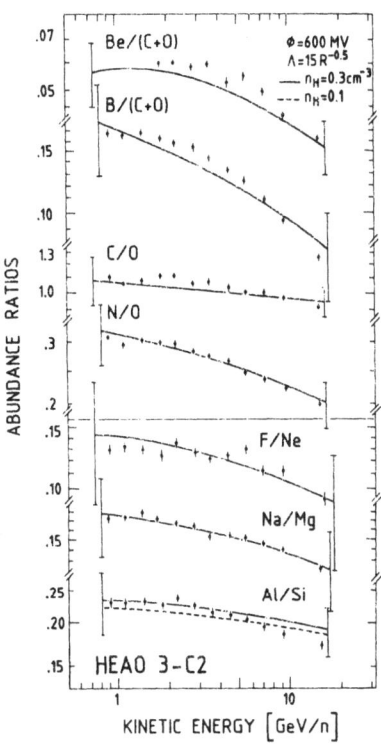

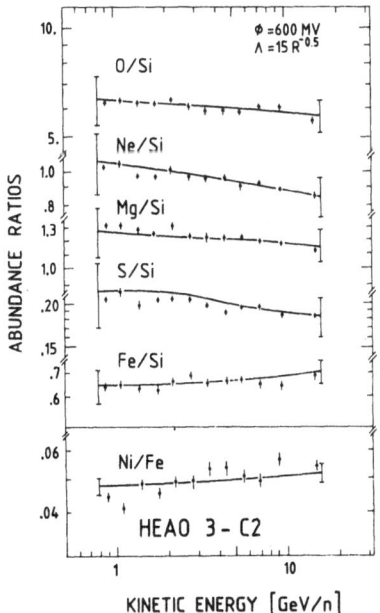

**Fig. 5, 6, 7.** Comparison of experimental and calculated relative elemental abundances as a function of energy. The calculated curves correspond to fits in the framework of the leaky-box diffusion model with escape length $\propto R^{-0.5}$ and solar modulation deceleration parameter $\emptyset = 600$ MV; the error bars at both ends of the calculated curves correspond to nuclear cross-section uncertainties.

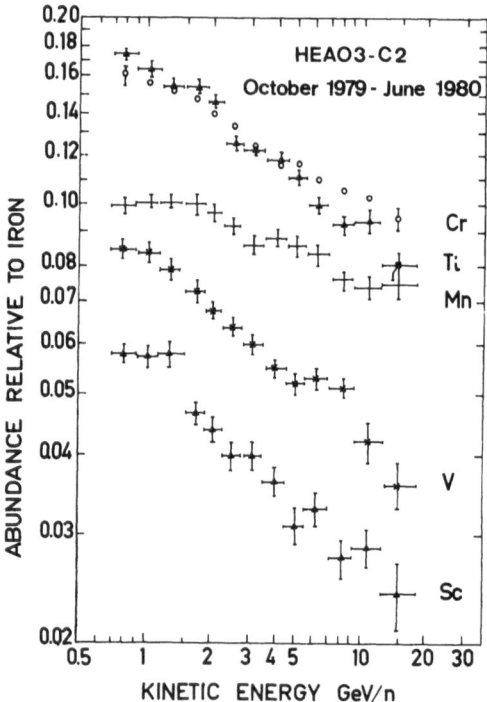

**Fig.8.** Observed abundances relative to Fe of the elements between Sc and Mn as a function of kinetic energy.

where the pathlength distribution of cosmic rays has a simple exponential form $P(x)$ $\propto \exp(-x/\Lambda_e)$. We have taken all primary nuclei to have the same source spectrum energy dependence, a power law in total energy with an index of - 2.2 and we have adjusted the source abundance ratios to give a best fit to the observations. Detailed calculations are presented in Koch-Miramond et al. (1981b) and Perron et al. (1981). The rigidity ($R$) dependence of the mean pathlength $\Lambda_e$ has been determined from the study of the sub-iron to iron abundance ratios : $\Lambda_e = 15 \, R^{-0.5} \, g.cm^{-2}$ of pure hydrogen (Perron and Koch-Miramond, 1981).

The solar modulation was taken into account using the force field approximation (Gleeson, 1968) with a mean energy loss in the heliosphere of 300 MeV/n corresponding to a deceleration parameter $\emptyset = 600$ MV. For nuclear cross sections we generally relied on Silberberg and Tsao's formulae (Silberberg and Tsao, 1977), with a few exceptions based on existing measurements. A detailed examination of nuclear cross sections was necessary to evaluate the uncertainties in the propagation calculations due to cross section errors. The corresponding error bars are shown at the two ends of calculated curves, in figures 5, 6 and 7. They are very large compared to the statistical errors on measured abundances ratios. However the shape of the calculated curves is better known than its absolute position because, in this energy range, the variations of cross-sections with energy are small and better known than their absolute values.

Our main conclusions are the following :

(i) Mostly secondary to primary ratios are well accounted for with the same simple exponential pathlength distribution with mean $\Lambda_e \propto R^{-0.5}$ from sub-iron down to Na and even N (figure 5). The agreement is poorer for Be/(C+O) and B/(C+O) considering the very accurate energy dependence given by our data : the low energy points seem to indicate that $\Lambda_e$ stops increasing below 2 GeV/n as proposed by other authors (Garcia-Munoz et al. 1979, Protheroe et al. 1981). The higher energy points seem to require a faster decrease of $\Lambda_e$ with energy than we assumed. Anyhow the corrections will be small. Work is now in progress to determine whether a pathlength distribution accounting for the B/(C+O) ratio can also explain the other secondary to primary ratios as well. In any case, neither Be nor B is overproduced by our calculations, which means that no truncation of the pathlength distribution is needed at low pathlengths.

(ii) Calculated ratios of mostly primary species also agree fairly well with the HEAO 3 data, confirming the assumption of an energy independent source composition in the energy range of our experiment : 0.7 to 20 GeV/n. The input source spectrum, a power law in total energy, seem quite acceptable but no study has been completed so far on the sensitivity of the data to the parameters defining the input energy spectrum. It will be easier to answer this question if we succeed in deriving reliable absolute energy spectra for the primary nuclei ; work is in progress in this direction.

## The radio-active cosmic ray clocks : $^{54}$Mn, $^{36}$Cl and $^{26}$Al.

In figure 8 one sees that the energy spectrum of the Mn abundance relative to Fe in the energy range 0.7 to 18 GeV/n is significantly flatter than those of other iron secondaries. In Koch-Miramond et al. (1981a) this difference is interpreted as being due to the survival of $^{54}$Mn above 15 GeV/n and to its progressive decay towards lower energies along the following arguments. Manganese has 3 isotopes : $^{55}$Mn is stable, $^{53}$Mn and $^{54}$Mn decay by electron capture with respective half-lives of 3.7 $10^6$ years and 312 days in the laboratory. In cosmic rays they can be considered as nearly stable in our energy range (Raisbeck et al. 1975). However $^{54}$Mn may also undergo beta decay as noted by Cassé (1973). But its $\beta^-$ partial half-life T has not been measured and Cassé (1973) has roughly estimated it to be

$T \sim 2.10^6$ yr. In the leaky-box model, the surviving fraction is then a function of the product of T and of the mean density $n_H$ of the propagation medium. We made calculations with different values for the product $(n_H . T)$. The results obtained with $\Lambda_e = 14 \ R^{-0.5}$ and $\emptyset = 600$ MV are compared with the data on fig.9. The use of a lower value for $\emptyset$ would not affect the comparison because of the relative flatness of the Mn/Fe ratio which results in a very small influence of modulation (Mn/Fe is only 2.5 per cent higher at 1 GeV/n with $\emptyset = 0$). Given that the errors on the formation and destruction cross-sections of Mn isotopes vary between 5% ($^{54}$Mn at 0.6 GeV/n) and $\sim$ 25% ($^{55}$Mn at high energy) it would appear at first sight that our calculated values are uncertain by 14% at low energy and 25% at high energy, so that all the curves of fig.9 are acceptable. However, one cannot change the cross-sections independently of one another, and, over all, independently of energy, within their error bounds : the shape of the excitation functions should not be distorted beyond permissible limits. For instance the two extreme curves could be brought into agreement with the data, without changing the source abundance ratio Mn/Fe (taken here equal to $(Mn/Fe)_{LG} = 0.9\%$), only by adopting excitation functions that increase with energy, which is unrealistic on the basis of nuclear physics. A change in the adopted $\Lambda_e(R)$ would have little effect on the low energy part of the Mn/Fe curves. The uppermost curve $(n_H . T = \infty)$ could be lowered by assuming that Mn is absent from the source, but the slope of the Mn/Fe curve would then be even steeper and thus more at variance with the data. Consequently we estimate that there are only two possible explanations to the observed flatness of the Mn/Fe curve :

(i) Partial decay of $^{54}$Mn (our preferred explanation), in which case the best agreement with the data is obtained for the central value $n_H . T = 0.3 \ 10^6$ yr.cm$^{-3}$ but values differing by a factor of $\sim$ 3 are still acceptable. (ii) Complete decay of $^{54}$Mn on the whole energy range combined with an increase of the source abundance of Mn by a factor of $\sim$ 3, to $\sim$ 0.03 of iron. This would correspond to shifting upward and flattening the lowest curve of fig.9 thus bringing it in reasonable agreement with the experimental points.

The value $n_H . T \sim 0.3 \ 10^6$ yr .cm$^{-3}$ gives a mean density $n_H$ consistent with that deduced by Wiedenbeck et al. (1980) from their $^{10}$Be measurements ($n_H \sim 0.3$ cm$^{-3}$) if $^{54}$Mn beta half-life is $\sim 10^6$ yr. With this value of $n_H . T$ we calculate that 45 % of $^{54}$Mn survive at 1 GeV/n and 93% at 20 GeV/n. An unambiguous evaluation of $n_H$ (and cosmic ray age) and of Mn source abundance must await the measurement of the $^{54}$Mn $\beta^-$ branching ratio, and the determination of Mn isotopic composition in cosmic rays. The difficult measurement of the $\beta^-$ branching

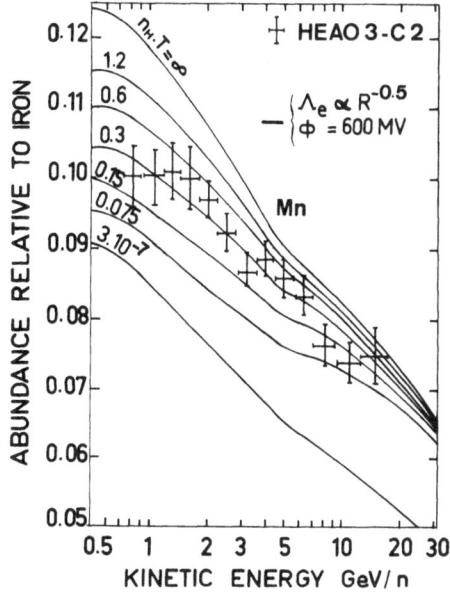

Fig.9. Comparison of experimental and calculated abundances of Mn relative to Fe as a function of kinetic energy. The calculated curves are obtained with $\Lambda_e \propto E^{-0.48}$ and different values of the product $n_H \cdot T$, as indicated. The upper curve corresponds to complete survival and the lower curve to complete decay of $^{54}$Mn.

ratio is now in progress in collaboration with the Niels Bohr Institute and the Laboratoire de Spectrométrie de Masse, Faculté d'Orsay.

Turning to Cl and Al the interpretation of the results on Cl/Ar and Al/Si abundance ratios in terms of cosmic ray lifetime is, as shown in Perron et al. (1981), out of reach of the present results owing to the great uncertainties in the interaction cross-sections. Here again the astrophysical interpretation of the data is limited by our lack of knowledge on nuclear cross-sections and not by the observational uncertainties. One hopes that a measurement of $^{26}$Al/Si ratio at high energy with HEAO 3 can help to solve this problem.

## 5. Propagation including a halo

The determination of the dimension of the confinement region of cosmic rays, the cosmic ray halo, is a very important but still unsolved problem. Data on $^{10}$Be abundance at low energy (Wiedenbeck 1980, Garcia-Munoz 1981), if interpreted in the framework of the leaky-box model, imply a mean density in the cosmic ray confinement volume of $\sim 0.3$ at.cm$^{-3}$, considerably below the average density of the galactic disk inferred from the CO millimeter wave observations (Gordon and Burton, 1976) even taking into account difficulties involved in the interpretation of the data (Paul, 1981). As shown by Prischep and Ptuskin (1974), the propagation of radioactive cosmic ray nuclei should be interpreted in the framework of more realistic diffusion models than the simple homogeneous leaky box model : the heterogeneous distributions of cosmic ray sources and of gas density must be taken into account. Cesarsky et al. (1981) consider one-dimensional models, where the cosmic ray sources are embedded in a gas disk of uniform density $n_o$ and of height h ; cosmic rays diffuse outward through a halo of density $n = 0$ and height $H \gg h$. The diffusion coefficient K is assumed to be constant in space. As shown by Ginzburg et al. (1980), the abundances of stable secondary nuclei, in the case considered here, depend only on the mean pathlength of cosmic rays before escape $\Lambda_e$. With the relation $\Lambda_e \propto R^{-0.5}$ deduced from the HEAO 3 data on stable secondary abundances, one shows that these static diffusion models imply either an energy dependent diffusion process where $K \propto R^{0.5}.v$ (v being the cosmic ray velocity ), or rigidity dependent size of the confinement region : $H \propto R^{-0.5}$. It follows that at high energy where ionization losses are negligible, the abundance of radioactive nuclei are determined by a combination of the parameters $(n_o T)$ and $(H/h)$, T being the half-life of the radioactive nucleus. When ionization losses are no longer negligible $n_o$ must be specified. The variation with energy of the ratios $^{54}Mn/Fe$ and Mn/Fe has been studied for various choices of the parameters and it was found that :

(i) The energy range of HEAO 3 is well suited for this analysis, but while the behaviour of the variation of the isotopic ratios is rather different in the various models, these effects are less conspicuous on a plot of the elemental Mn/Fe ratio where they can be confronted with the present HEAO 3 data (figures 10 and 11 and Table 1).

(ii) With the cross-section values used throughout, and assuming $T \sim 10^6$ yr, the HEAO 3 observations appear to favor relatively flat haloes with $H/h \sim 10$.

(iii) For a given value of $(n_o T)$ heterogeneous models with energy dependent size of the confinement region imply at low energy (1 -2 GeV/n) a larger confinement region than the heterogeneous energy dependent diffusion models.

This study shows that measurements at high energy with isotopes of known half-life may provide useful answers on halo size and confinement times of cosmic rays in the Galaxy.

## 6. Models with simultaneous acceleration and propagation of cosmic rays

Cesarsky et al. (1981, this conference) examine the constraints put by the HEAO 3 observations on the amount of reacceleration of cosmic rays in the Galaxy.

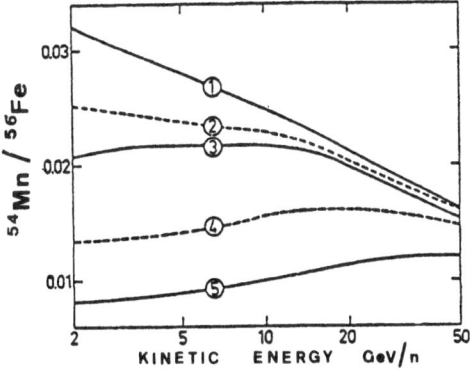

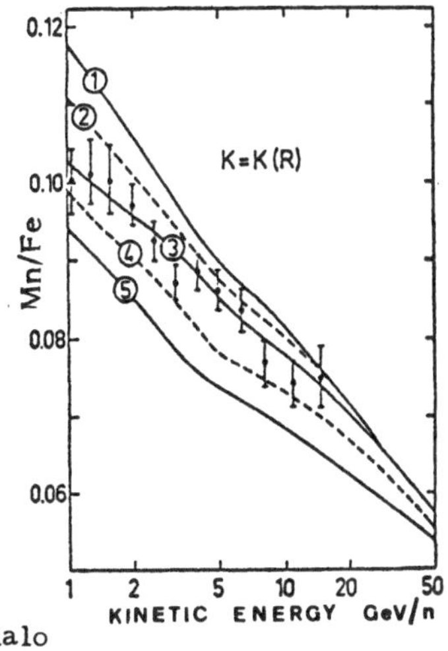

Fig.10. Variation with kinetic energy (in GeV/n) of the isotopic ratio ($^{54}$Mn$/^{56}$Fe) for the various diffusion models with a halo defined in Table 1, when the diffusion coefficient K is assumed to be constant in space and function of cosmic ray rigidity R and the height of the halo H = constant.

Fig.11 Same as fig.10 for the elemental ratio (Mn/ Fe) ; the observations are from HEAO 3-C2.

TABLE 1. Parameters of the models with a halo

| Model | $n_0 \cdot T$ cm$^{-3}$. $10^6$yr | H/h | $< n >$ cm$^{-3}$ | T Myr |
|-------|------|-----|------|-----|
| 1 | 30 | 50 | 0.6 | 10 |
| 2 | 10 | 75 | 0.13 | 5 |
| 3 | 1 | 10 | 0.1 | 1 |
| 4 | 1 | 50 | 0.02 | 1 |
| 5 | 0.3 | 50 | 0.006 | 0.3 |

The fifth column gives the value of T assumed to calculate the corrections due to ionization losses in interstellar space.

Blandford and Ostriker (1980) proposed that cosmic rays are continuously accelerated, through their galactic life, by successive encounters with supernovae remnant shock-waves. This process implies that at high energies, even if the escape time is very short, secondary particles are always present (Eichler 1980). As a consequence the secondary/primary ratios predicted under these hypotheses, tend to be too flat compared to the observations. Fransson and Epstein (1980) have shown that the observed L/M ratio at E < 100 GeV/n implies that the mean time between reacceleration is larger than the mean escape time $t_e$, so reacceleration can only be a minor effect. The HEAO 3 data on sub-iron elements cover a narrower energy range, and set less stringent conditions on the ratio $t_r/t_e$ ; still, we have found that it is not possible to fit these data at E > 10 GeV/n if $t_r < 0.2 \, t_e$.

Cesarsky et al. (1981) examine also the effect of reacceleration of cosmic rays in the Galaxy on the abundance of radioactive isotopes. The prediction of leaky box models or diffusion models with sudden acceleration, especially when the escape time is energy dependent, is that at high energies the unstable isotopes do not have enough time to decay, and behave like stable isotopes. If acceleration and propagation occur simultaneously the nuclei have a chance to decay while their energy is low, so that a substantial population of decayed particles is transferred to high energies. One shows that even when the reacceleration time $t_r$ is equal to the mean escape time at 1 GeV/n a radioactive isotope, say $^{54}$Mn appear underabundant at all energies. A more quantitative discussion does not seem warranted until $T(^{54}Mn)$ is measured.

## 7. Source abundances of cosmic rays

The source contributions in the figures 5, 6 and 7 affect mainly the absolute level of the observed ratios and to a lesser extent the slope of these ratios as discussed in sect. 3. In Perron et al. (1981), the source abundances of 16 elements are derived (see Table 2). The procedure followed, which takes into account the propagation errors, is illustrated in figure 12. The elemental N/O ratio observed by HEAO 3 is shown together with the calculated secondary component and its 1 $\sigma$ error associated with uncertainties on the formation and destruction cross-sections and on the propagation model. The difference represents the surviving primary N/O ratio. This graph shows that since the production of secondary nuclei decreases with increasing energy, the primary component can be better estimated towards higher energies : at 0.8 GeV/n only an upper limit of the N/O ratio at the source can be obtained, with only 20% of N arriving at Earth being primary, but this proportion rises to 40% at 15 GeV/n. The high energy values were therefore given more weight in deriving the final abundances and uncertainties presented in the first column of Table 2. In several cases our knowledge of the cosmic ray source elemental abundances is significantly improved with respect to previous data. But it is still impossible to obtain useful information on source abundances of the rare elements F, P, Cl, K, Sc, Ti, V, Cr and Mn, due mainly to the uncertainties on spallation cross-sections (and not on cosmic ray measurements).

Note that we get a source abundance ratio N/O = 9 $\pm$ 3%, fully consistent with the local galactic ratio N/O = 10 $^{+5}_{-3.5}$ % deduced from solar photosphere and HII region data (Meyer 1979a, b). The source N/O ratio in galactic cosmic rays has a considerable importance to pinpoint the nucleosynthetic birthmarks of cosmic rays. It is a matter of debate because it is difficult, as shown by Gusik (1981), to

TABLE 2

| ELEM. | GCR SOURCES HEAO3-C2 | | LOCAL GALACTIC (LG) Meyer 1979 | | OVERABUNDANCE GCRS/LG |
|---|---|---|---|---|---|
| C | 420. | ± 32. | 1300. | ± 300. | .32 $^{+ .10}_{- .06}$ |
| N | 46. | ± 15. | 240. | ± 80. | .19 $^{+ .12}_{- .07}$ |
| O | 505. | ± 20. | 2300. | ± 500. | .22 $^{+ .06}_{- .04}$ |
| F | < 2.5 | | 0.093 | $^{+ 0.056}_{- 0.035}$ | < 43. |
| Ne | 63. | ± 8. | 270. | $^{+ 190.}_{- 110.}$ | .23 $^{+ .17}_{- .10}$ |
| Na | 5.6 | ± 3.2 | 5.6 | ± 0.9 | 1.00 $^{+ .65}_{- .57}$ |
| Mg | 105. | ± 6. | 105. | ± 3. | 1.00 ± .07 |
| Al | 10.9 | ± 3.9 | 8.4 | ± 0.4 | 1.30 ± .47 |
| Si | ≡ 100. | ± 6. | ≡ 100. | ± 3. | ≡ 1.00 ± .07 |
| P | < 2.5 | | 0.96 | ± 0.20 | < 3.3 |
| S | 14.3 | ± 2.5 | 45. | ± 13. | .32 $^{+ .15}_{- .09}$ |
| Cl | < 1.6 | | 0.47 | $^{+ 0.28}_{- 0.18}$ | < 5.5 |
| Ar | 3.6 | ± 0.8 | 9.0 | $^{+ 6.3}_{- 3.7}$ | .40 $^{+ .29}_{- .18}$ |
| K | < 1.9 | | 0.36 | ± 0.12 | < 8.0 |
| Ca | 7.1 | ± 2.0 | 6.2 | ± 0.8 | 1.15 ± .37 |
| Sc | < 0.8 | | 0.0035 | ± 0.0005 | <270. |
| Ti | < 2.4 | | 0.27 | ± 0.04 | < 11. |
| V | < 1.1 | | 0.026 | ± 0.005 | < 52. |
| Cr | < 2.9 | | 1.30 | ± 0.12 | < 2.5 |
| Mn | < 3.7 | | 0.79 | ± 0.17 | < 4.7 |
| Fe | 91. | ± 5. | 88. | ± 6. | 1.03 ± .09 |
| Co | 0.31 | ± 0.13 | 0.21 | ± 0.03 | 1.48 ± .64 |
| Ni | 4.8 | ± 0.7 | 4.8 | ± 0.6 | 1.00 ± .20 |
| Cu | 0.061 | ± 0.011 | 0.052 | $^{+ 0.031}_{- 0.020}$ | 1.17 $^{+ .74}_{- .48}$ |
| Zn | 0.059 | ± 0.011 | 0.100 | ± 0.020 | .59 $^{+ .18}_{- .14}$ |

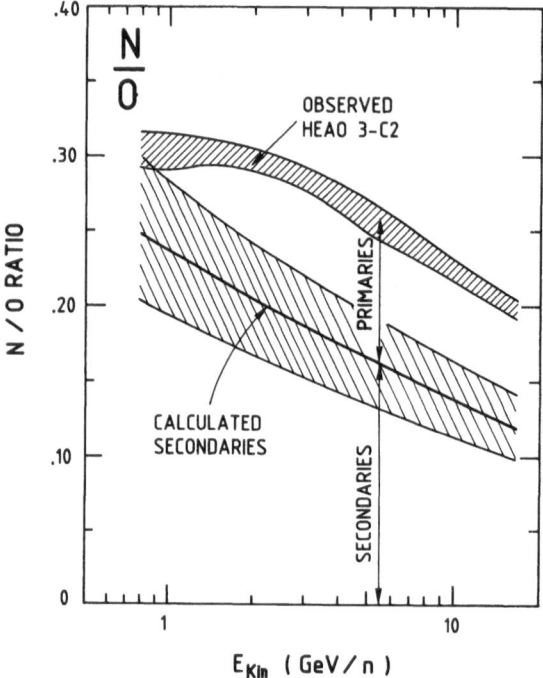

Fig.12. Elemental N/O ratio at high energy : range of observed ratios with HEAO 3-C2 and calculated secondary yield, with its range of uncertainty. The difference represents the surviving primaries seen at Earth.

reconcile the N/O source ratios deduced from both elemental and isotopic ratios obtained at low energy by Wiedenbeck et al. (1979), Mewalt et al. (1981) and Webber (1981).

Goret et al. (1981) reexamined this point in the light of the excellent elemental ratio N/O obtained by HEAO 3 (figure 6) and the good but unique proton induced partial cross-section measurements for the production of $^{14}$N and $^{15}$N from $^{16}$O spallation at 2.1 GeV/n of Lindstrom et al. (1975). The preliminary isotope ratio $^{15}$N/$^{14}$N at 3 GeV/n (Juliusson et al. 1981) obtained by HEAO 3 and discussed later on, is not yet accurate enough to be used in this context. The conclusion of our analysis is that with the best estimated cross-sections at low energy it is difficult to fit the combined $^{15}$N/$^{14}$N and N/O data at low energy but that within the $\pm$ 20% cross-section errors any source N/O abundance ratio between 3 and 8% could not be excluded. Therefore only a marginal disagreement exists between the N/O source abundances deduced from high and low energy observations. Once again our lack of knowledge on nuclear physics limits our astrophysics.

## 8. Comparison of elemental abundances in cosmic ray sources and in our local galactic environment

In figure 13 the ratio of cosmic ray source abundance (GCRS) deduced from HEAO 3 (Goret et al. 1981) to the local galactic elemental abundances (LG : Meyer 1979a, b) is shown as a function of first ionization potential for all elements between C and Zn. H(R) and He are from Meyer (1981c), where H(R) is the H abundance derived from the observed H/He ratio at a given rigidity. Error bars on local galactic and cosmic ray source abundances have been drawn separately in the

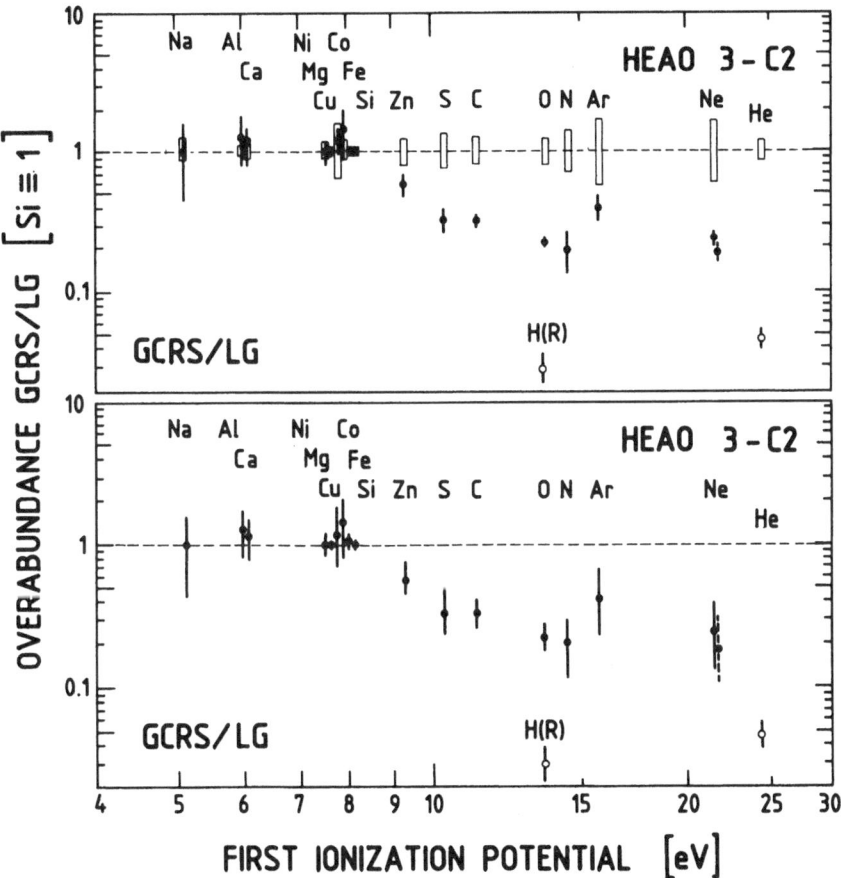

**Fig.13.** Ratios of Galactic Cosmic Ray Source to Local Galactic abundances normalized to Si derived from the HEAO 3-C2 data (except for H and He) vs. first ionization potential. H is derived from H/He ratio at a given rigidity. Dashed bar for Ne : "anomalous" $^{22}$Ne component subtracted.
Top : errors on GCRS and on the LG standard (boxes) plotted separately. Bottom : both errors are combined quadratically as in last column of Table 2.

upper part of the figure and combined quadratically in the lower part. The cosmic ray source overabundances relative to local galactic standard appear to be well correlated with the first ionization potential of the elements. HEAO 3 data on Na, Co, Cu, Zn and Ar confirm the validity of the correlation which appears mainly limited by the large uncertainties on local galactic abundances. For instance, abundances of Ar and Ne in the local interstellar medium are uncertain within a factor of 1.7 but their underabundances reach a factor of 2.5 to 4.

This correlation with the first ionization potential (or at least some related atomic parameter) has been known for some time (Bradt and Peters 1950, Kristiansson 1971, Cassé and Goret 1978). With the HEAO 3 data, only either H and He or H, Ne and Ar, cannot be smoothly ordered. It is most probably H and He that do not follow the general trend since He (and not Ne and Ar) behave differently from other elements in solar energetic particles (Meyer 1981a). Before discussing the significance of this correlation let us consider the relative elemental abundances in solar energetic particles.

## 9. Comparison with solar energetic particle composition

In figure 14 the ratio of our values for the cosmic ray source abundances to the Solar Energetic Particle (SEP) abundances (as derived by Meyer 1981a, b from the "mass unbiased baseline composition") is shown as a function of the first ionization potential. If we do not consider He, for which a mass unbiased baseline SEP abundance cannot be properly defined, a unique horizontal line can be drawn through the error bars associated with all elements but C, which is two times more abundant in galactic cosmic ray sources.

## 10. Implications for the galactic cosmic ray sources

The overabundances of elements heavier than He in GCRS and in SEP with respect to L.G. are nearly the same (except for C), and are correlated with the first ionization potential (FIP). The shape of this correlation suggests a structure with two plateaux, one for "low FIP" elements (5 to 8.5 eV), possibly not completely flat, one for "high FIP" elements (10 to 22 eV), with Zn behaving in an intermediate fashion. It has been suggested by Meyer (1981b, c) that the high FIP plateau corresponds to elements selected as neutrals before acceleration, in contrast with the low FIP elements which are commonly thought to be selected as ions in a plasma at $\sim 10^4$ °K.

On the other hand, the similarity of composition of SEP and Solar Corona suggests that the SEP composition is primarily governed by the transport of thermal neutral particles and ions from the relatively cool chromosphere into the

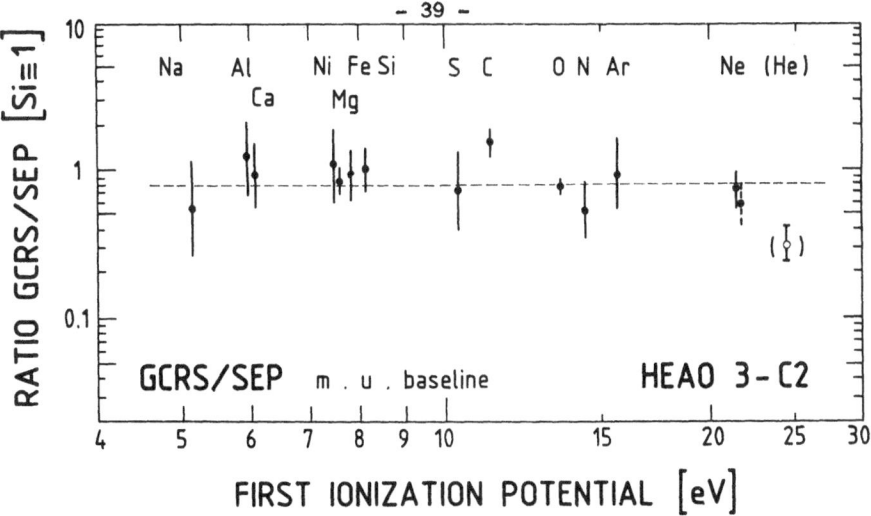

Fig.14. Ratio of GCRS to solar energetic particle (SEP) abundances vs. first ionization potential. GCRS abundances (except He) are from HEAO 3-C2 with SEP (mass unbiased baseline) from Meyer (1981a).

Corona from which SEP are later on accelerated (Meyer 1981b).

Finally the similarity between the elemental compositions of GCRS, SEP and solar corona suggests that the GCR composition might be governed by an injection by stellar flares, similar to solar flares and hence ultimately by transport of matter from chromospheres to coronae of lower mass stars.

Thus the major component of GCR might be due to an injection by stellar flares, similar to solar flares, the acceleration taking place in a two-stage process. As Gorenstein (1981) pointed out, the most prolific source of X-ray flares in the galaxy seem to be the dwarf M stars, the characteristics of these flare events being very similar to those of the sun and probably similar too in particle emission. In addition X-ray flares a hundred times larger seem a very frequent phenomena in pre-main sequence stars seen in the giant clouds close to the sun (Montmerle et al. 1981, 1982) and energetic particle emission might be present as well. This implies solar like particle emission throughout the galaxy. However for pre-main sequence objects the problem of the propagation of these injected particles in the surrounding rather dense medium is not yet solved. The problem of keeping the "coronal" composition of cosmic rays after acceleration remains anyway, even in the framework of the suggestion by Axford (1981) of prompt acceleration by shock-waves associated with supernova remnants in the hot interstellar medium (see Cassé, 1981). Other models like injection of GCR from grain-destruction products (Cesarsky and Bibring, 1980 ; Bibring and Cesarsky, 1981) although not particularly favoured by the HEAO 3 results cannot be ruled out (Goret et al. 1981, Israel et al. 1981).

When comparing the composition of GCRS with that of SEP a striking anomaly is the overabundance of C nuclei, which has been confirmed by the HEAO 3 data. It could be related to the $^{22}$Ne excess observed in GCRS, if this isotopic overabundance is due to a minor component recently processed by standard He burning in massive stars (Meyer 1981c). This component may originate from Carbon rich Wolf-Rayet stars, as discussed by Cassé and Paul (1981), and Audouze et al. (1981).

Another interesting constraint on galactic cosmic ray sources is the time delay δt between nucleosynthesis and acceleration of cosmic rays. From our measured Fe/Co/Ni abundance ratios we can derive a lower limit of δt using the analysis made by Soutoul et al. (1978). In figure 15 we compare the source abundance ratios Co/Fe and Ni/Fe deduced from the HEAO 3-C2 data, found "normal" with the expected source composition, as a function of the time δt between e-process nucleosynthesis (Hainebach et al. 1974) and cosmic ray acceleration. The fast decrease of Co source abundance after one year is due to $^{57}$Co decaying to $^{57}$Fe by electron capture and the subsequent rise of Co abundance after $10^4$ yr is due to the electron-capture decay of $^{59}$Ni to $^{59}$Co. For Co, spallation "en route" is important, and we compare also its predicted "arriving" abundance with that observed near Earth. As can be seen, the observations are inconsistent with δt = 3 yr to $10^5$ yr and most probably imply δt > $10^5$ yr, thus extending the lower bound of δt far beyond the previous limit of 2 yr (Tueller et al. 1979). The acceleration of Fe, Co, Ni follows their nucleosynthesis in massive explosive stars after more than $10^5$ yr, a phase where the old remnant ejecta composition is dominated by swept up interstellar material.

## 11. Analysis of the isotopic composition of relativistic cosmic rays

The idea of isotope analysis of galactic cosmic rays by using the filtering effect of the earth magnetic field was first suggested by Balasubrahmanyan et al. (1963) in an attempt to measure the isotopic composition of cosmic ray helium with a balloon-borne instrument. In the past decade, an extensive development of the analysis methods was led by B. Peters who described the differential method in Peters (1974) ; meanwhile the results from several balloon experiments instigated by Lund et al. (1970), were reported by Dwyer and Meyer (1975), Lund and Sorgen (1977), Dwyer (1978), Dwyer and Meyer (1979). The data collected on HEAO 3 is potentially able to yield accurate and statistically significant information on the isotopic composition of abundant elements in high energy cosmic rays from Be to

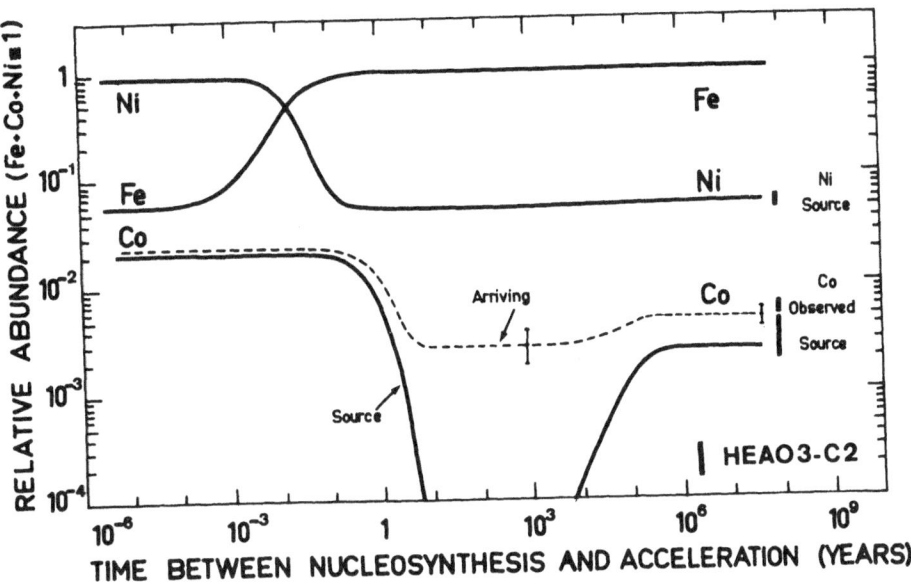

**Fig.15.** Curves : expected relative abundance of Fe, Co, Ni at the galactic cosmic ray source vs. time δt elapsed between e-process nucleosynthesis and GCR acceleration based on Soutoul et al. (1978). For Co we give also (dashed) the predicted abundances of Co at Earth ("arriving") with its uncertainty. Bars : observed and source abundances derived from HEAO 3-C2 data.

Ni. The problems encountered in the course of the data analysis are different in many respects from those of balloon experiments. A major feature is that, owing to the satellite orbit and spin, observations are made in all directions corresponding to all rigidity cut-offs from a few to several tens of GV, therefore an ordering of the data according to rigidity cut-off is mandatory. This is achieved with the concept of <u>transmission function</u> as explained in Byrnak et al. (1981) and Juliusson et al. (1981). The relevant parameter to describe the transmission function is the ratio of the measured momentum P to the rigidity cut-off $R_c$ associated with the arrival location and direction of the incoming particle, an ordering parameter which was first proposed by Hubert (1974). The transmission function is the point by point ratio of the filtered to the unfiltered $P/R_c$ distributions as shown in fig.16 and 17 for oxygen events near 3 GeV/n. The filtered distribution is readily measured with the instrument. The unfiltered distribution is obtained by selecting particles arriving at locations and from directions with sufficiently low cut-off rigidity, i.e. not affected by the geomagnetic field, and by combining this distribution with the relative time exposure at each rigidity cut-off which is obtained independently. The isotopic information is fully preserved in this method.

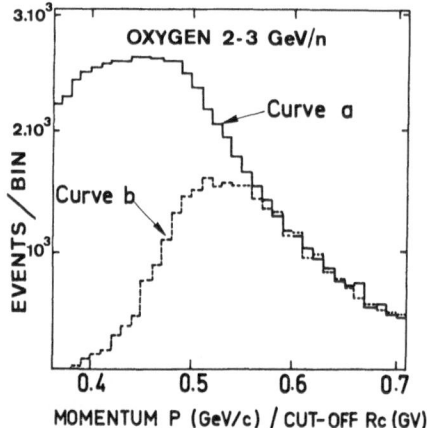

Fig.16. Filtering effect of the geomagnetic field on the distribution of ratios of momentum to rigidity cut-off for oxygen nuclei collected by HEAO 3-C2 in the energy range 2 to 3 GeV/n. Curve a : the unfiltered momentum to rigidity cut-off ratio distribution. Curve b : the filtered $P/R_c$ ratio distribution.

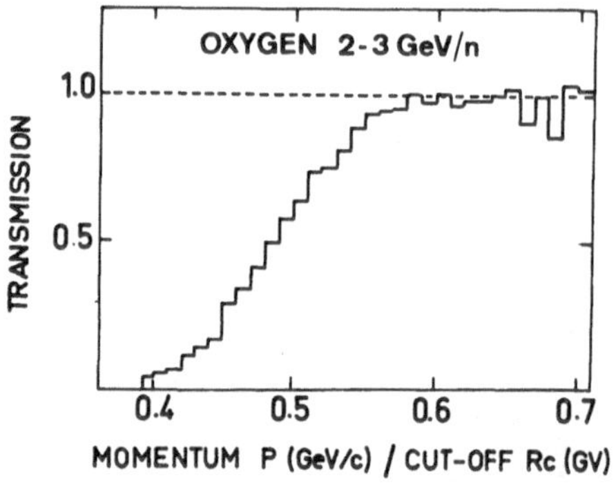

Fig.17. The transmission function with interpolated geomagnetic cut-offs for oxygen nuclei in the energy range 2 to 3 GeV/n. It is obtained taking the point to point ratio of curve ( b) to curve (a) from fig.16.

Furthermore spectral differences between the unfiltered momentum distribution of different elements and some possible systematic biases will cancel out so that systematic errors are minimized. Considering an element possessing 2 isotopes with mass $A_1$ and $A_2$ the observed transmission function will be a linear combination of the transmission functions corresponding to $A_1$ and $A_2$. They are identical in shape but usually of different strength and displaced with respect to each other on a log $P/R_c$ scale by an amount equal to log $A_1/A_2$.

Due to the presence of complex structures in the geomagnetic cut-off pattern (penumbra width) and to errors on momentum and on rigidity cut-off assignments the observed transmission functions are not sharp. Therefore individual isotopes are not seen at this stage of the analysis and only mean masses can be obtained, except for $^{22}$Ne which shows up when analysed with calculated geomagnetic cut-offs as seen in Byrnak et al. (1981) and Soutoul et al. (1981).

(i) Mean masses obtained with integral transmission functions using interpolated geomagnetic cut-offs (integral method)

In a first approach we used interpolated geomagnetic cut-offs deduced from a world-wide 5-dimensional grid and computed versus altitude, zenith, azimuth, latitude and longitude (Smart et al. 1979). In Juliusson et al. (1981) the mean masses of 8 elements are derived from integral transmission functions using a reference element with "known" isotopic composition, namely oxygen with < A > = 16.06 or < A/Z > = 2.008. This mean mass corresponds to cosmic ray source oxygen with solar system isotopic composition propagated at 3 GeV/n through $\Lambda_e = 15$ $R^{-0.5}$ g.cm$^{-2}$ of hydrogen. Since for a given (Z, A) isotope the transmission function is expected to extend down to $P/R_c = Z/A$, the mass scale is built by shifting the oxygen transmission function by a factor f = 2.008.Z/A. The observed integral transmission function for each element is then converted into a mean mass < A > relative to oxygen. The mean masses obtained for 8 elements from C to Fe at an energy of 3 GeV/n are displayed in the column 1 of Table 3. The error bars on the observed values are obtained from a convolution of the statistical error on the number of particles used in the integral transmission function computation with the sensitivity of the method. In column 2 of table 3 the mean masses obtained in the same energy range using a simplified momentum calculation with two reference elements O and Fe are shown. The mean masses obtained by both approaches agree quite well within the error bars.

TABLE 3. Observed mean masses of cosmic rays
at 3 GeV/n (HEAO 3-C2)

| Element | Column 1 Integral Method | Column 2 Integral Method | Column 3 Differential Method | Solar source propagated |
|---------|--------------------------|--------------------------|------------------------------|-------------------------|
| C  | 12. 08 $\pm$ 0. 03 | 12. 04 $\pm$ 0. 04 |              | 12. 06 $\pm$ 0. 02  |
| N  | 14. 50 $\pm$ 0. 05 | 14. 42 $\pm$ 0. 06 |              | 14. 41 $\pm$ 0. 04  |
| O  | = 16. 06           | = 16. 06           | = 16. 06     | .16. 06 $\pm$ 0. 02 |
| Ne | 20. 77 $\pm$ 0. 07 | 20. 76 $\pm$ 0. 07 | 20. 58 $\pm$ 0. 05 | 20. 48 $\pm$ 0. 08  |
| Mg | 24. 54 $\pm$ 0. 08 | 24. 46 $\pm$ 0. 07 | 24. 40 $\pm$ 0. 05 | 24. 42 $\pm$ 0. 04  |
| Si | 28. 38 $\pm$ 0. 08 | 28. 44 $\pm$ 0. 07 | 28. 26 $\pm$ 0. 05 | 28. 19 $\pm$ 0. 035 |
| S  | 32. 60 $\pm$ 0. 24 | 32. 69 $\pm$ 0. 20 | 32. 70 $\pm$ 0. 15 | 32. 53 $\pm$ 0. 16  |
| Ca | 42. 28 $\pm$ 0. 40 | 42. 12 $\pm$ 0. 30 | 41. 51 $\pm$ 0. 25 | 41. 68 $\pm$ 0. 22  |
| Fe | 55. 35 $\pm$ 0. 29 | = 55. 82           | 55. 45 $\pm$ 0. 13 | 55. 82 $\pm$ 0. 03  |

(ii) Mean masses obtained with differential transmission functions computed by trajectory tracing in the geomagnetic field (differential method)

Preliminary isotopic ratios were also obtained using the differential method which relies on the computation of the detailed penumbra pattern for each incoming particle as described by Byrnak et al. (1981). The geomagnetic cut-off is computed for each event via the trajectory tracing method developped by McCracken (1962) : for each arrival direction and geographical position a set of trajectories is computed in a model geomagnetic field for a sequence of equally spaced rigidities, each identified transition rigidity value from an allowed to a forbidden trajectory being calculated with higher precision through an iterative process. Moreover in

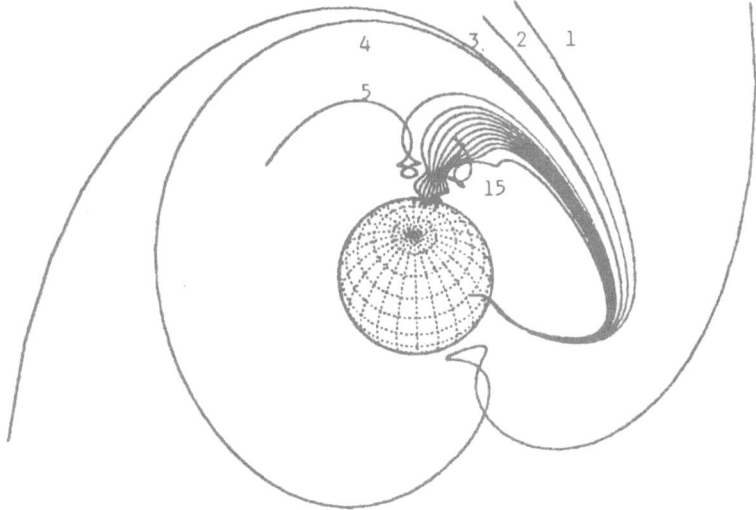

**Fig.18.** Investigation of the penumbra pattern in a particular direction of observation defined by the coordinates : 35°N, 315°E, altitude 500 km, zenith vertical. A sequence of trajectories 1 to 15 are calculated for assumed negatively charged particles travelling away form the observation point with a sequence of decreasing rigidity values. When the rigidity is very high (trajectories 1 to 3) the particles travel away from the Earth in a smooth curve without altitude oscillations ; at 4.11 GV (trajectory 4) one sees the onset of altitude oscillations ; between 4.01 and 3.92 GV (trajectories 6 to 14) the particles are no longer travelling away from the Earth but impinging the solid Earth, the so-called primary forbidden band occurs.

order to further clean up the data set only those directions with a peculiar cut-off feature, as illustrated in figure 18 are retained. This feature called primary forbidden band is defined by a sequence of trajectories with only one altitude oscillation corresponding to trajectories 6 to 14 on figure 18. The resulting transmission functions for oxygen, neon and iron are shown in figures 19 to 21.Comparing with figure 17 the sharpness of the transmission cut-off for oxygen has been much improved. One can see on fig.20 the excess due to $^{22}$Ne on the sharp edge of the transmission function of Neon. Using oxygen as reference element the preliminary isotopic results converted into mean mass values are shown in column 3 of Table 2 where they can be compared to the results of the integral method. When the residual charge dependent errors on our momentum assignments are corrected for, we should be able to use a least square technique to unfold the contribution of individual isotopes to the transmission function of the most abundant elements each isotope showing up in the "window" feature defined above and seen on figure 19.

Individual isotopes can also show up on the edge of the transmission function if sufficiently sharp and accurate. In Soutoul et al.(1981a) a detailed comparison of the computed and observed transmission functions is made using only those locations and directions with "sharp" cut-offs. It shows a systematic shift of about 3 percent between computed and observed momentum over cut-off scales of oxygen. Assuming a similar systematic shift in the calculated neon transmission function one obtains a preliminary estimate of the $^{22}$Ne/Ne ratio of 0.30 $\pm$ 0.05 (assuming $^{21}$Ne/Ne = 0.10). The corresponding mean mass of neon is : 20.70 $\pm$ 0.10. It is rather insensitive to the assumption on $^{21}$Ne abundance.

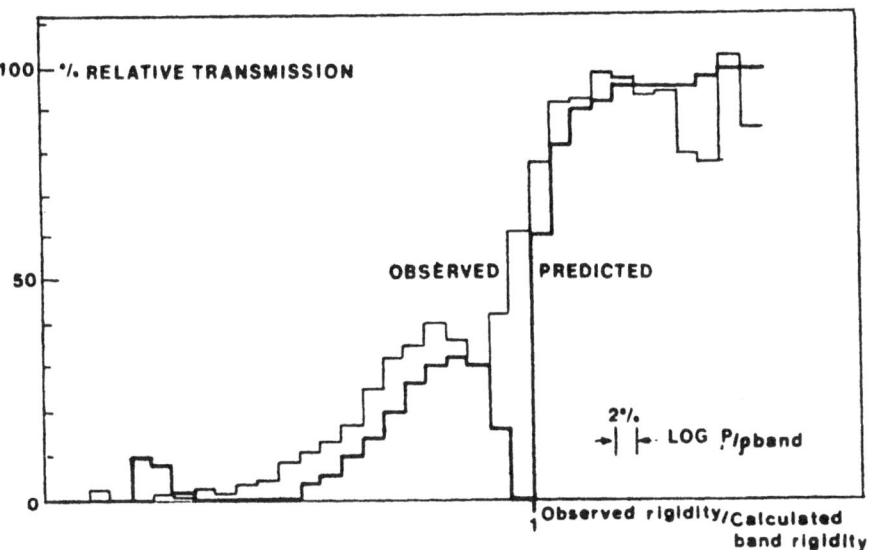

Fig.19. The transmission function for oxygen in a log $P/R_c$ scale (2% bins), in the energy range 2 to 2.5 GeV/n using calculated rigidity cut-offs with the full expansion of the MAGSAT field model and a selected set of trajectories (see text). The structure in the geomagnetic cut-off predicted by trajectory calculations can be seen also in the HEAO 3 data.

(iii) The mean mass of Neon, Magnesium and Silicon at 3 GeV/n near Earth.

Combining the results of the different approaches described above, we get the preliminary mean masses shown on the first column of Table 4. They are converted in the isotopic ratios : $^{22}$Ne/$^{20}$Ne (assuming $^{21}$Ne/$^{20}$Ne = 0.10), $^{26}$Mg/$^{24}$Mg, $^{30}$Si/$^{28}$S (assuming equal abundances of the 2 heavy isotopes of Mg and Si), in the column 2 of Table 4. They are compared with other isotopic ratio measurements on the same table.

Table 4 - Observed isotope ratios of Ne, Mg, Si in "arriving" cosmic rays

| Isotope Ratio | HEA03 -Preliminary 3 GeV/n (Mean) (mass) | Isotope ratio[1] | ISEE - 3 WG 100 - 300 MeV/n | ISEE -3 M 30 - 170 MeV/n | IMP7 G - M 80 - 250 MeV/n | Balloon W 600 - 800 MeV/n | Balloon F 390 - 530 MeV/n |
|---|---|---|---|---|---|---|---|
| $^{22}Ne/^{20}Ne$ | $(20.70_{+}.10)$ | $0.50_{+}.12$ | $0.67^{+.10}_{-.07}$ | $0.49^{+.39}_{-.11}$ | $0.54_{+}.07$ | $0.48^{+.11}_{-.07}$ | $0.49_{+}.16$ |
| $^{26}Mg/^{24}Mg$ | $(24.47_{+}.12)$ | $0.23_{+}.07$ | $0.30^{+.04}_{-.03}$ | $0.36^{+.20}_{-.07}$ | $0.22_{+}.07$ | $0.25_{+}.05$ | $0.34_{+}.10$ |
| $^{30}Si/^{28}Si$ | $(28.36_{+}.12)$ | $0.16_{+}.07$ | $0.084^{+.02}_{-.014}$ | $0.08^{+.10}_{-.03}$ | - | $0.06_{+}.02$ [1] | - |

(1) Isotope ratios : $^{22}Ne/^{20}Ne$ assuming $^{21}Ne/^{20}Ne = 0.10$ ; $^{26}Mg/^{24}Mg$ and $^{30}Si/^{28}Si$ assuming equal abundances of the 2 heavy isotopes.
WG : Wiedenbeek and Greiner (1981) ; M:Mewalt et al. (1980, 1981) ; G-M : Garcia-Munoz et al. (1979) ; W : Webber (1981) ; F : Freir et al. (1980).

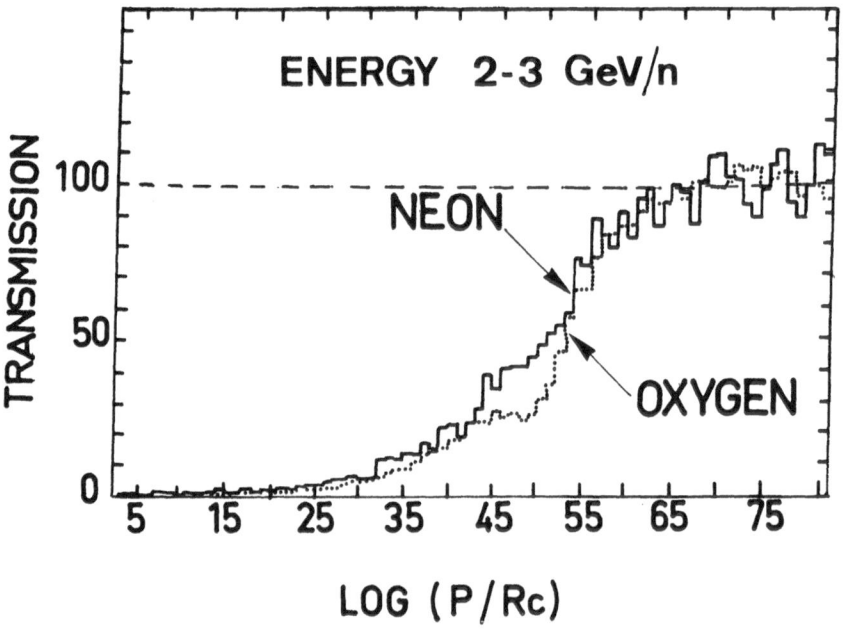

Fig.20. The transmision functions of Oxygen (dotted curves) and Neon (full curve) in a log $P/R_c$ scale (1% bins) with calculated cut-offs using a reduced expansion of the MAGSAT field model and the same selection than in fig.19. The structure in the Oxygen transmission function has been damped out due to the lower accuracy of the calculated cut-offs. The channel 54 on the $P/R_c$ scale corresponds to $P/R_c = 0.5$.

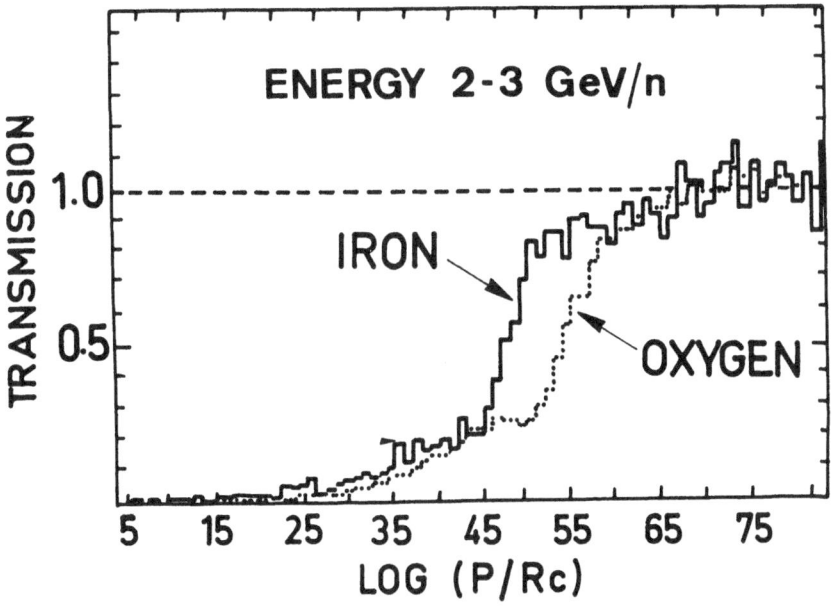

**Fig.21.** Same as figure 20 for Oxygen and Iron.

(iv) <u>Isotopic abundances at the source as deduced from the preliminary mean</u>
<u>masses observed by HEAO 3-C2</u>

Using the propagation program described in Perron et al. (1981) we have computed
the mean masses of the arriving Ne, Mg and Si at 3 GeV/n as a function of the
enhancement factor at the source of the heavier isotopes relative to the lighter
ones as compared to the same ratio in local galactic matter i.e. :

$$F_{Ne} = \frac{[\,(^{22}Ne + ^{21}Ne)/^{20}Ne]\,CRS}{[\,(^{22}Ne + ^{21}Ne)/^{20}Ne]\,LG} \qquad\qquad F_{Si} = \frac{[\,(^{29}Si + ^{30}Si)/^{28}Si]\,CRS}{[\,(^{29}Si + ^{30}Si)/^{28}Si\,]\,LG}$$

$$F_{Mg} = \frac{[\,(^{25}Mg + ^{26}Mg)/^{24}Mg]\,CRS}{[\,(^{25}Mg + ^{26}Mg)/^{24}Mg]\,LG}$$

Figure 22 shows the mean masses computed at 3 GeV/n , near Earth as a
function of the isotopic enhancement factors F at the source. The mean masses
corresponding to F = 1 are shown in column 4 of Table 3 for 9 elements. The area
defined by the dotted curves of figure 22 corresponds to the propagation errors,
mainly due to nuclear cross-section uncertainties. The relative weight of the
observational and propagation errors on the determination of F is shown on figure
22 around the central value of < A > given by the integral method (column 1 of
Table 3). One sees that even with a perfect mean mass measurement the

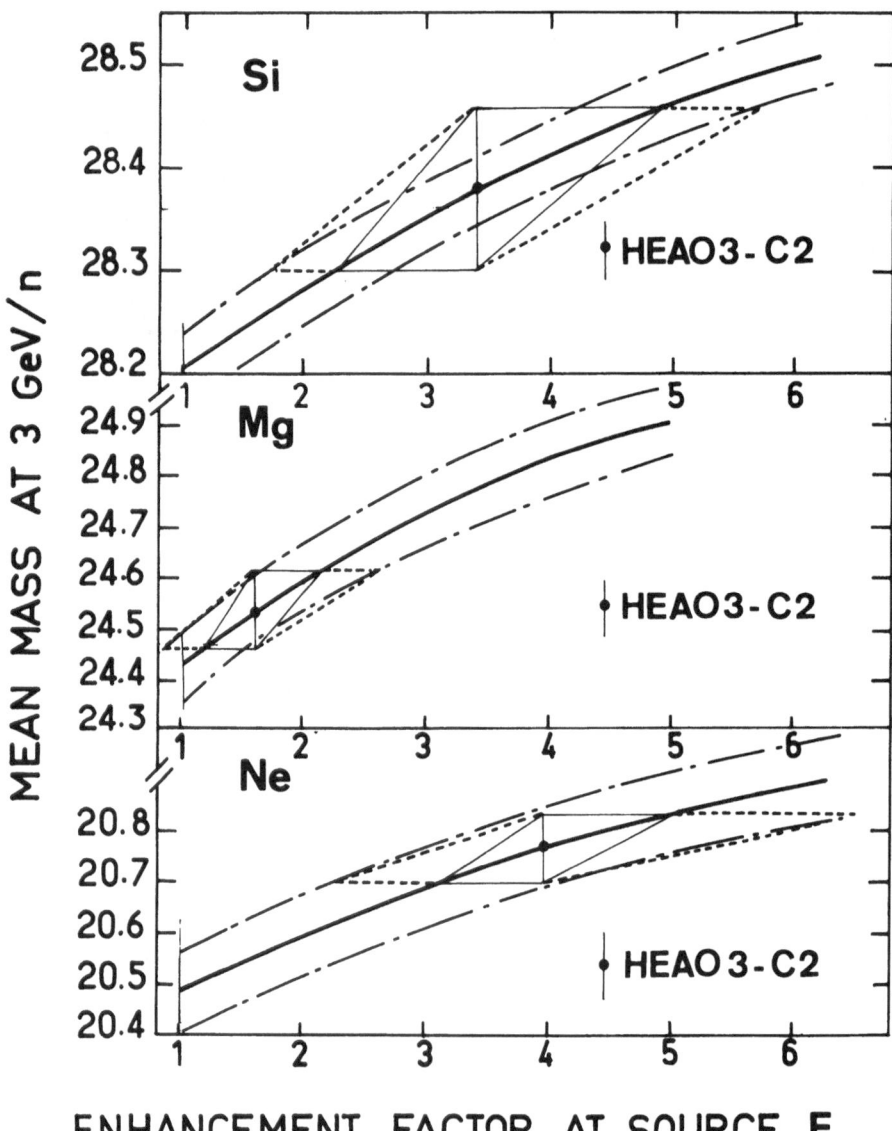

**Fig.22.** Calculated and observed (HEAO 3-C2 integral method) mean masses of Ne, Mg, Si at 3 GeV/n as a function of cosmic ray source isotopic enhancement factors F for the isotope ratios $^{21,22}Ne/^{20}Ne$, $^{25,26}Mg/^{24}Mg$ and $^{29,30}Si/^{28}Si$. The dotted curves show the range of uncertainty due to nuclear cross-sections error (at 1 σ for Ne and Si and 2 σ for Mg).

uncertainty on the enhancement factor at the source due to propagation errors would still be of 1 unit for Ne and 0.5 unit for Mg and Si. Our preliminary isotopic enhancement factor at source as deduced from our combined mean mass measurements at 3 GeV/n are shown in first column of Table 5 ; the first error bar including only the observational errors, the error bars in brackets including also the propagation errors.

Table 5 - Isotope enhancement factors F at the cosmic ray source

| Isotope Ratio | HEAO3 -C2 Preliminary 3 GeV/n | ISEE 3 WG 100 - 300 MeV/n | ISEE M 30 - 170 MeV/n | IMP7 GM 80 - 250 MeV/n | Balloon W 600 - 800 MeV/n | Balloon F 390 - 530 MeV/n |
|---|---|---|---|---|---|---|
| $^{21,22}$Ne/$^{20}$Ne | $3.1^{+2.2(+2.7)}_{-1.1(-1.8)}$ | $4.1^{+0.9(+1.0)}_{-0.6(-0.7)}$ | $2.8^{+3.0}_{-0.8}$ | $3.1{+}0.05$ | $3.4{+}.9$ | $3.5{+}.9$ |
| $^{25,26}$Mg/$^{24}$Mg | $1.2^{+0.7(+1.0)}_{-0.6(-0.8)}$ | $1.6{+}0.2(0.3)$ | $2.0^{+1.2}_{-0.7}$ | $1.1{+}0.05$ | $1.27{+}.25$ | $2.0{+}.4$ |
| $^{29,30}$Si/$^{28}$Si | $3.1^{+2.2(+3.1)}_{-1.7(-2.1)}$ | $1.6{+}0.2(0.3)$ | $1.4^{+3.6}_{-0.8}$ | - | $0.9 {+} .6$[1] | - |

( 1 ); WG, M, GM, W, F : See note of Table 4.

These results must be considered as preliminary due to the residual systematic errors mentioned above. They are compared to results obtained at lower energy by several authors on the same Table.

## 12. Search for cosmic ray antiparticles in the HEAO 3-C2 data

In Lund et al.(1981b) it is suggested that the ability of our HEAO 3 experiment to determine the arrival direction of the individual cosmic ray nuclei allows the search for events arriving from directions for which the Earth and its magnetic field prevents particles with a positive nuclear charge to come, whereas negatively charged particles -antinuclei- are allowed. Preliminary investigations are underway and the first conclusions are the following :

(i) The error rate in the arrival direction determination made by our HEAO 3 time of flight system is lower than $10^{-4}$ for $Z \geq 14$ nuclei going in the "down" direction in the instrument (Petrou et al. 1981).

(ii) Lund et al. (1981) analysed only $Z \geq 14$ "good" events and made the trajectory tracing in the MAGSAT geomagnetic field model for all possible isotopes with lifetimes greater than 1 s (possibly created in the atmosphere) ; the range of rigidities corresponding to the momentum uncertainty determination is scanned in 1 percent steps and only trajectories impinging the solid Earth after less than an Earth radii trajectory length are considered in order to avoid ambiguities due to trajectory oscillations for large distances. They found 32000 positively charged nuclei candidates and 94 antimatter candidates corresponding to an upper limit of $\sim 3.10^{-3}$ for antimatter to matter ratio of $Z \geq 14$ nuclei in the 1 to 10 GeV/n energy range. With subsequent analysis we expect to obtain a much lower limit.

## 13. Conclusion

Only a small part of the astrophysical information contained in the $10^7$ good events gathered by our experiment during the HEAO 3 lifetime has been extracted so far. Very accurate measurement of the relative abundances of 27 elements between Be and Zn as a function of energy in the range 0.7 to 20 GeV/n have allowed us to study the propagation of cosmic rays in the framework of galactic diffusion models, either homogeneous or with a halo. The radioactive cosmic ray clock $^{54}$Mn has been used to derive the galactic halo size as seen by cosmic rays in the GeV/n range : if the half-life of $^{54}$Mn is $\sim 10^6$ yr the HEAO 3-C2 data tend to favour a rather flat halo only $\sim 10$ times thicker than the galactic disk. We are presently working on the measurement of the $^{54}$Mn beta decay half-life in the laboratory in order to be able to infer from our data a cosmic ray lifetime at high energy. The abundances of 16 elements have been derived at the cosmic ray source with an accuracy only limited by our knowledge on formation and destruction cross sections of cosmic ray nuclei in the interstellar medium. There are differences between the source composition and the local galactic and solar system composition. However the overabundances in the cosmic ray source are correlated with the first ionization potential of the elements. It is remarkable that the same correlation holds for the solar nuclei accelerated during flares. All nuclei between C and Ni in cosmic ray sources and in solar energetic particles have identical overabundances, with the remarkable exception of C which is two times more abundant in cosmic ray sources.

Our data are compatible with the suggestion of an injection of cosmic ray particles by stellar flares in a two-stage acceleration process. The acceleration to high energies has to be prompt, i.e. must take place on a timescale short with respect to the cosmic ray mean escape time from the galaxy. The overabundance of C mentioned above may be related to the anomalous neon isotopic composition of cosmic rays already known at low energy and seen also at 3 GeV/n in our preliminary results, if a specific minor component of cosmic rays is accelerated from material having recently undergone a nucleosynthetic process, such as quiescent helium burning in massive stars. We find that the time elapsed between nucleosynthesis of Fe, Co and Ni nuclei and their acceleration at a few GeV/n is likely larger than $\sim 10^5$ yr ; this presumably excludes supernova explosions as <u>direct</u> accelerators of these nuclei.

Our isotope analysis has just begun and we have derived preliminary mean masses for 8 elements. The separation of the heaviest neon isotope, $^{22}$Ne, has been achieved at 3 GeV/n : one finds that $(30 \pm 5)\%$ of arriving cosmic ray neon is $^{22}$Ne. This preliminary result corresponds to a ratio $^{22}$Ne/$^{20}$Ne $\sim$ 3 times larger at the cosmic ray source than in meteoritic Neon A or solar energetic particles. When compared to the results obtained by several groups at lower energy, it shows no evidence of an energy dependence of the enrichment in $^{22}$Ne at the cosmic ray source (Mewalt 1981). A very precise knowledge of our instrument in flight has been obtained during this first phase of our analysis and we have found several residual systematic errors that our on-going work will hopefully correct. We have also realized that the present knowledge of the geomagnetic field is far better than we had expected : the uncertainty on the particle rigidity cut-off assignment is at most 3% and may be further improved. Since this isotopic separation at high energy relies upon a detailed analysis of the geomagnetic penumbra pattern for each event, it implies a very lengthy numerical simulation, and until now our computing facilities were the limiting factor. Their expected increase by a factor 30 in 1982 will certainly help us in reaching the intimate knowledge of the composition of cosmic rays at their source, at a level which will allow significant advances in the questions of origin and propagation of cosmic rays in the galaxy.

The progress recently achieved in cosmic ray astrophysics both in observations -thanks partly to HEAO 3- and in theory has opened a new exciting period of cosmic ray research. It is a great encouragement to cosmic ray physicists to prepare a new generation of large instruments to be hopefully installed for years on space platforms.

Acknowledgements. The HEAO 3-C2 experiment team would like to thank the NASA personnel at Headquarters, Marshall and Goddard Space Flight Centers, the TRW Systems Group, the Bevatron engineers and physicists. Our experiment depended on the excellent work of the entire team of the GERES Group at Saclay, of the Centre Spatial de Toulouse (CNES), and of the support personnel at Saclay and Copenhagen. We are very grateful to all these people. I am personally endebted to all my colleagues for enlightening discussions when preparing this presentation of our results. This work was supported in part by CNES contracts and by NASA with regard to the entire mission as well as Bevatron calibrations and balloon flights.

REFERENCES

Audouze, J., Cassé, M., Chièze J.P., Malinie, G., Paul, J.A., 1981, 17th ICRC, Paris, 2, 296.

Axford, W.I., 1981, 17th ICRC, Paris, 12, 155.

Balasubrahmanyan, V.K., Damle, S.V., Gokhale, G.S., Menon, M.G.K., Roy, S.K., 1963, 8th ICRC, Jaipur, 3, 110.

Bibring, J.P., Cesarsky, C.J., 1981, 17th ICRC, Paris, 2, 289.

Blandford, R.D., Ostriker, J.P., 1980, Ap.J. 237, 793.

Bouffard, M., Engelmann, J.J., Koch-Miramond, L., Lund, N., Rasmussen, I.L., Peters, B., Soutoul, A., and the Saclay-Copenhagen Collaboration, 1981, preprint.

Bradt, H.L., Peters, B., 1950, Phys. Rev. 80, 943.

Byrnak, B., Lund, N., Rasmussen, I.L., Peters, B., Risbo, T., Rotenberg, M., Westergaard, N.J., Petrou, N., 1981, 17th ICRC, Paris, 2, 8.

Caldwell, J.H., 1977, Ap.J. 218, 269.

Cantin, M., Goret, P., Jorrand, J., Jouan, R., Juliusson, E., Koch, L., Maubras,Y., Mestreau, P., Petrou, N., Rio, Y., Soutoul, A., Cawood, P., Linney, A., 1975, 14th ICRC, Munchen, 9, 3211.

Cantin, M., Cassé, M., Koch, L,, Jouan, R., Mestreau, P., Roussel, D., 1974, Nucl. Instrum. and Meth. 118, 177.

Cantin, M., Engelmann, J.J., Masse, P., Rotenberg, M., 1981, 17th ICRC, Paris, 8, 59.

Cassé, M., 1973, Ap.J. 180, 623.

Cassé, M., Goret, P., 1978, Ap.J. 221, 703.

Cassé, M., Paul, J., 1981, 17th ICRC, Paris, 2, 281.

Cassé, M., 1981, 17th ICRC, Paris, 13, 111.

Cesarsky, C.J., Bibring, J.P., 1980, IAU Symp. n°94, Bologna, 361.

Cesarsky, C.J., Koch-Miramond, L., Perron, C., 1981,17th ICRC, Paris, 2,22.

Corydon-Petersen, O., Dayton, B., Lund, N., Melgaard, K., Omoe, K., Peters, B., Risbo, T., 1970, Nucl. Instrum. and Meth. 81, 1.

Cowsik, R., Yash Pal, Tandon, S.N., Verma, R.P., 1967, Phys. Rev. 140, B1147.

Davis, Jr. L., 1959, 6th Int. Cosmic Ray Conf., Moscow, 3, 220.

Dwyer, R., Meyer, P., 1975, Phys. Rev. Lett. 35, 601.

Dwyer, R., 1978, Ap.J. 224, 691.

Dwyer, R., Meyer, P., 1979, 16th ICRC, Kyoto, 1, 461.

Eichler, D., 1980, Ap.J. 237, 809.

Engelmann, J.J., Cantin, M., 1978, Journ. Phys. 39, C357.

Engelmann, J.J., Goret, P., Juliusson, E., Koch-Miramond, L., Masse, P., Petrou, N., Rio, Y., Soutoul, A., Byrnak, B., Jakobsen H., Lund, N., Peters, B., Rasmussen, I.L., Rotenberg, M., Westergaard, N., 1981, 17th ICRC, Paris, 9, 97.

Fisher, A.J., Hagen, F.A., Maehl, R.C., Ormes, J.F., Arens, J.F., 1976, Ap.J. 205, 938.

Fransson, C., Eostein, P.I., 1980, Ap.J. 242, 411.

Garcia-Munoz, M., Margolis, S.H., Simpson, J.A., Wefel, J.P., 1979, 16th ICRC, Kyoto, 1, 310.

Garcia-Munoz, M., Simpson, J.A., Wefel, J.P., 1981, 17th ICRC, Paris, 2, 72.

Garcia-Munoz, M., Simpson, J.A., Wefel, J.P., 1979, Ap.J. Letters, 232, L95.

Ginzburg, V.L., Khazan, Y.M., Ptuskin, V.S., 1980, Ap. Sp. Sci., 68, 295.

Gleeson, L.J., Axford, W.I., 1968, Ap.J. 154, 1011.

Gordon, H.A., Burton, N.B., 1976, Ap.J. 208, 346.

Gorenstein, P., 1981, 17th Int. Cosmic Ray Conf., Paris, 12, 99.

Goret, P., Engelmann, J.J., Koch-Miramond, L., Meyer, J.P., Lund, N., Rasmussen, I.L., Perron, C., 1981, 17th ICRC, Paris, 9, 122.

Guzik, T.G., 1981, Ap.J. 224, 695.

Hainebach, K.L., Clayton, D.D.,Arnett, W.D., Woosley, S.E., 1974, Ap.J. 193, 157.

Hubert, P., 1974, Nucl. Inst. and Meth., 119, 183.

Israel, M.H., Klarmann, J., Binns, W.R., Fickle, R.K., Waddington, C.J., Garrard, T.L., Stone, E.C., 1981, 17th ICRC, Paris, 2, 36.

Juliusson, E., 1974, Ap.J. 191, 331.

Juliusson, E., Engelmann, J.J., Goret, P., Koch-Miramond, L., Masse, P., Petrou, N., Rio, Y., Soutoul, A., Byrnak, B., Lund, N., Peters, B., Rasmussen, I.L., Rotenberg, M., Westergaard, N., 1981, 17th ICRC, Paris, 9, 101.

Koch-Miramond, L., Engelmann,J.J., Goret, P., Juliusson, E., Petrou, N., Rio, Y., Soutoul, A., Byrnak, B., Lund, N., Peters, B., Rasmussen, I.L., Rotenberg, M., Westergaard, N., 1981a, Astron. and Astrophys. 102, L9.

Koch-Miramond, L., Perron, C., Goret, P., Cesarsky, C.J., Juliusson, E., Soutoul, A., Rasmussen, I.L., 1981b, 17th ICRC, Paris, 2, 18.

Kristiansson, K., 1971, Ap. Space Sci. 14, 485.

Lezniak, J.A., Webber, W.R., 1978, Ap.J. 223, 676.

Lindstrom, P.J., Greiner, D.E., Heckman, N.H., Cork, B., 1975, preprint LBL 3650.

Linney, A., Peters, B., 1972, Nucl.Instrum. and Meth. 100, 545.

Lund, N., Rasmussen, I.L., Peters, B., Westergaard, N.J., 1975, 14th ICRC, Munchen, 1, 2527.

Lund, N., Peters, B., Cowsik, R., Pal, Y., 1970, Phys. Lett. 318, 553.

Lund, N., Sorgen, A., 1977, 15th ICRC, Plovdiv, 11, 292.

Lund, N., Rasmussen, I.L., Rotenberg, M., Masse, P., Engelmann, J.J., Jorrand, J., Petrou, N., 1981a, 17th ICRC, Paris, 8, 67.

Lund, N., Risbo, T., Petrou, N., 1981b, 17th ICRC, Paris, 2, 12.

Lund, N., Westergaard, N.J., Engelmann, J.J., Goret, P., Juliusson, E., 1981c, 17th ICRC, Paris, 8, 63.

Mac Cracken, K.G., Rao, U.R., Shea, M.A., 1962, MIT Technical report 77 NYO 2670.

Mewalt, R.A., Spalding, J.D., Stone, E.C., Vogt, R.E., 1980, Ap.J. (Letters) 235, L95.

Mewalt, R.A., Spalding, J.D., Stone, E.C., Vogt, R.E., 1981, 17th ICRC, Paris, 2, 68.

Meyer, J.P., 1979a, 22nd Liège Int. Ap. Symp. pp.153, 465, 477, 489 ; 1979b, 16th ICRC, Kyoto, 2, 115.

Meyer, J.P., 1981a, 17th ICRC, Paris, 3, 145 ; 1981b, ib., 3, 149 ; 1981c, ib., 2, 265.

Montmerle, T., Koch-Miramond, L., Falgarone, E., Grindlay, J., 1982,in preparation.

Montmerle, T., Koch-Miramond, L., Grindlay, J., 1981, 17th ICRC, Paris, 1,166.

Paul, J., 1981, 17th ICRC, Paris, 12, 79.

Perron, C., 1975, Thesis Univ. of Paris.

Perron, C., and Koch-Miramond, L., 1981, 17th ICRC, Paris, 2, 27.

Perron, C., Engelmann, J.J., Juliusson, E., Koch-Miramond, L., Meyer, J.P., Soutoul, A., Lund, N., Rasmussen, I.L., Westergaard, N., 1981, 17th ICRC, Paris, 9, 118.

Peters, B., 1974, Nucl. Instrum. and Meth. 121, 205.

Petrou, N., Jorrand, J., Westergaard, N., 1981, 17th ICRC, Paris, 4, 217.

Prishchep, V.L.,Ptuskin, V.S., 1975, Ap. Sp. Sci. 32, 265.

Protheroe, R.J., Ormes, J.F., Comstock, G.M., 1981, Ap.J. 247, 362.

Raisbeck, G.M.,Comstock, G., Perron, C., Yiou, F., 1975, 14th ICRC, Paris, 2, 560.

Rotenberg, M., Rasmussen, I.L., Engelmann, J.J., Masse, P., Rio, Y., 1981, 17th ICRC, Paris, 8, 112.

Silberberg, R.,Tsao, C.H., 1977, Ap.J. suppl. 35, 137.

Simon, M., Spiegelhauer, H., Schmidt, W.K.H., Stohan, F., Ormes, J.F., Balasubrahmanyan, V.K., Arens, J.F., 1980, Ap.J., 239, 712.

Smart, D.F., Shea, M.A., Lund, N., Rasmussen, I.L., Byrnak, B., Westergaard, N.J., 1979, 16th ICRC, Kyoto, 12, 237.

Soutoul, A., Cassé, M., Juliusson, E.,1978, Ap.J. 219, 753.

Soutoul, A., Engelmann, J.J., Goret, P., Juliusson, E., Koch-Miramond, L., Masse, P., Petrou, N., Rio, Y., Risbo, T., 1981, 17th ICRC, Paris, 9, 105.

Tueller, J., Love, P.L.,Israel, M.H., Klarmann, J., 1979, Ap.J., 228, 582.

Webber, W.R., 1981, 17th ICRC, Paris, 2, 261 and preprint.

Wiedenbeck, M.E., Greiner, D.E., Bieser, F.S., Crawford, H.J., Heckman, H.H., Lindstrom, P.J., 1979, 16th ICRC, Kyoto, 1, 412.

Wiedenbeck, M.E., Greiner, D.E., 1980, Ap.J. 239, L139.

Wiedenbeck, M.E., Greiner, D.E., 1981, Phys. Rev. Letters 46, 682.

ULTRAHEAVY COSMIC RAYS — HEAO-3 RESULTS

Martin H. Israel
Department of Physics and the
McDonnell Center for the Space Sciences
Washington University
St. Louis, Missouri 63130, U.S.A.

I have been invited to report on results from the experiment on HEAO-3 which was designed to measure the elemental composition of the Ultraheavy (UH) cosmic rays, i.e., nuclei with atomic number (or nuclear charge, Z) greater than 30. Before I do that, I want to take a few minutes to remind you why we designed this experiment.

## Early History

At the 1965 International Cosmic Ray Conference in London, Bob Walker presented a paper [1] in which he described the first evidence for the existence of cosmic rays significantly heavier than iron. He and his colleagues had been studying tracks in meteorites due to radiation damage from heavily ionizing particles. They found many tracks which could convincingly be attributed to cosmic-ray iron nuclei near the end of their range, where they are most heavily ionizing; and they found a few tracks which were significantly longer, attributed to higher-charge nuclei whose ionization rate is above the track threshold over a longer distance. They estimated a flux of these heavier nuclei about $2 \times 10^{-4}$ of Fe, which we know today to be the correct value for $Z \gtrsim 34$.

By the next International Cosmic Ray Conference, at Calgary in 1967, Peter Fowler and his co-workers [2] had flown a 4.5 m$^2$ stack of nuclear emulsions on a 14.5-hour stratospheric balloon flight and found 9 tracks attributed to nuclei of $Z \gtrsim 40$, with the charge estimate based on the grain density along the track relative to that along an iron track. The overall abundance of these nine nuclei relative to iron, $\sim 5 \times 10^{-5}$, was about the same as in the condensed bodies of the solar-system. But the striking feature was that two of the nine events were attributed to nuclei with charge near 90, which implied that actinide abundance in the cosmic rays was much higher than in the solar system, by perhaps 2 orders of magnitude. As I shall report later, we now know that the actinide abundance in the cosmic rays is rather similar to the solar system and nowhere nearly as enriched as those first 2 events would indicate.

Regardless of the details of those early results, they established unequivocally that the cosmic rays include nuclei substantially heavier than iron, although those elements are extremely rare in the cosmic rays just as they are rare in the bodies of the solar system. It was apparent that measurements of the composition of these UH cosmic rays promised to provide important clues to the nature of the cosmic-ray source.

## Significance of Ultraheavy Cosmic Rays

The nucleosynthesis of elements beyond atomic number 30 is understood to occur by neutron capture. For most of these heavy nuclides,
the observed abundances in the solar system are generally understood as
a combination of two neutron-capture processes, the slow s-process and
the rapid r-process. The two processes produce the various elements
with characteristic relative-abundance features [3]. For example, the
r-process produces very little $_{38}$Sr compared to $_{34}$Se and $_{36}$Kr, while
the s-process produces a lot of Sr. Similarly in the r-process very
little $_{56}$Ba is produced compared with relatively abundant $_{52}$Te and
$_{56}$Xe, while in the s-process Ba is much more abundant. Likewise, $_{78}$Pt
is mainly produced by r-process while $_{82}$Pb by s-process.

The r-process has conventionally been attributed to supernova
explosions. Since supernovae have long been thought to be the source
of cosmic rays, the hope after discovery of these ultraheavy cosmic
rays was that they would show the elemental abundance pattern characteristic of the r-process and thus confirm the supernova origin of
cosmic rays. The two actinides which Fowler identified in his first
flight whetted everyone's appetite, since actinides are only made in
the r-process.

Of course life is not really that simple. Recent models of cosmic-
ray acceleration, by supernova shocks in the interstellar medium, derive
the cosmic-ray energy from supernovae, but the material which gets
accelerated does not necessarily originate inside the supernova. And
the HEAO data, which I shall present shortly, show some interesting
differences from the solar-system abundances, but do not give any clear
evidence of an r-process enhancement.

In addition to their interest as probes of the cosmic ray source,
these ultraheavy nuclei are probes of the galactic confinement of cosmic
rays. Nuclei with atomic number around 50 have a mean-free-path against
nuclear fragmentation in the interstellar medium of less than 2 g/cm$^2$
and by atomic number 90 the mean free path is only 1 g/cm$^2$. Abundances
of secondary nuclei resulting from fragmentation of lighter cosmic rays
are understood by an exponential distribution of galactic path lengths
with mean around 6 g/cm$^2$; however; the shape of the distribution at
short path lengths has been debated. Secondary fragments of the ultra-
heavies will principally reflect the distribution of short path lengths

## More Background

It was thus apparent that these discoveries of ultraheavy cosmic
rays deserved to be followed by detailed studies. It was also apparent
that these studies were difficult. Detectors of very large area were
necessary to collect significant numbers of these rare nuclei. Table 1
gives approximate event rates in a 5 m$^2$sr detector at 5 GV geomagnetic
cutoff for nuclei of various atomic numbers [4].

Table 1

|  |  |
|---|---|
| Z = 26 | 1/sec |
| Z = 30 | 2/hour |
| $32 \leq Z < 40$ | 1/hour |
| $40 \leq Z < 60$ | 5/day |
| $60 \leq Z \leq 90$ | 2/day |
| $Z \geq 90$ | 2/year |

In the late 1960's and early 1970's experimenters from Bristol, Washington University, University of California (Berkeley), General Electric, Dublin, and Johnson Space Center accumulated an exposure of about 1.5 $m^2$ster years on a series of balloon flights using very large areas of nuclear emulsion and/or plastic track detectors, seeing a total of about 120 events attributed to nuclei with atomic number greater than 70. Then in 1973 the Skylab provided an eight-month exposure of 1.2 $m^2$ of plastic track detector. The Skylab results [5], like the balloon data before, showed significant actinide fluxes and suggested more Pt than Pb, taken as further evidence of significant r-process enhancement in the cosmic rays.

The emulsion and plastic-track detectors succeeded in delineating the general fluxes of UH cosmic rays, but they did not have the charge resolution required for a definitive measurement. Good resolution is required in the sense that individual elements must be identified. Abundance peaks of the r-process and of the s-process are two to four charge units apart. For example, a signature of the r-process would be a lack of $_{56}$Ba compared to $_{52}$Te and $_{54}$Xe, but individual element resolution is required to make the distinction. Similarly, individual element resolution is required if we are to look for other features of the cosmic-ray source, such as effects of ionization potentials of elements.

Good resolution is essential in another sense. When relative abundances change by orders of magnitude over a few charge units, it is not sufficient to have an instrument whose resolution is characterized by a narrow full-width-at-half-maximum. It is also essential that there be a very small probability of mis-identifying an event by several charge units. The cosmic-ray abundances, like the solar-system abundances, fall by a factor of $10^4$ from $_{26}$Fe to $_{32}$Ge. Obviously we cannot measure the abundance of Ge if there is even $10^{-4}$ probability of an Fe being mis-identified by six charge units. Similarly, if the abundance of $_{90}$Th and $_{92}$U in the cosmic rays is about 1% relative to $_{78}$Pt and $_{82}$Pb, as it is in the solar system, then an experiment will give the completely wrong picture if it has a several percent chance of misidentifying a Pb nucleus as Th.

## The HEAO-3 Heavy Nuclei Experiment

On 20 September 1979 the third High Energy Astronomy Observatory (HEAO-3) was launched, carrying our Heavy Nuclei Experiment and the other two instruments which have been described earlier in this HEAO-3 Symposium. Our experiment is a collaboration involving several scientists

from various institutions in the U.S. [6]. The instrument (Figure 1)
is composed of six independently analysed dual-gap ionization chambers,
each of which measures dE/dx, the rate of energy loss of the cosmic ray;
and a Cerenkov counter in which eight independently analysed photo-
multipliers view two sheets of Pilot 425 plastic in a white box [7].
The trajectory of each cosmic-ray nucleus is given by the multi-wire
ionization hodoscopes. Both the ionization chambers and the Cerenkov
counter give signals which to first order vary with the square of the
nuclear charge, but are different functions of velocity.

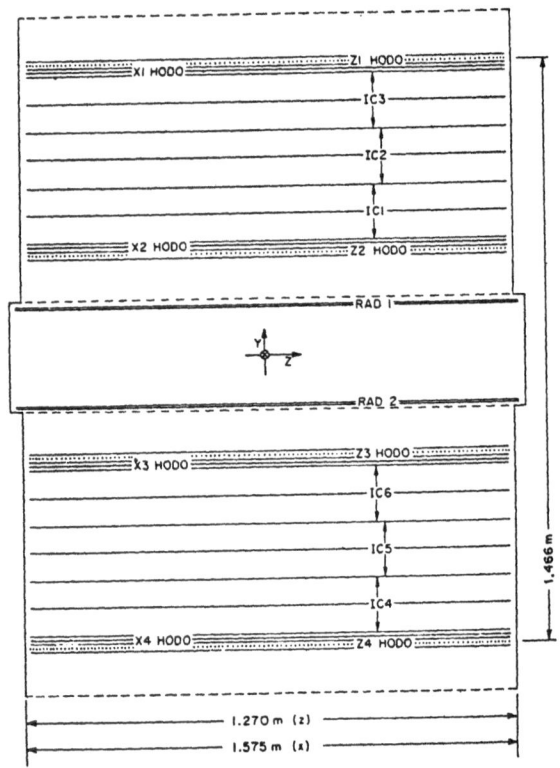

Figure 1

Figure 2 displays well-defined peaks at even charges in the region
from Fe to Ge where the abundance falls by four orders of magnitude, and
Figure 3 shows that we have resolved peaks at even charges and can
determine abundances for each even-charged element at least to $_{42}$Mo. In
particular note the prominent peak at Sr. This is the first experiment
ever to combine adequate size to give good statistics and adequate
resolution to provide well defined peaks above charge 30 [8].

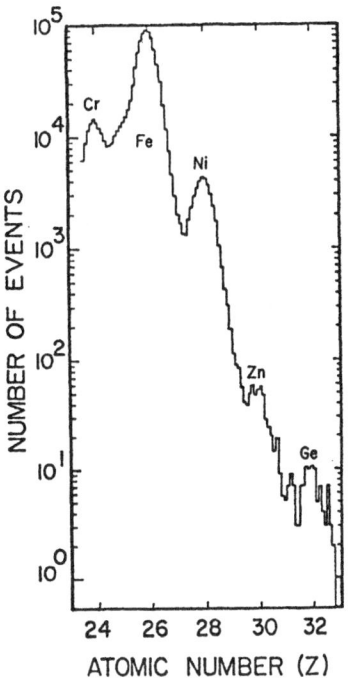

Figure 2

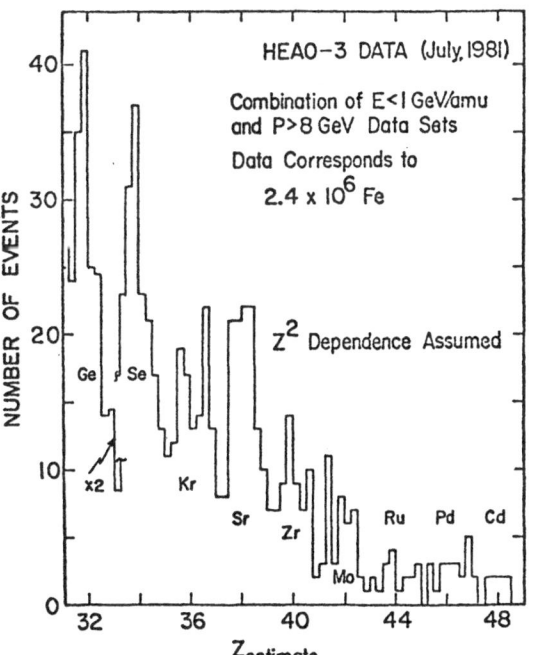

Figure 3

Figure 4 shows as open circles the observed abundances inferred from the data of Figures 2 and 3. In this particular charge interval the generally falling abundance with increasing charge means that at each even-charge element the nuclei we see near earth are mainly surviving primaries with relatively small contributions of secondary fragments from interstellar nuclear interactions of heavier elements. As a result the source abundance can be inferred from our observations and the result is relatively insensitive to the propagation model. The squares give these inferred abundances at the cosmic-ray source.

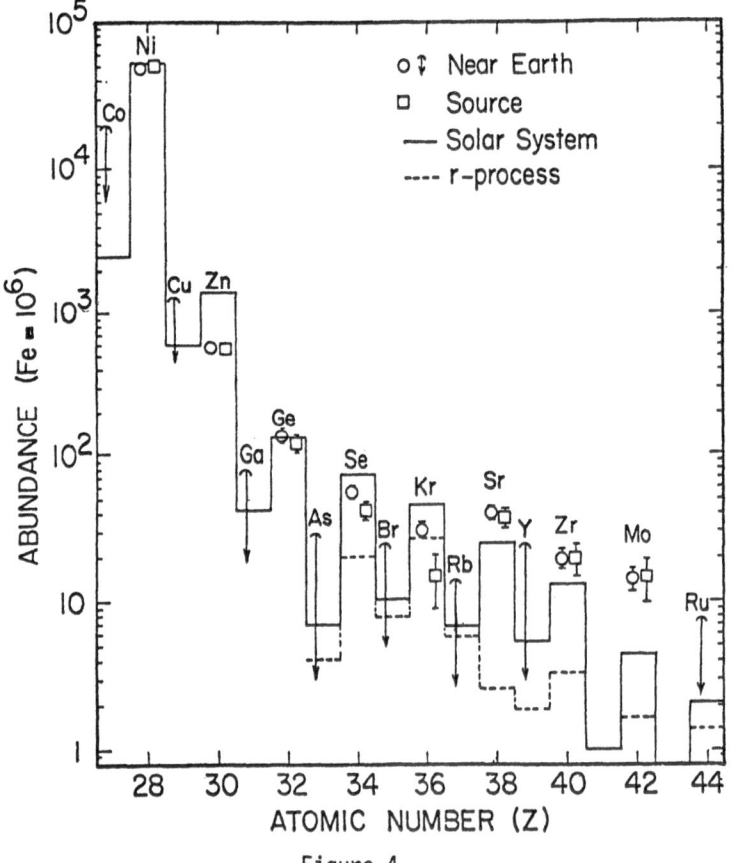

Figure 4

The source abundances are generally quite similar to those of the solar system. And the prominent peak at Sr is clear evidence that in this charge interval the r-process is not dominating the abundances.

Cassé and Goret [9] have pointed out that cosmic-ray abundances of iron and lower charges relative to solar system abundances, exhibit an anti-correlation with first ionization potential; elements which are harder to ionize are underabundant in the cosmic rays. Figure 5 demonstrates that our source abundances for elements with Z ≥ 26 in the cosmic rays, relative to solar system abundances, show the same general organization with first-ionization potential as do the abundances of elements with Z ≤ 26. (In this figure the squares, for Z ≥ 26 use cosmic-ray data from our experiment and the circles, for Z ≤ 26, use cosmic-ray data presented at this conference from the Danish-French HEAO-3 cosmic-ray experiment. Solar system abundances for all points are from Cameron [10] and their uncertainties are not included in the plotted error bars. Relative abundances are normalized to unity at $_{26}$Fe.) At the same time, we note that variations in abundances are not perfectly ordered by first-ionization potential. Significant differences between elements of similar first-ionization potential can be seen between Zr and Mo and among Zn, Se, and S.

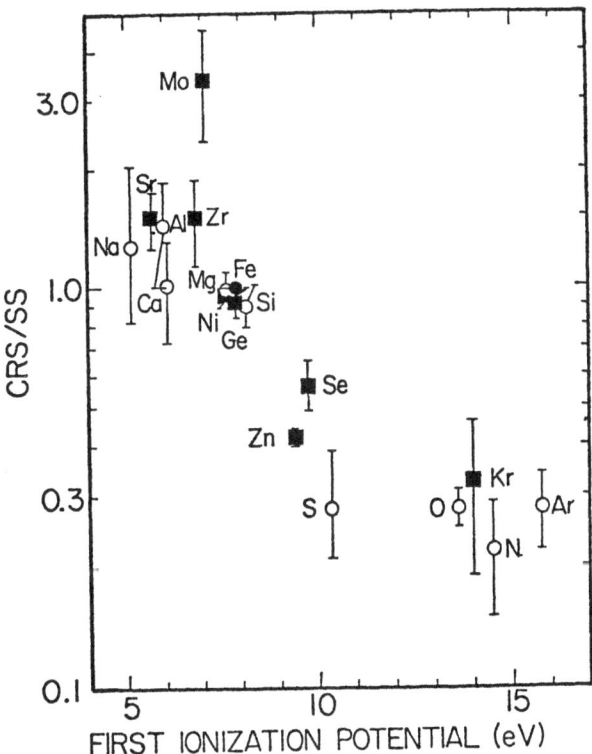

Figure 5

One suggestion that has been made is that the volatility rather than the ionization potential is the controlling atomic property. Most of the elements on this figure which are volatile have ionization potential above 9 eV and the refractory elements are below. One notable exception is Ge which is volatile but has the same first ionization potential as Fe; it seems to indicate that volatility is not the controlling factor [11].

At the current stage of our analysis we have confined our results for $Z \leq 42$ to the high resolution subset of our data, shown in Figures 2 and 3. In order to have adequate statistics for the still rarer elements at higher charge we must be less selective. Figure 6 shows that a slight relaxation of selection criteria still gives adequate charge resolution for $Z \leq 42$ and is beginning to give adequate statistics for discerning peaks in the region from $_{50}$Sn to $_{56}$Ba. The lower histogram in this figure assumes the first-order $Z^2$ response for both the Cerenkov and ionization signals. The upper histogram displays the same data but with an approximate application of higher order corrections to the charge assignments [12]; these corrections are between half and one charge unit in the Sn to Ba region. Results here are still preliminary, and conclusions drawn here today are less certain than those in the region up to charge 42; but it is important to note that peaks appear to be emerging at both Sn and Ba, and these elements do not appear to be depleted compared with Te and Xe. Since Sn and Ba are made mainly in the s-process, while Te and Xe are made mainly in the r-process, the individual element abundances in this charge interval again suggest a cosmic-ray source which is not dominated by r-process nucleosynthesis.

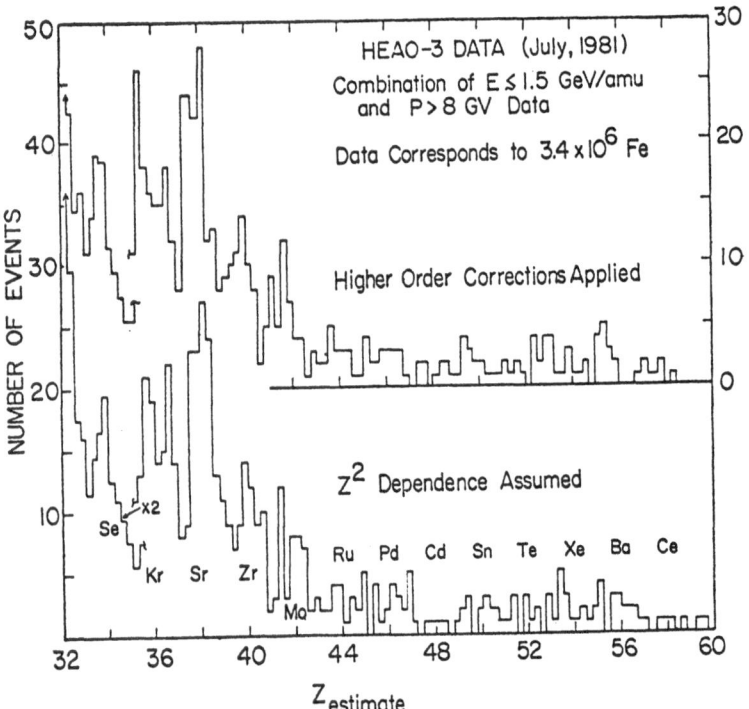

Figure 6

At higher charges, we have not yet extracted data with individual element resolution, and we have looked at only a very preliminary correction for higher-order (non $Z^2$) effects on the detector responses [13]. A preliminary look at our higher-charge results, combined with those at lower charge, is shown in Figure 7. Data have been grouped into bins two charge units wide centered at even charges. At this stage of analysis, our data do not warrant any conclusions about relative abundances of nearby elements; for example, we would not attempt to infer a Pt/Pb ratio from our data at this time. These data do show the expected sharp drop in abundance just above $_{56}$Ba, and another sharp fall off above $_{82}$Pb. The plotted upper limits at $Z = 86$ and $88$ are one-sigma limits based on seeing no events of these charges.

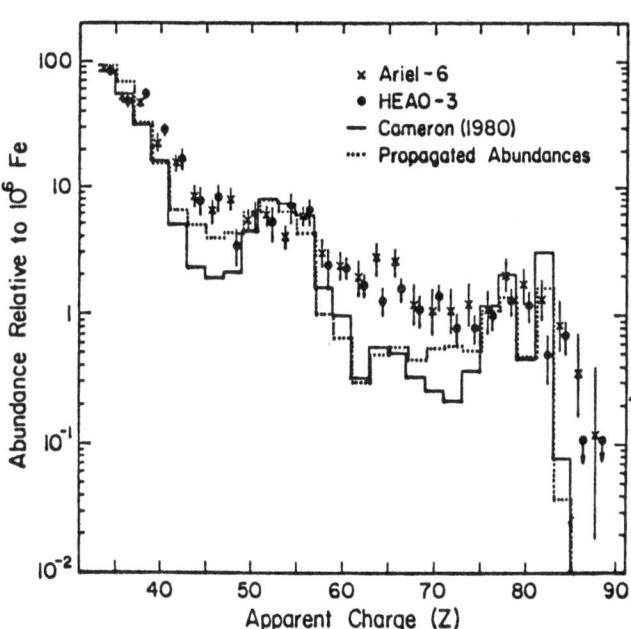

Figure 7

For comparison, data from the United Kingdom spacecraft Ariel-6 are also shown in Figure 7 [14]. The broad abundance features observed by Ariel and by HEAO are in general agreement.

Also shown in Figure 7 are the solar-system abundances [10] and those expected at earth as a result of propagating these abundances through an exponential path-length distribution with mean of 5.5 g/cm$^2$ (interstellar medium). [15]. The observed flux of secondary nuclei in the interval $60 \leq Z \leq 74$ appears to be higher than expected from this propagation. At least qualitatively, a path-length distribution truncated at short path lengths would increase the secondary production in this manner. If one does interpret this excess in the 60 to 74 interval as excess secondaries from fragmentation in the 76 to 83 ("Pt-Pb") region, then one is lead to conclude that the cosmic-ray source is enriched in "Pt-Pb" relative to higher elements [16]. The significance of these apparent effects should become clearer if we can continue our analysis to extract individual-element resolution in this interval.

Finally, turning to the flux of actinides, in about 15 months of data we have observed only one event which may be an actinide; in the same time we have 106 events assigned to the "Pt-Pb" region of $74 \leq Z \leq 84$ [13]. The resulting ratio of about 1% is shown in the lower-right corner of Figure 8. Ariel has observed two apparent actinides and 69 "Pt-Pb" [14]. Thus their value for the same ratio, 3%, is higher than, but statistically consistent with, our ratio. (Assymetric error bars in this figure give "one-sigma", i.e., 84% confidence, limits based on Poisson statistics of small numbers.) Our HEAO result is not consistent with the much higher actinide abundances inferred from nuclear emulsions and plastic track detectors on balloons [17] or from published results from the plastic track detectors on Skylab [5]. However, Price (private communication) points out that five of the seven events identified as actinides in the Skylab data were at very low energy; if one looks only at Skylab data above about 500 MeV/amu, a lower ratio is obtained (open circle in Figure 8).

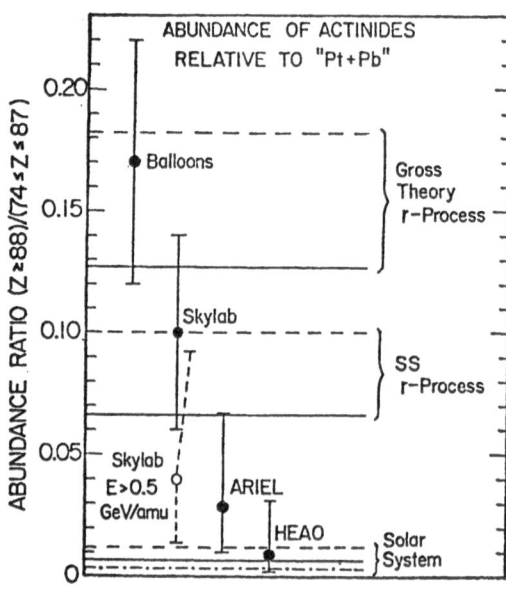

Figure 8

For comparison, the lines show what this ratio would be at earth if the cosmic-ray source composition were that of the solar system at its formation (lowest solid line); that of the implied r-process contribution to the solar system, without decay since nucleosynthesis (solid line labeled SS-r-process); or that of a theoretical r-process based on a gross theory of β-decay rates (upper solid line) [18]. The dashed lines give results for the same sources modified by an assumed first-ionization-potential correlation [18]. The lowest, dot-dash, line shows the expected ratio for a source with the relative abundances of the solar system today, instead of those of $4.6 \times 10^9$ years ago [19].

Our low observed actinide abundance, is clearly inconsistent with a cosmic-ray source composed principally of fresh r-process material. However, we cannot rule out a cosmic-ray source which is either enhanced or depleted in actinides, relative to the early solar system, by a factor as great as three. And HEAO-3 will not significantly narrow this range of possible values, even when all twenty months of data have been fully analyzed.

## Summary

In summary, our HEAO-3 instrument has resolved individual even-charge elements at least to $_{42}$Mo, and peaks are emerging from further analysis, at least to $_{56}$Ba. The implied source composition is not identical with that of the solar system; deviations from solar abundances show a similar first-order anti-correlation with elemental first-ionization potential, but there are distinct deviations from a simple smooth anti-correlation. Finally, the cosmic-ray source does not appear to be dominated by r-process nucleosynthesis. In the cosmic-rays Sr is abundant; Sn and Ba also appear to be abundant; Th and U are not greatly enhanced. (However, lack of individual-element resolution near charge 80 leaves us at this time with no conclusion concerning the ·r-process from Pt/Pb.)

With further data analysis, now underway, we expect to increase the statistics and have clear individual-element resolution in the Sn to Ba region. We also hope to resolve elements in the Pt-Pb region. Furthermore, we are planning to calibrate a prototype of our instrument with the Bevalac at Lawrence Berkeley Laboratory next year, when they expect to produce relativistic nuclei at least as heavy as Pb. This calibration will provide a direct measure of deviations from first order, $Z^2$, instrument response for energies up to about 1 GeV/amu, and as a result may give us improved charge resolution, especially at the highest charges.

Acknowledgments: The HEAO results presented here have been the product of the efforst of our whole experiment team [6], and the publications of our team [3], [7], [8], [13] should be referenced when quoting these results. The HEAO project was made possible by the dedicated efforts of personnel in the program office at NASA headquarters, in the project office at NASA-MSFC, at TRW Systems Group who built the spacecraft and operated the operation control center, and at NASA-GSFC in operation control, and data and tracking. Our experiment depended on the excellent work of people at Ball Aerospace Systems Division who designed and fabricated major portions of the instrument and integrated it, and to support personnel at Caltech, University of Minnesota, and Washington University. We are grateful to all these people. This work was supported in part by NASA under contracts NAS 8-27976,7,8 and grant NGR 26-008-001.

## References

[1]  This post-deadline talk was presented orally but did not appear in the printed proceedings. These results were subsequently published by R. L. Fleischer, P. B. Price, R. M. Walker, M. Maurette, and G. Morgan, J. Geophys. Res. 72, 355 (1967).

[2]  P. H. Fowler, R. A. Adams, V. G. Cowen, and J. M. Kidd, Can. J. Phys. 46, S601 (1968); and Proc. Roy Soc. A301, 39 (1967).

[3]  For example see the first section of M. H. Israel, J. Klarmann, W. R. Binns, R. K. Fickle, C. J. Waddington, T. L. Garrard, E. C. Stone, this conference Vol. 2, 36 (1981).

[4]  Rates in this table are based on our HEAO results. The actual HEAO event rate, averaged over many orbits, is about half the rate shown in this table.

[5]  E. K. Shirk and P. B. Price, Astrophys. J. 220, 719 (1978).

[6]  California Institute of Technology
     E. C. Stone and T. L. Garrard
     University of Minnesota
     C. J. Waddington, R. K. Fickle, W. R. Scarlett
     Washington University
     W. R. Binns (formerly at McDonnell Douglas Research Laboratories) M. H. Israel, and J. Klarmann.

[7]  W. R. Binns, M. H. Israel, J. Klarmann, W. R. Scarlett, E. C. Stone, and C. J. Waddington, Nucl. Instr. and Meth. 185, 415 (1981).

[8]  The data analysis and selection leading to these figures are described by W. R. Binns, R. K. Fickle, T. L. Garrard, M. H. Israel, J. Klarmann, E. C. Stone, and C. J. Waddington, Astrophys. J. 247, L115 (1981).
     Figures 3, 4, and 5 of this talk are updated versions of corresponding figures in that paper, based on analysis of nearly twice as many days of data. See also this conference, Vol. 2, 13; and Vol. 2, 36. (Normalizations relative to Fe, for $Z \geq 32$, may have systematic error as great as ±20%.)

[9]  M. Cassé and P. Goret, Astrophys. J. 221, 703 (1978).

[10]  A. G. W. Cameron, Center for Astrophysics preprint no. 1357 (1980).

[11]  In a comment after of this talk, C. J. Cesarsky pointed out that although meteorite abundances indicate that Ge is a volatile element, in fact its condensation temperature indicates that it is only marginally volatile; in interstellar grains it could act refractory. See also J. P. Meyer in this conference Vol. 2, page 281. For discussion of grains as sources of cosmic ray elements see J. P. Bibring and C. J. Cesarsky, this conference Vol. 2, page 289; and R. I. Epstein, Mon. Not. R. Astron. Soc. 193, 723 (1980).

[12]  J. H. Derrickson, P. B. Eby, and J. W. Watts, this conference Vol. 8, 83 (1981); and S. P. Ahlen, Phys. Rev. A 17, 1236 (1978).

[13] C. J. Waddington, R. K. Fickle, T. L. Garrard, E. C. Stone, W. R. Binns, M. H. Israel, and J. Klarmann, this conference, paper OGH.2-9, to be published in late-paper volume.

[14] P. H. Fowler, M. R. W. Masheder, et al; this conference, paper OGH.2-10, to be published in late paper volume; see also P. H. Fowler, R. N. F. Walker, M. R. W. Masheder, R. T. Moses, and A. Woorley, Nature 291, 45 (1981).

[15] S. H. Margolis, Washington University, St. Louis, private communication.

[16] The apparent excesses of ultraheavy secondaries and of "Pt-Pb" are discussed more quantitatively by R. A. Mewaldt in his OG rapporteur talk, elsewhere in this volume.

[17] P. H. Fowler, C. Alexandre, V. M. Clapham, D. L. Henshaw, D. O'Sullivan, and A. Thompson, 15th International Cosmic Ray Conference (Plovdiv) 11, 165 (1977).

[18] J. B. Blake, K. L. Hainebach, D. N. Schramm, and J. D. Anglin, Astrophys. J. 221, 694 (1978).

[19] N. R. Brewster, P. S. Freier, and C. J. Waddington, private communication.

```
*******************
*                 *
* INVITED PAPERS  *
*                 *
*******************
```

THE NATURE OF THE INTERSTELLAR MEDIUM

Christopher F. McKee
Departments of Physics and Astronomy
University of California, Berkeley

## 1. Introduction

The physical properties of the interstellar medium (ISM) are determined by its interactions with stars in the galaxy and by its lack of interaction with matter and radiation outside the galaxy. Stars inject energy into the ISM in the form of starlight, stellar winds, and, in some cases, as supernovae; some of this energy is injected as cosmic rays. Stars inject mass into the ISM via stellar winds, planetary nebulae, and supernovae. The ISM recycles this mass by the process of star formation in regions where enough gas accumulates to become gravitationally unstable. External influences on the ISM are relatively small, however: the effective temperature of the ambient radiation (~ 3K) is much less than the typical temperature of interstellar gas, and since our galaxy is not located in a rich cluster of galaxies, there is no evidence that the ambient intergalactic medium has a significant effect on the ISM.

The lifetime, isotropy, and composition of cosmic rays are determined to a significant extent by the structure of the ISM. Recent observational and theoretical work has led to a picture of the ISM in which most of the mass is concentrated in cold (T $\lesssim 10^2$ K) clouds, whereas most of the volume is occupied by warm (T ~ $10^4$ K) and hot (T ~ $10^{5.7}$ K) gas. The gas in the galactic disk, which has a thickness of order 200 pc, exchanges mass and energy with that in the halo relatively efficiently. This three phase model is consistent with much of the existing data on the ISM, but further observational confirmation is required to establish its validity. Here I shall summarize three phase models of the ISM, focussing on the role of supernova remnants (SNRs) in generating hot gas in the galaxy and the mechanisms by which a steady state can be maintained.

## 2. Development of Inhomogeneities in the ISM

Optical and radio observations some time ago established that the ISM is inhomogeneous on many length scales. These inhomogeneities can arise due to self-gravity, to thermal instability, or to mechanical effects. Gravitational effects are important on a large scale, such as in spiral structure, or in small regions of high density in which star formation is occurring, and I shall not consider these effects in detail here.

The role of thermal instability in creating inhomogeneities can be analyzed in terms of the idealized problem in which an initially uniform gas is steadily heated at a rate $n\Gamma$ per unit volume, where n is the gas density. In diffuse regions of the ISM the radiation emitted by the gas can escape freely, so it is impossible for the gas to be in complete thermodynamic equilibrium. However, it can be in thermal equilibrium when the radiative cooling rate $n^2\Lambda$ equals the heating rate $n\Gamma$; here $\Lambda$ is the cooling function in erg cm$^3$ s$^{-1}$ (Dalgarno and McCray 1972; Raymond, Cox and Smith 1976). If the energy balance equation $n\Gamma = n^2\Lambda$ is solved together with the ionization equilibrium equation for the case of constant heating rate $\Gamma$, an interesting result emerges: the pressure p is a multi-valued function of the temperature T, p $\propto$ $\Gamma T/\Lambda$ (Field, Goldsmith, and Habing 1969). Schematically, the p-T curve is shaped like the letter W. Cooling exceeds heating above the curve, and vice versa. Below a few hundred K the cooling function $\Lambda$ varies as $T^m$ with $m > 1$, so equilibrium can be maintained at progressively lower densities as T increases; hence p falls as T rises. A second region of declining pressure occurs in the temperature range $10^4$-$10^{4.7}$ K, where collisional excitation of hydrogen and helium results in efficient cooling. On the other hand, in the regions $10^{2.5}$-$10^4$ K and above $10^{4.7}$ K, the cooling rate increases slowly with temperature, if at all, and p and T rise together.

Now consider the stability of this equilibrium against isobaric perturbations (Field 1965). Regions in which p falls as T rises are thermally stable: an increase in T at constant pressure moves the gas into a region where cooling exceeds heating, so the temperature drops back to its equilibrium value. However, if p rises with T the gas is thermally unstable since an increase in T moves the gas into a region in which heating is dominant, so the temperature runs away. Over the range of pressure in which the equilibrium pressure is multi-valued, we expect the gas to be partially in the thermally stable cold phase (T $\lesssim 10^{2.5}$ K) and partially in the stable warm phase (T $\sim 10^4$ K). Since the pressure and density of the gas are normally controlled by several different processes, it is unlikely that they will conspire to place all the gas in just one phase in this multi-valued region (Krolik, McKee and Tarter 1981).

The idea of two thermally stable phases coexisting at the same pressure forms the basis of the two-phase model of the ISM (Field et al. 1969). The cold phase is identified with the cold neutral hydrogen clouds seen in absorption in 21 cm studies, whereas the warm, partially ionized phase corresponds to the broad 21 cm emission component. The model predicts that the ISM lies in a narrow range of pressure (the value depending on $\Gamma$) over a wide range of mean density. Originally the heating was ascribed to low energy (2 MeV) cosmic rays, but now it is believed that the heating is due to some other mechanism, such as photoelectric heating (e.g., deJong 1980) or soft X-rays (McKee and Ostriker 1977).

An important question which is not addressed in this model is what happens when the heating is time dependent. If dynamical effects are ignored, a modified two phase model results (Bottcher et al. 1970). However, time dependent heating due to SNRs can produce inhomogeneities in the ISM by mechanically displacing the gas, and we turn to this topic now.

## 3. SNRs Generate a Hot Phase in the ISM

The dominant energy sources for the ISM are starlight and supernovae. The luminosity of the Galaxy is about $10^{43.5}$ erg s$^{-1}$, but almost all of that is radiated away into intergalactic space, either directly or indirectly after being absorbed by interstellar dust. The supernova energy injection rate is of order $10^{42}$ erg s$^{-1}$ in the Galaxy, corresponding to one supernova of $10^{51}$ erg occurring every 30 years.

This energy is not injected uniformly as assumed in the above discussion of thermal instability, but rather at random points in the disk. After an initial stage in which the SN ejecta act as a piston driving a shock into the surrounding ISM, the SNR evolves into an adiabatic blast wave. In a uniform medium, the mean pressure and temperature inside the SNR fall off as $R^{-3}$, where R is the radius of the SNR. Young SNRs fill a tiny fraction of the volume of the ISM with very hot gas (T ~ $10^8$ K), whereas older remnants fill a larger fraction of the volume with gas at a lower temperature. When the temperature falls below about $10^6$ K, radiative cooling becomes important and a dense, radiative shell forms at the periphery of the SNR. The shell encloses a hot bubble of low density gas at T ~ $10^6$ K. This radiative stage commences at R ≃ 20 $E_{51}^{0.3}$ $n_0^{-0.4}$ pc, where $E_{51}$ is the SNR energy in units of $10^{51}$ erg. It ends when the SNR has expanded to the point that it is in pressure equilibrium with the ambient gas, at

$$R_E = \frac{50\ E_{51}^{0.3}}{n_0^{0.16}} \left(\frac{10^4\ K\ cm^{-3}}{p_0/k}\right)^{0.2} pc \tag{1}$$

A useful description of the effect of SNRs on the ISM is in terms of their role as entropy generators (Kahn 1975). The entropy of a parcel of gas is increased when it passes through the shock front at the perimeter of the SNR. Thereafter, it is conserved except that high entropy gas is destroyed by thermal conduction and low entropy gas by radiation. Let

$$s^* = T^{3/2}/n \tag{2}$$

so that the entropy per unit mass is $(k/\mu)\ln s^*$, where $\mu$ is the mean mass per particle. Then the mass of gas with an entropy exceeding that corresponding to $s^*$ is

$$M(>s^*) = \frac{80\ E_{51}}{(n_0 s_{10}^*)^{2/3}}\ M_\Theta \tag{3}$$

where $s_{10}^* = s^*/(10^{10}\ K^{3/2}\ cm^3)$. Between $10^5$ K and $10^{7.5}$ K, the radiative cooling time inferred from the results of Raymond, Cox, and Smith (1976) is about

$$t_{cool} = 6.3 \times 10^5\ s_{10}^*\ yr. \tag{4}$$

Since the time interval between SNRs at a given point is about $4 \times 10^5$ yr (McKee and Ostriker 1977), gas with $s_{10}^* \gtrsim 1$ cannot cool. The fate of this gas will be discussed in §IV below.

Cox and Smith (1974) were the first to point out that this hot gas in the interiors of old SNRs must occupy a substantial fraction of the volume of the ISM. Consider a random point in the ISM. Let Q equal the mean number of SNRs in which the point is embedded; Q is termed the porosity by Cox and Smith. If interactions between different SNRs are neglected, then

$$Q = \frac{V_{SNR}}{V_{Gal}} \cdot \frac{t_{SNR}}{\tau_{SN}} \tag{5}$$

where $V_{SNR}$ is the volume of the SNR, $t_{SNR}$ its age, $V_{Gal}$ the volume of the Galaxy, and $\tau_{SN}$ the mean time interval between SN in the Galaxy. The volume of an old SNR is $4\pi R_E^3/3$. Its lifetime is less certain; we adopt $t_{SNR} = R_E/$ (sound speed in the ambient gas). Then it is simple to show that (McKee and Ostriker 1977)

$$Q \simeq \frac{E_{51}^{1.3}}{n_0^{0.14}} \left(\frac{10^4 \text{ K cm}^{-3}}{p_0/k}\right)^{1.3} \left(\frac{30 \text{ yr}}{\tau_{SN}}\right) . \tag{6}$$

Observations show $n_0 < 1 \text{ cm}^{-3}$ and $p_0/k \sim 4000 \text{ K cm}^{-3}$ (Jura 1975). Hence, $Q \gtrsim$ 1, and the SNR filling factor ($= 1 - \exp(-Q)$ in this non-interacting model) is close to unity.

This argument for a large porosity relies on the observational value for the ambient interstellar pressure $p_0$. Alternatively, one can argue that if SNRs are the dominant energy source for the ISM, then they must expand to fill the ISM so that they can determine the pressure, and $p_0$ in equation (6) will necessarily be low enough that $Q > 1$.

Although these arguments are reasonably convincing, they are not conclusive. The principal potential flaw is that an SNR contains many embedded clouds; if the interior can cool so that the internal pressure falls below the mean interstellar value, then the clouds can expand until they occupy a larger volume than normal, thereby reducing the volume left for the hot gas (McKee 1982). Thus, negative fluctuations in the SNR rate might produce regions of low porosity, which could take several million years at the average SNR rate to recover the typical porosity.

Observational support for pervasive hot gas comes from several sources: Analysis of the soft X-ray background (e.g., Fried et al. 1980) and of L$\alpha$ absorption toward nearby stars (Bohlin 1975) has established that the porosity within 100 pc or so of the sun is indeed of order unity. The sun appears to be in a somewhat unusual position, however; according to Tanaka and Bleeker (1977), less than 5% of the galactic disk could have the same X-ray emissivity as the solar neighborhood. The strongest evidence for the widespread nature of the hot gas comes from observation of ultraviolet absorption lines of the $O^{+5}$ ion seen in many stars, most of which are beyond 100 pc (Jenkins and Meloy 1974, York 1974). The ionization potential of $O^{+4}$ is 114 eV, and the temperature indicated by the presence of $O^{+5}$ is several hundred thousand K. It is thought that these absorption lines arise in the conductive interfaces of clouds embedded in the hot phase of the ISM (McKee and Cowie 1977, McKee and Ostriker 1977).

In our discussion, we have tacitly assumed that the SNRs occur at random points in the disk. This is a good assumption for the relatively low mass stars which produce Type I supernovae (about half the total) and for some of the high mass stars which produce Type II supernovae. Many massive stars are born in large associations, however, and when they explode they are but one of many supernovae occurring in a limited region of space. The evolution of such associations has been discussed by Bruhweiler et al. (1980), and it is quite possible that they produce the large neutral hydrogen structures - "super-shells" - with sizes ranging up to a kiloparsec which have been observed by Heiles (1979). These regions are estimated to occupy no more than about one tenth of the volume of the Galaxy, and we shall not consider them further here.

In §II we discussed how thermal instability in the ISM would lead to two distinct phases. Based on the discussion in this section, however, we see that supernovae will cause such a configuration to rapidly self-destruct and to evolve into a three phase ISM with a large volume of hot, low density gas.

According to McKee and Ostriker (1977), the hot intercloud medium (HIM) typically has a density n $\simeq 10^{-2.5}$ cm$^{-3}$, a temperature T $\simeq 10^{5.7}$ K, and a volume filling factor f $\simeq 0.7$. The cold neutral medium is in the form of clouds with $(n,T,f) = (40$ cm$^{-3}$, 80K, 0.024). These cold shells are surrounded by shells of warm gas (T $\sim$ 8000 K) which is either predominantly ionized by starlight (the WIM) or is mainly neutral (the WNM). In the absence of magnetic pressure (cf. Cox 1981) all three phases are at the same pressure. The neutral part of the warm phase - the WNM - is the analog of the second phase in the two phase model. An important consequence of the large porosity for the HIM is that SNRs can expand beyond 150 pc before cooling, and hence they are effective at connecting the gas in the disk with that in the halo.

## 4. Regulation of the Three Phase ISM

In a steady state, the energy injected into the ISM by supernovae must be balanced by that which escapes from the ISM. Furthermore, there must be mass balance as well (McKee and Ostriker 1977): the rate at which the HIM is created must be balanced by the rate at which it is destroyed.

The energy injected by supernovae initially goes primarily into the HIM; some of it may go into cosmic rays as well. The HIM may lose this energy by radiation, by compressing the embedded clouds, or by transferring it to the halo. The energy in the clouds is eventually radiated; that in the halo is either radiated or convected away in a galactic wind. Chevalier and Oegerle (1979) have shown that a galactic wind cannot carry away a substantial fraction of the energy injected by supernovae (see below). Hence, the energy injected by supernovae must eventually be radiated by the HIM (in the disk or in the halo) or by clouds. The typical pressure in the ISM is then of order

$$p \sim \left(\frac{t_{cool}}{\tau_{SN}}\right) \frac{E}{V_{Gal}} \ . \tag{7}$$

Next consider mass balance for the HIM. It is created by shocking clouds to sufficiently high entropy or by evaporating the clouds. The primary destruction mechanism is radiative cooling. In addition, however, there is the possibility of producing very hot gas (T $\gg 10^6$ K) from the HIM: just as SNRs create hot, low density bubbles in an ISM with n $\sim$ 0.1-1 cm$^{-3}$ (Cox and Smith 1974), so they will create very hot, very low density bubbles in the HIM unless restrained by conduction to the HIM, conduction to the clouds (evaporation), or escape to the halo. There is no good observational evidence that such a very hot medium exists in the disk.

Two principal models for energy balance in the three phase ISM have been proposed: One is the galactic fountain (Shapiro and Field 1976), in which high entropy gas rises above the plane into the halo, is stored there until it radiates, and then returns to the disk in the form of cold or warm clouds. The other is the three phase model of McKee and Ostriker (1977), in which the HIM is regulated by the evaporation of embedded clouds and most of the SNR energy is radiated away in the disk. The two models are not mutually exclusive; for example, SNRs in the McKee-Ostriker model expand to such large radii (R > 150 pc) that they drive a weak fountain. However, we shall argue that a strong fountain, in which most of the total SNR energy is radiated away in the halo, is inconsistent with mass balance.

## a) Galactic Fountain

Since the original suggestion by Shapiro and Field (1976), galactic fountains have been studied by Chevalier and Oegerle (1979), Bregman (1980), Habe and Ikeuchi (1980), Cox (1981), and Kahn (1981). Many of these authors were interested in the dynamics and appearance of the fountain rather than in constructing a model for the ISM, and they did not require that the fountain carry most of the galactic SNR energy. Cox (1981) did make this requirement, but he adopted an SNR energy injection rate $E/\tau_{SN}$ almost an order of magnitude less than the $10^{42}$ erg s$^{-1}$ estimated in §III. Interestingly enough, he found that the height of the fountain was comparable to the thickness of the disk, so his model is qualitatively similar to the McKee-Ostriker model.

Observational evidence in favor of a fountain comes from the high velocity neutral hydrogen clouds seen in 21 cm emission (Verschuur 1975), although much of this emission could be in the disk (H. Weaver, private communication), and from detection of UV absorption lines of $C^{+3}$ and $Si^{+3}$ ions in the halo (Savage and deBoer 1979). Cosmic rays would be convected into the halo as well, and the radio halo of the Galaxy (Webster 1978) is naturally accounted for. Note, however, that this evidence does not require that the fountain be strong in the sense defined above.

The difficulty with a strong fountain is that it does not satisfy mass balance. The amount of mass which must cool in order to dissipate the energy from a single supernova is

$$\Delta M \simeq \frac{E}{\left(\frac{5}{2}\frac{kT}{\mu}\right)} = 1500 \frac{E51}{T_6} M_\Theta \qquad (8)$$

where $T_6 = T/10^6$ K and where we made the conservative assumption that the cooling is isobaric. In a strong fountain, most of this mass is HIM. Since the cooling above $10^5$ K is thermally unstable, a substantial fraction of this gas will go into clouds (Schwarz, McCray, and Stein 1972). However, in the absence of cloud evaporation there is no way of generating such a large mass of HIM: the cloud mass shocked to an entropy high enough to become part of the HIM ($s_{10}^* \gtrsim 1$) is given by equation (3) reduced by the cloud filling factor, and is only about 100 $M_\Theta$/SNR. This imbalance between the rates of creation and destruction of HIM is characteristic of any model in which the injected energy is radiated away by the HIM, resulting in its destruction, without having cloud evaporation as a source of new HIM. A similar conclusion was reached by Habe, Ikeuchi, and Tanaka (1981) on the basis of numerical modelling of the ISM.

Another problem with a strong fountain model is also apparent from equation (8). With one supernova every 30 yr in the Galaxy, one expects a mass flow of order 50 $M_\Theta$ yr$^{-1}$ from the halo onto the disk, considerably higher than that normally estimated. Chevalier and Oegerle (1979) used equation (8) to show that energy balance could not be achieved by a galactic wind, since the mass loss rate would be catastrophic unless the temperature were so high ($>10^7$ K) that it violated observation.

## b) Evaporative Regulation

In a three phase ISM, large temperature gradients occur around the cold and warm clouds embedded in the HIM. If thermal conduction is not substantially impeded by magnetic fields or plasma instabilities, then the clouds will evaporate, provided they are not too large (Cowie and McKee 1977, McKee and Cowie 1977). The heat flux toward a cloud $\sim \kappa T/a$ heats the cloud surface and drives an outward energy flux $\sim (5/2)pv$, so that the evaporation rate is $\dot{m}$ = $4\pi a^2 \rho v \sim 8\pi a \kappa \mu/5k$; here $\kappa$ is the thermal conductivity, which we take equal to $\phi_c < 1$ times the classical value (Spitzer 1962), and a is the cloud radius. For spherical clouds, the exact value of $\dot{m}$ is (2/5) of this:

$$\dot{m} = 2.75 \times 10^4 \, T^{5/2} \, a_{pc} \phi_c \, g \, s^{-1} \tag{9}$$

where $a_{pc}$ = a/1 pc. The temperature sensitivity of this result means that cloud evaporation can act as a thermostat for the HIM; in particular, it is effective at lowering the entropy per unit mass of the hot gas in the SNRs down to the point that radiative cooling becomes important.

Cloud evaporation has a major effect on the evolution of SNRs in a three phase medium (McKee and Ostriker 1977): since the hot phase has a low density, it can be significantly modified by the evaporation of a small fraction of the cloud material. The mean density in the hot phase governs the rate of expansion of the SNR. The blast wave propagates through the hot gas, leaving behind the clouds to be compressed later by slower shocks, so that the kinetic energy is primarily in the hot gas and $v^2 \propto E/M \propto (\rho R^3)^{-1}$, where $\rho$ is the mean internal density. In the absence of cloud evaporation, $\rho = \rho_0$ = const and $T \propto v^2 \propto R^{-3}$; if cloud evaporation dominates the swept up gas, then $\rho \propto R^{-5/3}$ and $T \propto v^2 \propto R^{-4/3}$. These results have been confirmed by a similarity solution due to Chieze and Lazareff (1981) and in numerical calculations by Cowie, McKee, and Ostriker (1981). As an example, the latter authors found that for $(E_0, n_0, f[WIM])$ = (3 x $10^{50}$ erg, 2.4x$10^{-3}$ cm$^{-3}$, 0.27) the mean internal density dropped from 0.2 cm$^{-3}$ 5000 yr after the explosion to 5x$10^{-3}$ cm$^{-3}$ at 7x$10^5$ yr, and it was within a factor 2 of the analytic estimate throughout this period.

Despite the large conceptual difference between evaporative and non-evaporative models of SNR evolution, it has proved difficult to obtain clear observational evidence which would distinguish between them (McKee 1982). One potential test involves the rim of an evaporative SNR (Cowie and McKee, in preparation). As pointed out by Cox (1979), the density just behind the blast wave shock is four times the ambient value and may be small compared to the mean internal value. The temperature tends to be approximately uniform, so in contrast to a non-evaporative blast wave, the pressure peaks inside the SNR rather than at its edge. Physically, this pressure gradient serves to accelerate the evaporated gas up to the expansion velocity of the SNR. Possible evidence for this effect has been obtained in the Cygnus Loop (see McKee 1982).

In the three phase ISM model due to McKee and Ostriker, mass balance is explicitly included by equating the rate of cloud evaporation to the rate of cloud condensation due to thermally unstable radiative cooling. Energy balance is achieved in the galactic disk by direct radiation from the HIM, prior to the point at which the SNRs overlap.

The advantages of the model are that (1) it is the only three phase model consistent with both mass and energy balance; (2) the temperature sensitivity of the evaporation rate ($\dot{m} \propto T^{5/2}$) tends to make the ISM stable against perturbations; (3) it is consistent with observed features of SNR evolution; and (4) it is not inconsistent with the existence of a cosmic ray and radio halo, since SNRs in the disk expand into the lower halo.

There are two key assumptions (and thus potential problems) in the model. Theoretical and observational arguments can be advanced in favor of each, but they are not conclusive. The first is that the filling factor of the HIM is large (see discussion of §III). If, on the other hand, the filling factor were small, then the ISM would resemble the two phase model of Field et al. (1969). In fact, either model could be correct, depending on the local supernova rate, the location in the Galaxy, etc.

The second assumption is that thermal conduction is not strongly inhibited by plasma instabilities or magnetic fields. This requires that the field between the clouds and the HIM be well connected; in view of the rapid mass exhange between the HIM and the clouds, this assumption is self-consistent. Observationally, there is considerable evidence that conduction is efficient in the solar wind, on scales small compared to those of the ISM, whereas Binney and Cowie (1981) have argued that it is inefficient in clusters of galaxies, on scales large compared to those of the ISM. Direct optical or X-ray observations of evaporating clouds (cf. Dopita and Mathewson 1979) would be of great value in resolving this question.

## 5. Properties of the Three Phase ISM

The arguments presented above indicate that SNRs produce a hot component of the ISM ($T \sim 10^{5.7}$ K, $n \sim 10^{-2.5}$ cm$^{-3}$) which fills a significant fraction of the volume of the disk. A typical SNR expands to a radius well in excess of 100 pc. Two steady state models of the ISM incorporating such a hot phase have been discussed: the fountain model, in which much of the SNR energy is radiated by the hot gas after it bubbles up out of the disk into the halo, and the evaporative model, in which the density of the hot gas is regulated by cloud evaporation so that cooling occurs in the disk. I shall conclude by commenting on several properties of three phase models of the ISM, which by and large are common to both the fountain and evaporative models.

Whenever the filling factor of the hot gas exceeds 0.5 or so, the dense gas may be thought of as clouds embedded in a hot ambient medium. Ionizing starlight is not absorbed by this hot gas, but instead forms warm ($T \sim 10^4$ K) partially ionized envelopes around the clouds (McKee and Ostriker 1977). Most clouds have cold cores which are small compared to the radius of these ionized envelopes ($\sim 2$ pc), although some are a good deal larger. This ionized gas accounts for the observed pulsar dispersion measure and diffuse Hα.

The effects of magnetic fields in a three phase ISM have not been considered in detail. The field will produce a weak dynamical coupling between the different phases. The mean value $B \simeq 3\times10^{-6}$ G inferred from pulsar observations refers primarily to the warm ionized envelopes around the clouds. If the fields of the clouds and the HIM are well-connected, as they must be in the evaporative model, then the fact that the warm envelopes occupy about 1/4 of the volume of the disk implies that the field in the HIM cannot be much less than the cloud value: $3 \times 10^{-6}$ G $\gtrsim$ B(HIM) $\gtrsim 10^{-6}$ G. This implies an Alfven velocity in the HIM of 30-100 $\tilde{km}$ $s^{-1}$. Under these conditions cosmic rays spend most of their time in the HIM, but the time-averaged density they experience is the mean ISM density, including cloud material on well-connected field lines. A detailed discussion of cosmic rays in a three phase ISM has been given by Blandford and Ostriker (1980).

At the outset of my talk I remarked that gas is recycled between stars and the ISM. Processed material is ejected from stars in supernova explosions; gas is returned to stars by gravitational collapse of interstellar clouds. In fact, supernovae and their remnants may be involved in both halves of the cycle: SNRs can compress interstellar clouds and drive the more massive ones over the brink of instability, so that the death of one star leads to the birth of others.

## 6. Acknowledgements

I wish to thank the National Organizing Committee of the 17th ICRC for the privilege of presenting this inaugural lecture. Many of the ideas presented here were worked out in collaboration with L. Cowie and J.P. Ostriker. My research is supported in part by NSF grant AST 79 23243.

## References

Binney, J., and Cowie, L.L. 1981, Ap.J. 247, 464.
Blandford, R., and Ostriker, J.P., 1980, Ap. J. 237, 793.
Bohlin, R.C. 1975, Ap.J. 200, 402.
Bottcher, C., et al, 1970, Astrophys. Letters 6, 237.
Bregman, J., 1980, Ap. J. 236, 577.
Bruhweiler, F.C., Gull, T.R., Kafatos, M., and Sofia, M., 1980, Ap. J.
    (Letters) 238, L27.
Chevalier, R.A. and Oegerle, W.R., 1979, Ap. J. 227, 398.
Chieze, J.P., and Lazareff, B., 1981, Astron. Astrophys. 95, 194.
Cowie, L.L., and McKee, C.F., 1977, Ap. J. 211, 135.
Cowie, L.L., McKee, C.F., and Ostriker, J.P., 1981, Ap. J. 247, 908.
Cox, D.P., 1979, Ap. J. 234, 863.
Cox, D.P., 1981, Ap. J. 245, 534.
Cox, D.P., and Smith, B.W., 1974, Ap. J. (Letters) 189, L105.
Dalgarno, A., and McCray, R.A., 1972, Ann. Rev. Astron. Astrophys. 10, 375.
de Jong, T., 1980, in Highlights of Astronomy, Vol. 5, P. Wayman, ed.,
    (Reidel: Dordrecht), p. 301.
Dopita, M. and Mathewson, D.S. 1979, Ap.J. (Letters) 231, L147.
Field, G.B., 1965, Ap. J. 142, 531.
Field, G.B., Goldsmith, D.W., and Habing, H.J., 1969, Ap. J. (Letters) 155,
    L49.
Fried, P.M. et al., 1980, Ap. J. 242, 987.
Habe, A., and Ikeuchi, S., 1980, Progr. Theoret. Phys. 69, 1995.
Habe, A., Ikeuchi, S., and Tanaka, Y., 1981, Pub. Astr. Soc. Japan 33, 23.
Heiles, C. 1979, Ap.J., 229, 533.
Jenkins, E.B., and Meloy, D.A., 1974, Ap. J. (Letters) 193, L121.
Jura, M., 1975, Ap. J. 197, 581.
Kahn, F., 1981, Z. Kopal Memorial Volume (in press).
Krolik, J., McKee, C.F., and Tarter, C.B., 1981, Ap. J. (in press).

McKee, C.F., 1982, in Supernovae (Proceedings of a NATO ASI), M. Rees and R. Stoneham, eds. (Cambridge University Press: Cambridge), in press.

McKee, C.F., and Cowie, L.L., 1977, Ap. J. 215, 213.

McKee, C.F., and Ostriker, J.P., 1977, Ap. J. 218, 148.

Raymond, J.C., Cox, D.P., and Smith, B.W., 1976, Ap. J. 204, 290.

Savage, B.D., and deBoer, K.S., 1979, Ap. J. (Letters) 230, L77.

Schwarz, J., McCray, R.A., and Stein, R.F., 1972, Ap. J. 175, 673.

Shapiro, P.R., and Field, G.B., 1976, Ap. J. 205, 762.

Spitzer, L., 1962, Physics of Fully Ionized Gases, 2nd ed. (New York: Wiley).

Verschuur, G. 1975 Ann. Rev. Astron. Astrophys. 13, 257.

Webster, A., 1978, M.N.R.A.S. 185, 407.

York, D.G., 1974, Ap. J. (Letters) 192, L127.

GAMMA RAYS AND ASTRONOMY

J.A. PAUL

Section d'Astrophysique, Centre d'Etudes Nucléaires de Saclay, France

ABSTRACT

Nowadays, the studies of celestial gamma
rays are of great interest for several areas of
galactic astronomy.  On a large scale, an upper
limit of about a factor 3 is placed on the
possible increase in density of the interstellar
gas in the ring at 5 kpc from the galactic center with
respect to the solar neighbourhood. Also the
locations of enhanced gamma-ray emissivity seem
to be preferentially distributed in accordance
with a spiral pattern. Gamma rays from the local
galactic environment appear to be a valuable probe
of the content and structure of the local interstellar medium. On a smaller scale, the detection
of numerous localized gamma-ray sources forces
to explore some particular phases of clusters of
young and massive stars.

## 1. Introduction

For a long time, cosmic gamma rays have been considered as "the most
elusive species among the cosmic rays" (Marcello Ceccarelli). Nowadays,
cosmic gamma rays are regarded as the most energetic natural form of the
electromagnetic radiation aiming at revealing  celestial bodies where
the largest energy transfers occur. The detection of high-energy gamma
rays always points to the presence of concentrations of relativistic
particles and matter or photon fields. The astronomical objects where
such concentrations are seen constitute the scope of gamma-ray astronomy, the youngest branch of the new astronomies.

This paper will concentrate on several galactic bodies for which the gamma-ray studies are of particular interest. In each case, the attention will be focused on the particular aspects emphasized by the gamma-ray observations. Note that the deep physical interpretations of the gamma-ray emission from the celestial bodies are far beyond the scope of this paper, and should be found elsewhere, in particular in many of the contributed papers gathered in the XG sessions of the Paris conference.

Also it should be considered that the gamma-ray domain is not yet fully exploited. The high-energy photons are observed above a background of $10^4$ times as many charged particles. It is thus not surprising that most of the experimental effort was devoted to the range above a few tens of MeV, where the pair-production event can be unambiguously identified by pictorial-type detectors among the events caused by the interaction of unwanted cosmic-ray particles. However, at low energies, triangulation techniques involving several space probes have succeeded to locate some of these transient gamma-ray sources which emit short bursts of photons of energy $\lesssim 1$ MeV.

At this time, high-energy gamma-ray observations results mainly from three successive space missions : the NASA satellites OSO-3 and SAS-2 and the ESA satellite COS-B. This european experiment, which has been successfully operated since its launch in August 1975, has provided most of the results presented in the next chapters. The dramatic breakthrough of COS-B in gamma-ray astronomy is best depicted by figure 1 which shows, as a function of time, the number of high-energy photons ( $\gtrsim 100$ MeV) collected from the galactic disc by the three main high-energy gamma-ray astronomy missions flown so far.

2.  The large scale characteristics of the Milky Way from gamma-ray observations

The strongest feature of the high-energy gamma-ray sky is the over-

whelming emission of the galactic disc.[1] Even the radiation observed
away from the galactic plane appears to be predominantly galactic, on
the basis of its latitude dependence (Fichtel et al. 1977) and of its
spatial correlation  with galactic species (Lebrun et al. 1981).

In contrast with the optical panorama, the gamma-ray appearance of
the Milky Way (fig. 2) immediatly reveals the very large scale structure
or "grand design" of our Galaxy ; for a gamma-ray eye, there is no absorption able to mask the remote regions of the Galaxy. In addition, as
high-energy gamma rays relate directly to the product of cosmic-ray
intensity and gas density,[2] the galactic gamma-ray observations appear
as an efficient technique to trace and study the spatial distribution
of the interstellar gas, including in particular its molecular component.
In this respect, gamma-ray surveys are complementary of radio-wave measurements.

One often tries to derive the molecular content of the galactic
disc from millimeter-wave observations of CO. But going from the intensity of the CO lines to the molecular hydrogen density is a multi-step
procedure, with a possible bug hidden at nearly every step (for a detailed discussion see Lequeux 1981). In this context, gamma-ray observations yield a clear constraint : in the restrictive hypothesis that the
gamma-ray emissivity is only due to interactions of cosmic rays with
the interstellar gas and under the assumption of a uniform cosmic-ray

---

[1]Nevertheless, extragalactic gamma-ray astronomy is not hopeless. The
active galactic nuclei and the quasars seem to emit most of their
energy in the gamma-ray range as illustrated by the MISO observations
of the Seyfert galaxy NGC4151 at low energy (Perotti et al. 1979, 1981)
and by the COS-B observations of the quasar 3C273 (Swanenburg et al.
1978, Bignami et al. 1981a).

[2]Compton radiation, resulting from galactic cosmic-ray electrons interacting with the 3°K blackbody radiation, with optical photons
and infrared photons may be also important in the 10 to 100 MeV range
(Kniffen and Fichtel 1981), but is usually neglected at higher energies.

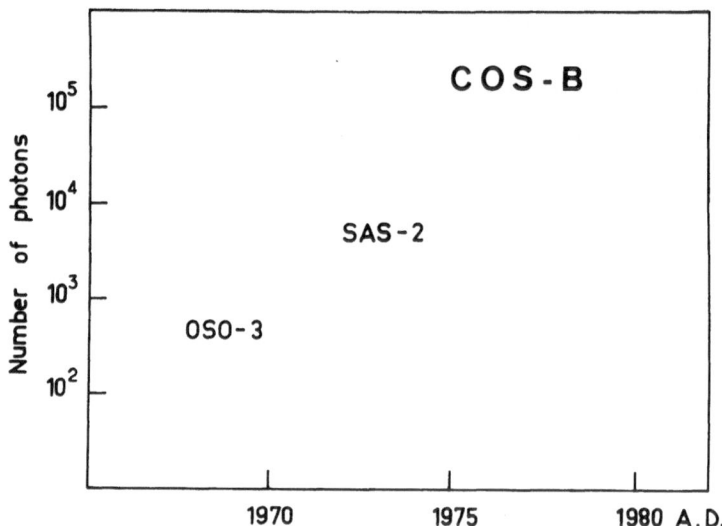

FIGURE 1. Total number of photons ( ≥ 100 MeV) from the Galaxy as a
function of the launch year for the major high-energy gamma-
ray missions flown so far (from Bignami, 1980).

FIGURE 2. (opposite side) map of the galactic gamma-ray intensity in
three energy intervals as observed by COS-B (from Mayer-
Hasselwander et al. 1981)

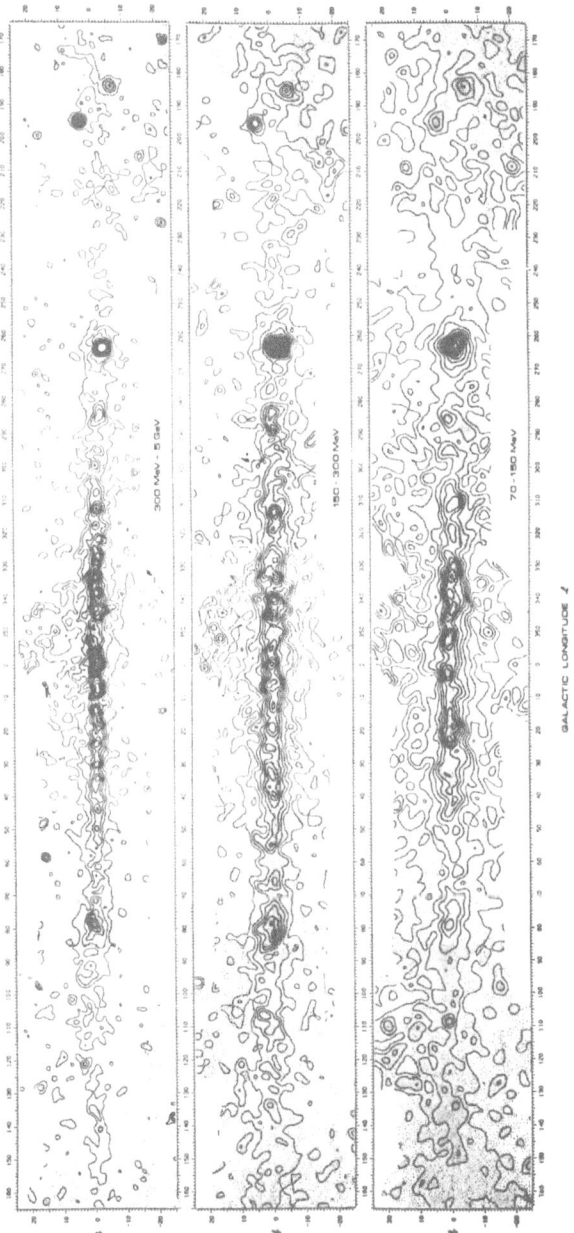

GALACTIC LONGITUDE ℓ

density, it is seen from the COS-B data that the total gas increase
cannot be more than a factor of 3 between 10 kpc and 5 kpc from the
galactic center (Mayer-Hasselwander et al. 1981). Since it is hard to
imagine that the cosmic-ray density is smaller in the 5 kpc galactic
ring of population I objects, the COS-B results tend to indicate that
the gas density in the 5 kpc ring cannot be more than 3 times higher
than locally.[3]

The large scale structure of the gamma-ray disc can be analyzed
without any assumption regarding the physical origin of the galactic
radiation. Such an approach has been attempted by Caraveo and Paul
(1979) using the unfolding technique described by Strong (1975), which
only implies the assumption of cylindrical symmetry (at least in one
half of the Galaxy). Figure 3 sketches the gamma-ray Galaxy, i.e. the
location of the regions of high gamma-ray emissivity, which contribute
significantly to the gamma-ray map. The regions of enhanced gamma-ray
emissivity appear to be preferentially distributed in accordance with
a spiral pattern. In a second step, in view of such an evidence for
spiral structure, a fixed  pattern can be imposed in the course of the
unfolding procedure (Kanbach and Beuerman 1979). The pitch angle of
the spiral is found to be $\sim 13°$, consistent with the consensus on the
galactic structure as known from radio data.

In addition to these model-independent derivations from the large
scale galactic gamma-ray observations, many attempts have been made to
yield quantitative informations on the total gaseous content of the
Galaxy from the gamma-ray data. But it is necessary to go through a
series of assumptions which are more or less questionable. First,
the point-source contribution to the galactic gamma-ray emission
should be taken as negligible. One should also consider that the gamma-
ray producing cosmic rays penetrate dense clouds. This is not unrea-

---

[3]The amount of gas derived by Solomon and Sanders (1980) for the inner
Galaxy cannot be reconciled  with the maximum tolerable amount of gas
compatible with the gamma-ray data, unless the cosmic-ray density is
    2 times lower than in the solar neighbourhood.

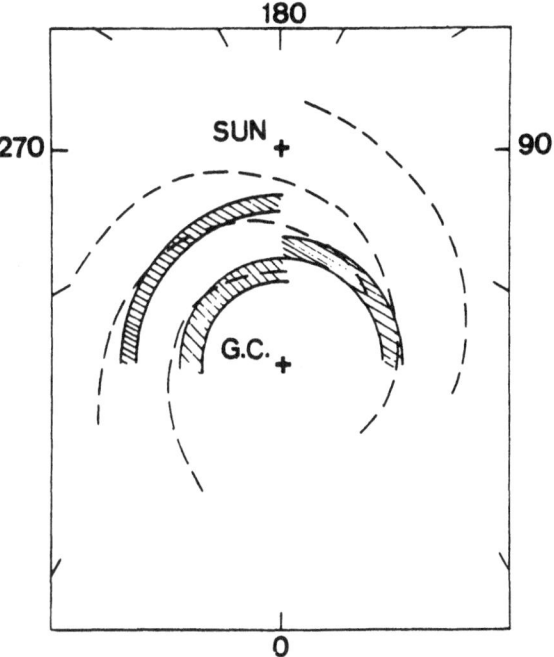

FIGURE 3. Repartition of high-energy gamma-ray emissivity regions
across the galactic plane (shaded area) unfolded from the
SAS-2 longitude profile, superimposed on the spiral pattern
derived by Georgelin and Georgelin (1976). The regions of
the galactic center and beyond are excluded from this
analysis of Caraveo and Paul (1979).

listic on the basis of the theoretical works of Skilling and Strong
(1976) and of Cesarsky and Völk (1978). Also it is shown in the case
of the Orion cloud complex that the simplest interpretation of the
gamma-ray observations is that the totality of the mass is involved
in the production of the gamma rays (Caraveo et al. 1980).

The most critical assumption concerns the large scale distribution
of the cosmic rays. For instance, taking advantage of the similarity
in the distribution of gamma rays and synchrotron radiation,Paul, Cassé,
and Cesarsky (1976) postulated that, on a large scale, the cosmic-ray
density is proportional to the gas density. The distribution of gas
thus predicted seems again inconsistent with that derived from radio
data (Fig. 4).

### 3. Gamma rays as a probe of the local interstellar medium

Also in the local galactic environment, the gamma-ray studies are
of great interest. Paradoxically, the local interstellar medium is not
well known, mainly because of the large size of the local structures
when compared with the very small fields of the optical and radio
telescopes. Large-scale studies, mandatory to depict the local galactic
environment, are often neglected ; large-scale surveys are not abundant
over the various domains of galactic astronomy. In spite of the limited
performance of the present generation of gamma-ray telescopes (angular
resolution, sensitivity,...), the observations of high-energy photons
provide a new insight into the local interstellar medium.

As a consequence of the thickness of the gaseous disc, the local
interstellar gas can be observed at medium galactic latitudes ( $|b| \geq 10°$)
i.e. in regions of the sky where there is no point-source contribution.
Moreover, in the local galactic environment, the cosmic-ray density can
reasonably be considered as uniform, and the intensity of the high-
energy gamma radiation observed at $|b| \geq 10°$ then gives a straightforward
estimate of the total gas column density. The local interstellar medium
contains an unknown amount of molecular hydrogen. The total gas (molecular + atomic) could be traced by gamma rays as well as by interstellar
absorption while the observation of the 21 cm transition line depicts
the distribution of the neutral atomic component HI only.

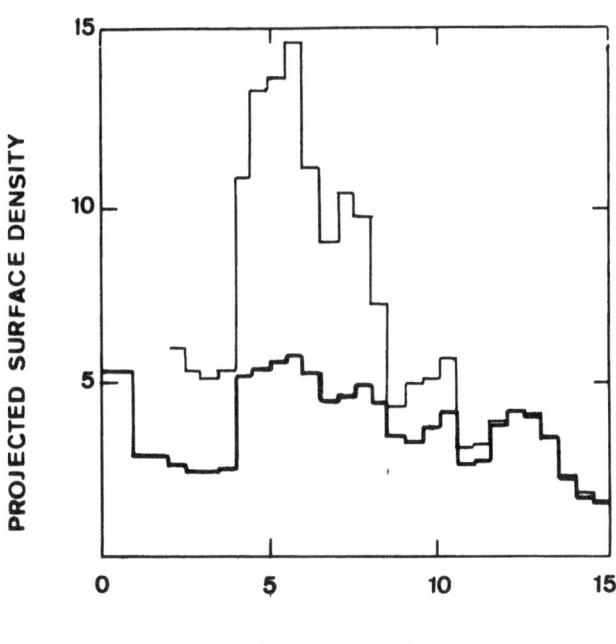

FIGURE 4. Radial distribution of the projected surface density of the
interstellar gas (in unit of $M_\odot$ per sq.pc.). Thin line : results of Gordon and Burton (1976). Thick line : derived in
Cesarsky, Cassé and Paul (1977) from the gamma-ray observations.

Since the work of Lilley (1955), it has been recognized that a relationship exists between gas and dust in the local interstellar medium.
The interstellar absorption, traced e.g. by the galaxy counts, has been
shown to be closely related to the total gas column density (see for exem-
ple Strong and Lebrun 1981 and references therein). On the basis of the
SAS-2 data, Lebrun and Paul (1979, 1981) have shown that at medium galactic latitudes, gamma rays are better correlated with the interstellar
absorption, derived from galaxy counts, than to the atomic hydrogen
column density $N_{HI}$.[4] This implies that, in addition to the atomic hydrogen,
a substantial fraction of the interstellar medium emits gamma rays and
contains dust ; this fraction is likely to be molecular hydrogen. The
same conclusion is supported by the recent COS-B data (Lebrun et al.
1981).

Such an analysis give the opportunity to map the regions where the
molecular gas is particularly conspicuous, i.e., the regions where the
observed gamma-ray intensity exceeds strongly the gamma-ray intensity
expected from the neutral atomic hydrogen alone. This map of the local
molecular gas, as derived by Lebrun and Paul (1979, 1981), reveals
several regions of interest : in addition to some well-known closeby
molecular clouds (as e.g. the Taurus and Rho Oph cloud complexes), a
wide (angular size $\sim$ 50°) concentration of molecular gas in apparent in
the Aquila-Ophiuchus-Sagittarius region. This nearly molecular complex,
discovered by the gamma-ray study, has attracted the attention of the
radio astronomers : a complete CO survey of this region has been undertaken at the Columbia-Goddard Institute for Space Studies millimeter-
wave telescope . Preliminary results of this survey testify that the
predicted interstellar molecules are indeed present in this region of
the sky (Lebrun and Huang 1981).

---

[4]On the basis of the same SAS-2 data, Fichtel, Simpson and Thompson
(1978) have found that gamma rays are well correlated to $N_{HI}$. But they
had performed their analysis in sky bins taken at different latitudes
so that the correlation they had observed results mainly from the fact
that the latitude distributions of both gamma rays and $N_{HI}$ follow a
cosec $|b|$ law and does not reflect an intrinsically similar distribution.

4. The gamma-ray source : an unforeseen aspect of clusters of young and massive stars

After having considered large-scale structures, I will turn now to smaller objects, whose angular extent is so small ( $\leq$ 2°) that they cannot be resolved by the present generation of gamma-ray instruments. Based on the data acquired during the first 32 COS-B observation periods, the second COS-B catalogue of gamma-ray sources has now been compiled (Swanenburg et al. 1981). The 25 sources contained in this catalogue are displayed in figure 5, which also illustrates the region of the sky which has been searched. Only four sources have been identified : with the Crab and Vela pulsar, with the quasar 3C273 and most probably with the Rho Oph cloud complex as well. All the remaining sources (except one) are located very close to the galactic plane : they are obviously galactic. It is beyond the scope of this paper to present a detailed review of the suggested models for gamma-ray sources. Also I will not attempt to describe the high-energy gamma-ray emission processes of pulsars ; such processes are reviewed in Cesarsky and Paul (1981). I will solely concentrate on several aspects of the possible relationship between gamma-ray sources and young and massive stars.

As a galactic population, the clusters of young and massive stars (the OB associations) present some similarities with the gamma-ray sources as e.g. the very limited latitude extent. But a detailed study shows that only a sub-class of the OB associations may be the site of a copious gamma-ray emission. The young and massive star clusters are generally associated with dense and massive clouds. Also, in some cases, star clusters could be the site of an efficient cosmic-ray energization. Only the vicinity of these "active" OB associations may turn out to be a powerful source of gamma radiation, provided that the gamma-ray producing cosmic rays are efficiently trapped in the dense surrounding medium. The SNOB concept (supernova remnants physically linked with OB associations ) proposed by Montmerle (1979), provides a plausible scenario to interpret the gamma-ray emission from active OB associations : in this model, low energy particles injected by active stars pertaining to an OB association are accelerated by the passing shock induced by a super-

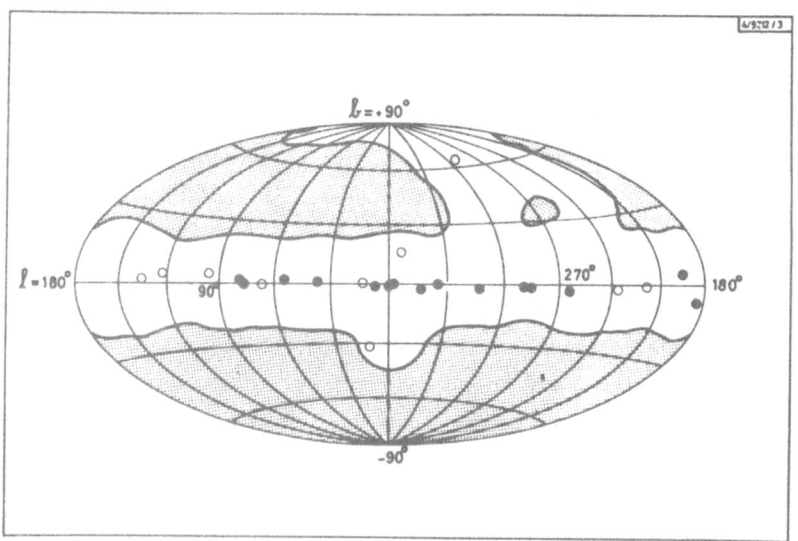

FIGURE 5. Gamma-ray sources contained in the 2CG catalog (Swanenburg
et al. 1981). Filled circle : sources with flux $\geq$ 1.3 $10^{-6}$
photons $cm^{-2}s^{-1}$. Open circles : weaker sources. Unshaded
area : region of the sky searched.

nova remnant expanding within the association. Thirty two SNOBs have been listed and 7 to 9 COS-B sources coincide with them (Montmerle and Cesarsky 1980). Several other SNOBs are situated in more extended regions of intense gamma-ray emission.

On the other hand, recent observations (see e.g. Snow and Morton 1976, Cassinelli 1979, Conti and Garmany 1980) have shown that O stars and B supergiants have strong stellar winds with typical velocities $V_w \sim 2000$ km s$^{-1}$ and mass-loss rate $\dot{M} \sim 10^{-7}$ to $10^{-5}$ $M_\odot$ yr$^{-1}$. The adjustment process of the stellar-wind flow to the environment should involve a shock transition. This has prompted Cassé and Paul (1980) to propose stellar-wind terminal shocks as a site of acceleration of charged particles, extending an early suggestion of Jokipii (1968) for the solar wind. Howevern most of the known OB associations do not contain stars emitting enough mechanical energy to be detectable as individual gamma-ray source by the COS-B satellite (Cassé, Montmerle and Paul 1981) On the contrary, the Carina Nebula, which comprises several OB associations and Wolf-Rayet stars - the most powerful mass losing massive stars - meets the visibility requirement. The presence of the gamma-ray source 2CG288-00 in the direction of the Carina Nebula forces to focus the attention on this rich concentration of young and massive stars (figure 6).

Montmerle, Cassé and Paul (1981) have proposed that a modest fraction ( < 10%) of the mechanical energy released by the stars in the form of supersonic stellar wind may be converted into cosmic-ray energy. The cosmic rays which are efficiently trapped by the resonant Alfèn-wave scattering (e.g. Cesarsky 1980) in the large and dense ionized HII region which surrounds the Carina star clusters, interact with the gas and produce the observed flux of gamma radiation. Some of the empirical parameters entering into this model, as e.g. the stellar-wind energy budget, are now substantiated by the recent UV observations of the O3 star HD93205 pertaining to the Carina Nebula (Laurent, Paul and Pettini 1981).

Note that cosmic-ray confinement is an essential ingredient in these models : for instance, in spite of the presence of numerous mass-

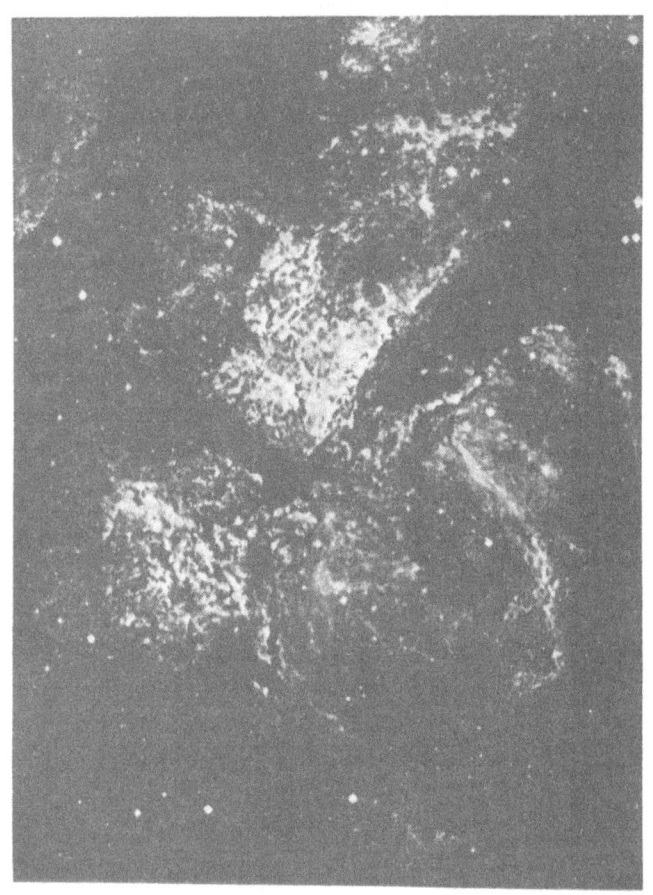

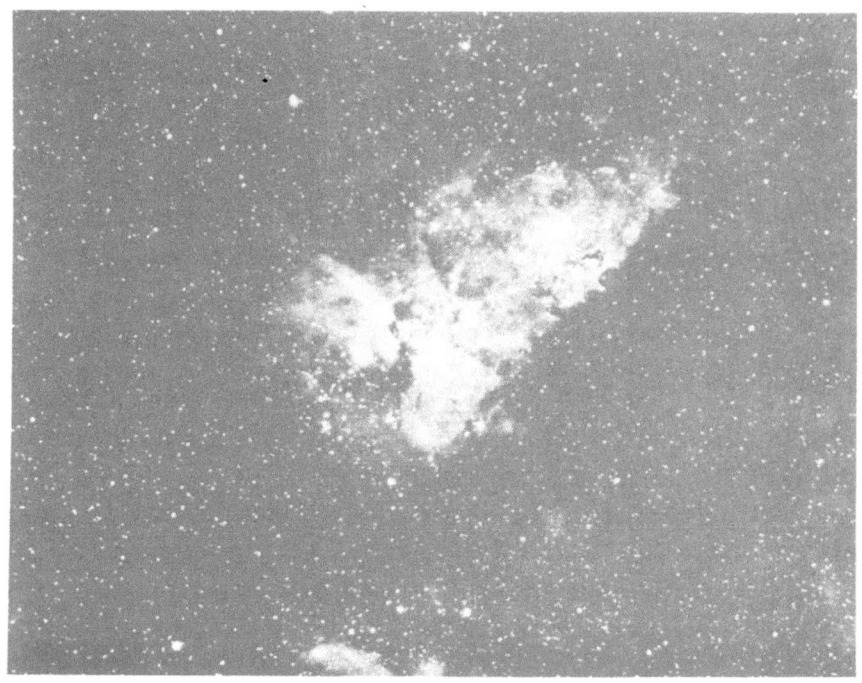

FIGURE 6. (Left) The Great Carina Nebula. (Right) The central region
of the Carina Nebula which contains the youngest and
brightest stars.

losing young stars in the Orion complex, together with a huge concentration of dense interstellar gas, confinement by resonant Alfèn-wave scattering is inefficient. No enhanced gamma-ray emissivity is expected within the Orion complex, as observed (Montmerle 1981).

The Carina complex, the richest concentration of young and massive stars can be considered as a candidate gamma-ray source. The same is for the closest aggregate of gas, dust and young stars : the Rho Oph cloud complex. The dense part of the cloud lies within the error circle of the COS-B source 2CG353+16. The observed flux is 4 to 10 times higher than the flux expected if, within the core of the cloud, the cosmic-ray intensity and energy spectrum are as observed in the solar vicinity (Simpson 1979, Cassé and Paul 1980, Bignami and Morfill 1980). A natural interpretation is that relativistic particles are accelerated within the cloud complex either by the supersonic stellar-wind mechanism (Cassé and Paul 1980, Paul, Cassé and Montmerle 1981) or by a close-by supernova shock - the North Polar Spur - (Morfill et al. 1981). However Issa, Strong and Wolfendale (1981) dispute the requirement of a higher cosmic-ray intensity within the cloud, noting that the surroundings of the cloud may also contribute to the observed flux of the gamma-ray source.

In spite of the arguments presented here in favour of the possible links between a class of gamma-ray sources and OB associations, it should be stressed that in several other cases the observational material does not substantiate the idea that the gamma-ray source results simply from a significant concentration of gas and cosmic rays. So is the first yet unidentified gamma-ray source, discovered by the SAS-2 satellite (Kniffen et al. 1975), and called "Geminga" by the COS-B workers. Searches for objects conspicuous at other wavelengths within the error box of Geminga and of other similar unidentified gamma-ray sources appear as a promising way to clarify their nature. For example, the peculiar radio star LSI+61°303 is located well inside the ∼ 3 sq. deg. error box of the gamma-ray source 2CG 135+01. This star is now recognized to be an X-ray source, whose emission seems to exceed that expected from the mere coronal emission (Bignami et al. 1981b). This result presses the identification of the star with the gamma-ray source, even if the suggested presence of a <u>SNOB</u> in the direction of 2CG135+01 (Montmerle 1979) makes the situation only more confusing.

## 5. Conclusions

Thanks to the efforts of the high-energy physicists, the study of the celestial gamma radiation can be now regarded as a promising branch of astronomy. But nowadays, gamma-ray astronomy is faced with the crucial problem of the very limited angular resolution of the gamma-ray telescope. Historically, major advances in all fields of astronomy had a clear correlation with the development of sensitive telescopes. A significant increase in the angular resolving power, in particular, has invariably led to major discoveries concerning the Universe. One may recall that the discovery of the quasars occured at the point when radio astronomers developed techniques capable of locating objects with sufficient accuracy for the isolation of optical counterparts (see also the results obtained with the VLA on the nuclei of extragalactic radiosources). The gamma-ray astronomers anticipate similar advances from their forthcoming experiments. Without neglecting the imperative exigency of factual observations in the low energy domain (0.1 to 30 MeV), it remains that the most vital goal for the next generation of gamma-ray telescopes is a striking improvement of the angular resolving power. Experiments with angular resolution $\sim 1$ arc min. are already planned. At the time of their completion, the links between gamma rays and astronomy, outlined in this paper, will set the framework for the emergence of an adult gamma-ray astronomy.

## Acknowledgements

I thank Thierry Montmerle for very fruitful discussions.

## REFERENCES

Bignami, G.F. 1980, in Non-Solar Gamma-Rays (COSPAR), ed. R. Cowsik and
R.D. Wills, Pergamon Press : Oxford, p. 227.
Bignami, G.F., Bennett, K., Buccheri, R., Caraveo, P.A., Hermsen, W.,
Kanbach, G., Lichti, G.G., Masnou, J.L., Mayer-Hasselwander, H.A.,
Paul, J.A., Sacco, B., Scarsi, L., Swanenburg, B.N., and Wills, R.D.,
1981a, Astron. Astrophys., 93, 71.
Bignami, G.F., Caraveo, P.A., Lamb, R.C., Markert, T.H., and Paul, J.A.,
1981b, Ap.J. (Letters), 247, L85.
Bignami, G.F., and Morfill, G.E., 1980, Astron. Astrophys., 48, 481.
Caraveo, P.A., Bennett, K., Bignami, G.F., Hermsen, W., Kanbach, G.,
Lebrun, F., Masnou, J.L., Mayer-Hasselwander, H.A., Paul, J.A., Sacco,
B., Scarsi, L., Strong, A.W., Swanenburg, B.N., and Wills, R.D. 1980,
Astron. Astrophys., 91, L3.
Caraveo, P.A., and Paul, J.A. 1979, 75, 344.
Cassé, M., Montmerle, T., and Paul, J.A. 1981, in Origin of Cosmic Rays,
IAU/IUPAP Symp. n° 94, ed. G. Setti, G. Spada, A.W. Wolfendale, p. 323
Cassé, M., and Paul, J.A., 1980, Ap.J., 237, 236.
Cesarsky, C.J. 1980, Ann. Rev. Astr. Ap., 18, 289.
Cesarsky, C.J., Cassé, M. and Paul, J.A., 1977, Astron. Astrophys., 60,
139.
Cesarsky, C.J., and Paul, J.A. 1981, Nucl. Sciences Appl.,in press.
Cesarsky, C.J., and Völk, H.J. 1978, Astron. Astrophys., 70, 367.
Conti, P.S., and Garmany, C.D. 1980, Ap.J., 238, 190.
Fichtel, C.E., Hartman, R.C., Kniffen, D.A., Thompson, D.J., Ogelman,
H.B., Ozel, M.E. and Tümer, T. 1977, Ap.J. (Letters), 217, L9.
Fichtel, C.E., Simpson, G.A. and Thompson, D.J. 1978, Ap.J.,222, 833.
Georgelin, Y.M., and Georgelin Y.P. 1976, Astron. Astrophys., 49, 57.
Gordon, M.A. and Burton, W.B. 1976, Ap.J., 208, 346.
Issa, M.R., Strong, A.W., and Wolfendale, A.W., 1981, J. Phys. G : Nucl.
Phys., 7, 565.
Jokipii, J.R. 1968, Ap.J., 152, 799.
Kanbach, G., and Beuermann, K. 1979, Proc. 16th Int. Cosmic Ray Conf.,
Kyoto, 1, 75.
Kniffen, D.A., Bignami, G.F., Fichtel, C.E., Hartman, R.C., Ogelman, H.,
Thompson, D.J., Ozel, M.E., and Tümer, T. 1975, Proc. 14th Int.
Cosmic Ray Conf., München, 1, 100.
Kniffen D.A., and Fichtel, C.E. 1981, Ap.J., in press.
Laurent, C., Paul, J.A., and Pettini, M. 1981, Ap.J., in press.
Lebrun, F., Bignami, G.F., Buccheri, R., Caraveo, P.A., Hermsen, W.,
Kanbach, G., Mayer-Hasselwander, H.A., Paul, J.A., Strong, A.W., and
Wills, R.D. 1981, Astron. Astrophys., in press.
Lebrun, F. and Huang, Y.L. 1981, Ap.J., submitted.
Lebrun, F., and Paul, J.A. 1979, Proc. of 16th Int. Cosmic Rays Conf.,
Kyoto, 12, 13.
Lebrun F. and Paul, J.A. 1981, Ap.J., submitted.
Lequeux, J. 1981, Comments on Astrophysics, in press.
Lilley, A.E. 1955, Ap.J., 121, 559.
Mayer-Hasselwander, H.A., Bennett, K., Bignami, G.F., Buccheri, R.,
Caraveo, P.A., Hermsen, W., Kanbach, G., Lebrun, F., Lichti, G.G.,

Masnou, J.L., Paul, J.A., Pinkau, K., Sacco, B., Scarsi, L.,
Swanenburg, B.N., and Wills, R.D., 1981, Astron. Astrophys., in press.
Montmerle, T. 1979, Ap.J., 231, 95.
Montmerle, T. 1981, Phil. Trans. R. Soc. Lond. A 301, 505.
Montmerle, T., Cassé, M. and Paul, J.A., 1981, Ap.J., submitted.
Montmerle, T., and Cesarsky, C.J. 1980, in Non Solar Gamma-Rays (COSPAR)
ed. R. Cowsik and R.D. Wills, Pergamon Press : Oxford, p. 61.
Morfill, G.E., Völk, H.J., Forman, M., Bignami, G.F., Caraveo, P.A., and
Drury, L. 1981, Ap.J., 246, 810.
Paul, J.A., Cassé, M. and Cesarsky, C.J. 1976, Ap.J., 207, 62.
Paul, J.A., Cassé, M. and Montmerle, T. 1981, in Origin of Cosmic Rays,
IAU/IUPAP Symp. n° 94, ed. G. Setti, G. Spada and A.W. Wolfendale,
p. 325.
Perotti, F., Della Ventura, A., Sechi, G., Villa, G., Di Cocco, G.,
Baker, R.E., Buther, R.C., Dean, A.J., Martin, S.J., and Ramsden, D.
1979, Nature, 282, 484.
Perotti, F., Della Ventura, A., Villa, F., Di Cocco, G., Bassani, L.,
Buther, R.C., Carter, J.N., and Dean, A.J., 1981, Ap.J., (Letters),
239, L49.
Simpson, G. 1979, in Particle Acceleration Mechanism in Astrophysics,
ed. J. Arons, C. Mc Kee and C. Max, American Institute of Physics :
New York, p. 289.
Skilling, J. and Strong, A.W., 1976, Astron. Astrophys., 53, 253.
Snow, T.P. and Morton, D.C. 1976, Astrophys. J. Suppl., 32, 429.
Solomon, P.M. and Sanders, D.B. 1980, in Giant Molecular Clouds in the
Galaxy, ed. P.M. Solomon and E.G. Edmunds, Pergamon Press : Oxford,
p. 41.
Strong, A.W., 1975, J. Phys. A. Math. Gen., 8, 617.
Strong, A.W. and Lebrun,F. 1981, to be published in Astronomy and Astro-
physics.
Swanenburg, B.N., Bennett, K., Bignami, G.F., Buccheri, R., Caraveo, P.,
Hermsen, W., Kanbach, G., Lichti, G.G., Masnou, J.L., Mayer-Hasselwan-
der, H.A., Paul, J.A., Sacco, B., Scarsi, L., and Wills,R.D. 1981,
Ap.J. (Letters), 243, L49.
Swanenburg, B.N., Bennett, K., Bignami, G.F., Caraveo, P., Hermsen, W.,
Kanbach, G., Masnou, J.L., Mayer-Hasselwander, H.A., Paul, J.A.,
Sacco, B., Scarsi, L. and Wills, R.D. 1978, Nature, 275, 298.

# X-RAYS ASSOCIATED WITH HIGH ENERGY PARTICLES

Paul Gorenstein
Harvard/Smithsonian Center for Astrophysics
60 Garden Street
Cambridge, MA 02138
United States

## ABSTRACT

A number of recent results in X-ray astronomy, particularly from the Einstein Observatory, are relevant to the acceleration or interaction of high energy particles. X-ray observations of M stars by the Einstein Observatory indicate that their frequent flares include X-ray emission. The correspondence of the flux and duration of M star X-ray flares to those of solar flares can be used to estimate the rate of >200 MeV proton injection into the galaxy. Given the uncertainties, there is no disagreement with the cosmic ray density. Thus, a two stage acceleration process may be viable at least for protons.

Supernova remnants (SNR's) are characterized by hot, turbulent shock waves with temperatures in excess of $10^6$ K to age $10^4$ years as seen, for example, in the Vela SNR. Thus, SNR's are good candidates as cosmic ray acceleration sites. X-ray jets due to particle beams are seen to emanate from SS433 and G109.1-1.0, two objects where SNR's are associated with compact X-ray sources.

Among extragalactic X-ray sources, the radio galaxies, Cen A and M87, have X-ray jet features emanating from their centers. M87 has more extended X-ray lobes in addition. These M87 features are due to synchrotron radiation of very high energy electrons and inverse Compton scattering of moderately high energy electrons against the 2.7K background.

## 1. Introduction

There are good reasons to expect an intimate connection between the various forms of high energy radiation. The acceleration, transport, and interaction of high energy particles is likely to occur in circumstances where they are accompanied by radiation of X-rays or gamma rays. It is interesting and useful to explore this connection both theoretically and through observations. In so doing, we can further our understanding of a wide variety of processes in the universe ranging from the active coronae of stars to the powerful particle accelerators in the nuclei of quasars. Cosmic ray measurements provide the number density, energy spectrum, mass composition, and charge composition of particles arriving at the solar system. Accounting for perturbing effects of the Sun, we can apply this information to other regions of the galaxy and beyond. However, cosmic ray measurements give very little information about their direction of origin. On the other hand, in the X-ray band with the use of imaging telescopes, we can precisely determine arrival directions and study morphology. Thus, we can identify sources where particle activity is likely to be involved but have virtually no information about the particles themselves. In this respect, cosmic ray measurements are complementary to the cosmic X-ray observations. Figure 1 illustrates the physical environments in which X-rays and particles are connected.

After two and a half years of observing with the Einstein Observatory (HEAO-2), a number of results have been obtained by consortium investigators and guest observers that are examples of processes shown in Fig. 1 and thus relevant to the presence, the acceleration, or the interaction of high energy particles. (The Einstein Observatory concluded operations in April 1981 when gas for the control system was exhausted.) I will discuss these results by referring mostly to observations on galactic sources. Outside of the galaxy, the main probes of high energy electrons are X-ray jets produced by synchrotron radiation and X-ray lobes produced by inverse Compton scattering of the 2.7°K microwave background.

---

Figure 1. Various processes in which X-rays are associated with particles. Paticle events on the Sun are accompanied by X-rays (insert 1). X-ray flare events from M stars with characteristics to Sun could be similar in particle emission. In insert 2, a rapidly rotating neutron star ("Rapid Pulsar") accelerates electrons to sufficiently high energies to produce X-rays by synchrotron radiation ("SYNC"). Low energy protons emitted by various compact objects and supernova explosions interact with ambient cloud medium to produce shock waves that accelerate electrons (and presumably protons and nuclei) to high energies. X-rays may be radiated thermally in the vicinity of the shocks (diagonally shaded region). Very high energy electrons produce X-rays by synchrotron radiation. Further out, moderately high energy electrons (few GeV) produce X-rays by inverse Compton scattering ("I.C.") against the 2.7K microwave background.

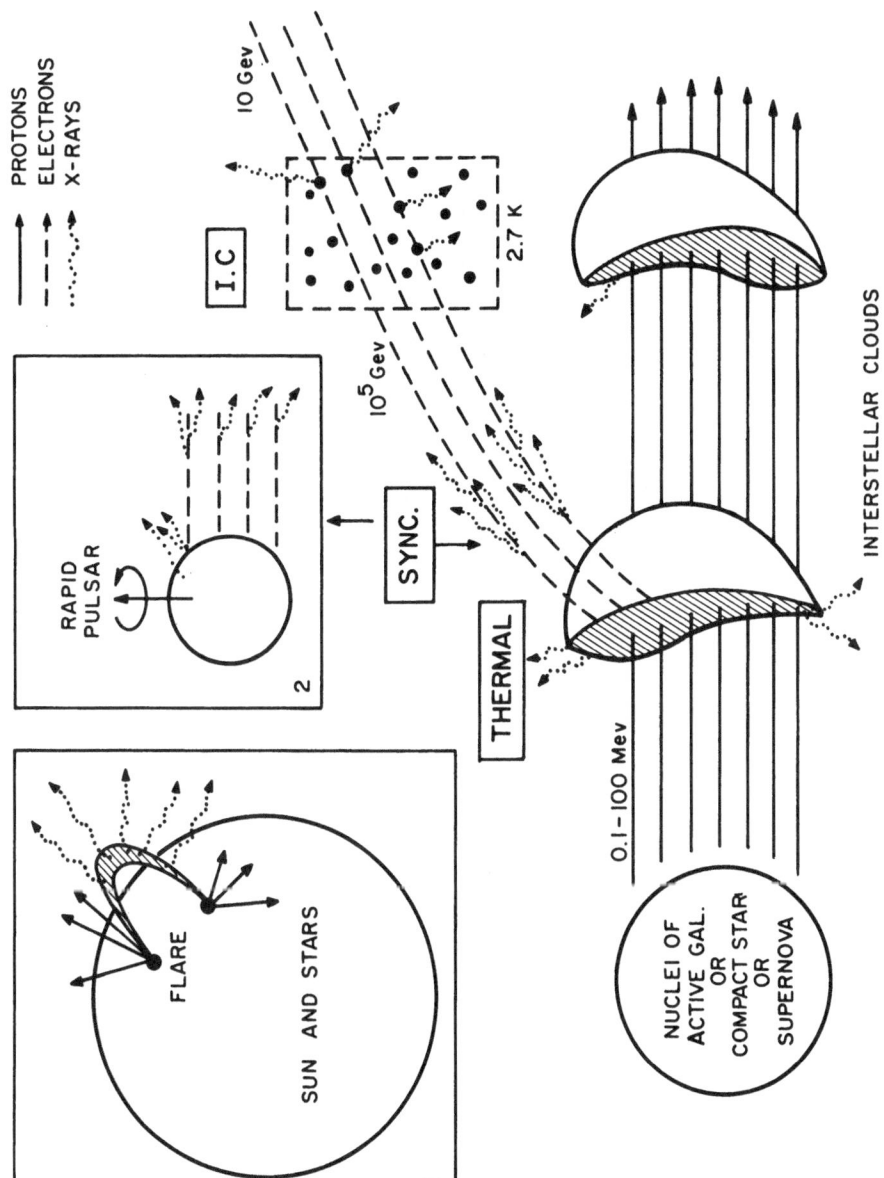

Figure 1

## 2.   The Galaxy

In the galaxy, four types of X-ray sources are of interest with respect to high energy particles. They are:
(1) Flare stars
(2) Pulsars and other active neutron stars
(3) Supernova remnants
(4) A possible non-thermal diffuse component

For each of the first three types of objects, it has been proposed that they are accelerators of charged particles and are possible sources of cosmic rays. The fourth type, if it is measurable at all, would be produced by the interaction of high energy electrons most likely through synchrotron radiation in a galactic magnetic field or by inverse compton radiation against the 3° background or stellar radiation. We do not yet have a means of discriminating between these sources of truly diffuse galactic X-rays and a large number of unresolved point sources. Hence, this possible source will not be discussed.

### 2.1   Flare Stars and the Injection of Particles, The Two Stage Acceleration Process

The Sun is known to be a source of particles. Particle streams are emitted episodically in flares which are characterized by strong X-ray emission. Because the solar particle number energy spectra declines rapidly above 100 MeV, solar-like particle acceleration is not likely to account for galactic cosmic rays. However, it is interesting to estimate the density of particles that could emanate from flare like phenomena as a primary source for a subsequent acceleration process. An increase in X-ray emission seems to be an excellent indicator of particle flare events on the Sun. X-ray emission is a common feature of stars. Figure 2 is a diagram of X-ray luminosity versus spectral type (Topka, 1980). The intrinsic X-ray luminosity stops decreasing as we examine stars further to the right. These are low mass, low temperature K and M stars. These are extremely numerous and are the majority of stars in the galaxy. In addition, the most prolific sources of X-ray flares in the galaxy seem to be dwarf M stars with emission lines in their optical spectrum or dMe stars. dMe stars emit relatively little visible light, yet their intrinsic X-ray flux is comparable to much brighter stars, like F and G spectra types. They are expected to flare about $10^2$ times as often as the Sun (Rosner & Vaiana, 1978). An X-ray flare

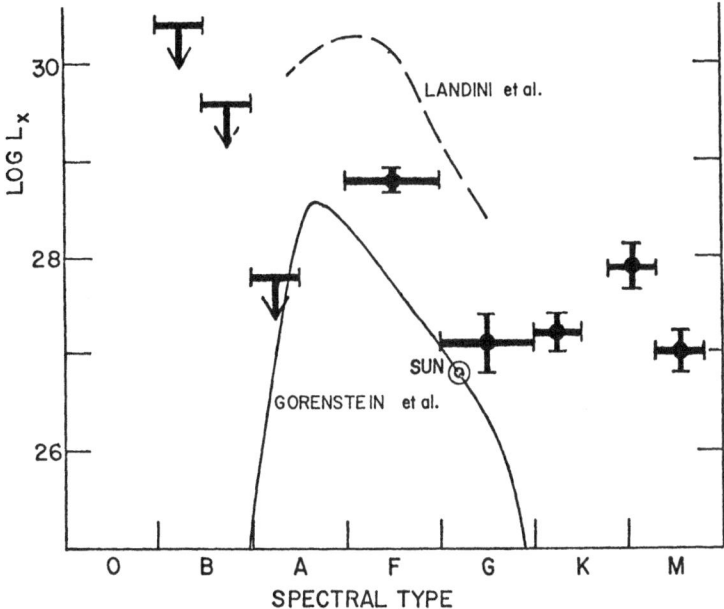

Figure 2. X-ray luminosity versus spectral type for stars. The X-ray luminosity remains high for "late" stars (K and M), which are the most numerous objects in the galaxy (Topka, 1980)

from the nearest M dwarf, Proxima Centauri, is shown in Figure 3 (Haisch et al., 1980).

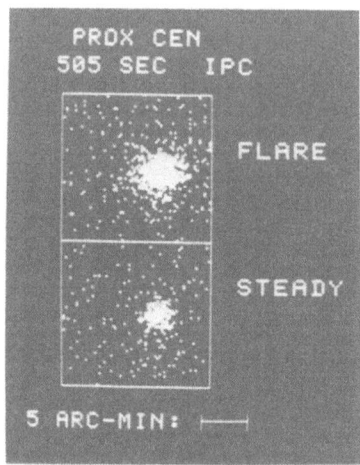

Figure 3. An X-ray flare in our nearest known stellar neighbor, Proxima Centauri, an M star (Haisch et al., 1980).

The X-ray energy radiated during this flare, ~$10^{32}$ ergs, is about equal to what might be seen in a large solar flare. To the extent we know about them, the soft X-ray characteristics of flares from dMe stars are similar to that of the Sun. On the basis of the direct resemblence between X-ray flare events on the Sun and dMe's, we could assume without scaling that their particle fluxes are similar. The physical connection between particle fluxes and X-ray emission is the internal magnetic field of the star and its rotation. Severe changes in the field configuration at the surface leads to concurrent X-ray flares and particle acceleration. The idea that flare stars are sources of particles is not new and even precedes the discovery of cosmic X-rays. The significance of their X-ray emission is that through the Sun, it is a quantitative estimator of the particle fluxes. M stars are the most common object in the galaxy, and dMe's are a large fraction of them. Their number density is several times larger than all the other stellar spectral types summed together. As discrete sources, they contribute more X-ray emission to the galaxy than all the other single stars (i.e., non-binary systems) combined. The total X-ray flux in the galaxy from dMe stars is about $10^{39}$ erg/sec. This implies solar-like particle emission throughout the galaxy. Because they are solar-like, the energy of these particles from M stars is, of course, far short of cosmic rays actually observed. Thus, dMe's cannot by themselves explain the acceleration of cosmic rays. However, the energy is high enough to satisfy one of the major requirements of the original classical Fermi acceleration mechanism. In this process, the particles need an initial boost in energy. Subsequently, particles are accelerated to relativistic energy by collisions with moderate or high velocity clouds. The Fermi mechanism becomes more efficient as the particle energy increases because ionization losses are reduced and the ratio of the number of head-on to overtaking encounters increases. Particles emitted from dMe stars could have the injection energy needed (the order of

200 MeV for protons) to possibly make this acceleration mechanism once again a viable idea.

More intense stellar X-ray flares have been seen at certain times from individual stars with the IPC of the Einstein Observatory. Stern, Underwood, and Antiochos (1981) detected a 2500 sec X-ray flare of $10^{34}$ ergs from a binary system of two normal stars in the Hyades cluster. This is a hundred times larger than a typical solar flare. The system contains a rapidly rotating K star. Because of the probable link between rapid rotation and flare activity, the K star is the more likely candidate. X-ray coverage of stars is still too sparse to know the rate of occurrence of giant stellar X-ray flares in the galaxy. Also, one does not know how to scale the flux and energy spectrum of particles associated with giant stellar flares to that of solar flares. In any case, giant stellar X-ray flares are another potential source of particles to add to the output of the dMe's.

With the similarity of X-ray flare events serving as a connection between dMe's and the Sun, we can estimate the output of particles with energies exceeding 200 MeV for the former from observations of the latter. Typical values for the solar particle emission fluxes can be estimated from the solar particle data reviewed by Lanzerotti (1977) (Fig. 4).

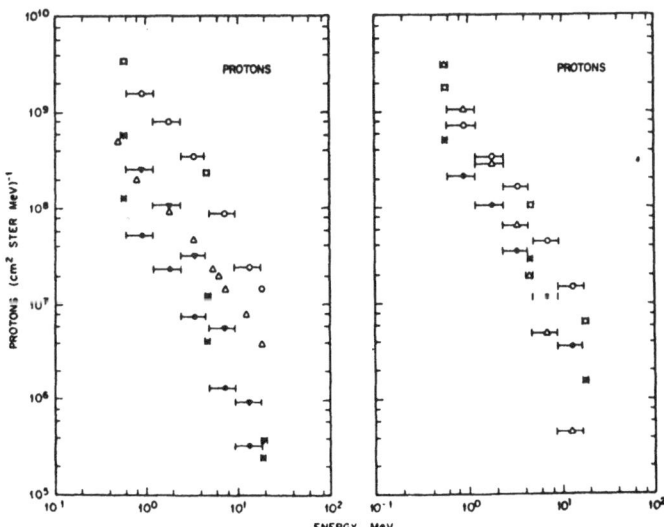

Figure 4. Proton emission from solar flares (Lanzerotti, 1977).

One must extrapolate the data shown, a not unreasonable amount, beyond the final measurements to estimate the flux of solar particles > 200 MeV. The density of dMe's in the galaxy is about 0.065 $pc^{-3}$, or each dMe is associated with a volume of 15 pc$^3$ .

Using these numbers, a typical dMe is expected to emit about $10^{31}$ particles/sec > 200 MeV. Measurements show that cosmic rays have traversed a path length of the order of 10 grm/sec. This implies residence times in the galaxy of about $10^7$ years. Thus, the typical 15 pc$^3$ cell around a dMe can accumulate a density of a few x $10^{-11}$ cm$^{-3}$ of particles > 200 MeV. This is somewhat short of the cosmic ray density; but in light of the various uncertainties and the extra contribution from the rarer giant flares, it may be compatible with it. Thus, given the number of protons that are given an initial boost to 200 MeV by dMe's, it may be worthwhile to reconsider the old Fermi process as the mechanism by which particles are accelerated in high energies. The problem of explaining the cosmic abundance of cosmic rays remains. The ejection spectrum may be near cosmic but subsequent acceleration is less efficient for heavier nuclei, so it is not evident how cosmic abundance is maintained (Eichler 1978).

### Compact X-ray Binaries

The X-ray emission from all the compact X-ray binaries in the galaxy exceeds the dMe total by far above 2 keV. This is illustrated in studies of our neighbor galaxy, M31 (Van Speybroeck et al., 1979), where some hundred discrete sources exceed by many times the total emission of M stars in M31. If we were merely scaling particle flux to X-ray emission, we would conclude that compact binaries are a bigger source of particles. However, with two important exceptions, there is no reason to believe that high energy particles are associated with the X-ray emission from compact binaries. The process of accretion onto a compact object which is effective in producing gas at $10^7$ K does not necessarily produce particles as high as 200 MeV. The full energy of gravitational infall onto a sphere of radius 10 km with 1 M$_\odot$ is only about an MeV. There is no evidence for a particle acceleration process in the hot gas environment of most compact binaries.

### 2.2 Supernova Remnants and Pulsars

In contrast to the classical Fermi mechanism where particle acceleration takes place over long times and very large distances, there are models for cosmic ray acceleration by discrete objects, such as pulsars and supernova remnants. The time averaged energy input to the galaxy by supernova explosions which occur at a rate of $3 \times 10^{-2}$ per year is several times $10^{42}$ ergs/sec. This is about what is needed to maintain the cosmic ray energy density in steady state. It is likely that supernovae play a role in the acceleration of cosmic rays either in the explosive phase, a late phase through the original Fermi process or during some intermediate phases marked by the existence of supernova remnants. It is a question of what stage in the evolution of a supernova remnant is the kinetic energy released by the explosion transformed into accelerated particles. For the Fermi process, it would be a relatively late stage in the evolution of a SNR when the remaining kinetic energy of the SNR resides in moderate velocity clouds and magnetic field energy density of the interstellar medium. If the acceleration takes place in young remnants where the expansion velocities are several hundred to several thousand km/sec, the mechanism would involve shock waves and stronger magnetic fields. The role of X-ray observations is important in this context for the existence of shock waves is reflected by the presence of gas at temperatures of $10^6$ -$10^7$ K. We do observe the presence of high energy electrons by the detection of radio and, in some cases, X-ray synchrotron radiation.

### 2.3  Pulsars and Other Active Neutron Stars

There is direct evidence that two pulsars are accelerating fast particles. These are the pulsars associated with the Crab Nebula and the Vela supernova remnant. Both are rapidly rotating neutron stars. In both cases there in an X-ray emitting nebula surrounding the pulsar which is produced by synchrotron radiation from high energy electrons. The detection of gamma rays is more indisputable evidence for the existence of high energy particles.

The Crab Nebula is the outstanding example of a particle acceleration center where a pulsar is the machine. Figure 5 is an X-ray image of the Crab showing the extent of

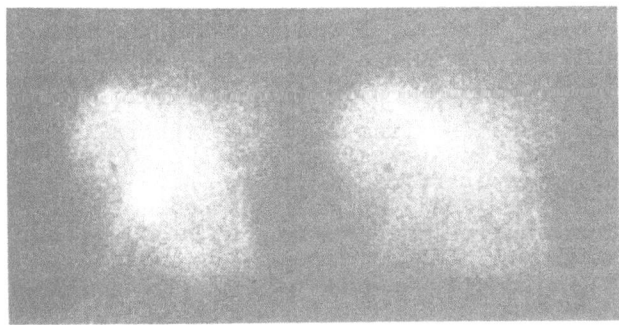

Figure 5.  X-ray image of the Crab Nebula for two phase intervals, the pulsar on and the pulsar off.

synchrotron radiation at ~1 keV. Images of the pulsar are shown for two phases of the cycle, minimum and maximum

light.   About 2% of the rotational energy loss of   the   Crab
pulsar   appears   as   synchrotron   radiation from the nebula.
There are no X-ray indications of shock waves in   the   Crab;
the X-ray spectrum is strictly non-thermal.
     The Vela pulsar PSR833-45 appears to be a   less   active
version of the Crab pulsar.   At the rate of which its period
is increasing the loss of rotational energy is a   factor   of
50   smaller.   A local nebula exists around the pulsar which
is much smaller in size and a factor of $10^3$ lower   in   X-ray
intensity   than the Crab.   Particle fluxes and energies must
be lower as might be expected from the factor   of   3   slower
rotation.   A   point   X-ray   source   does   coincide with the
pulsar which is conspicuous as the strongest source   in   the
gamma   region.   However, X-ray pulsations are not detectable
and the point X-ray source is most likely the hot surface of
a   neutron   star.   For   most of the other cases where point
X-ray sources have   been   found   in   supernova   remnants   or
coincident   with   pulsars, there are reasons to believe that
the emission is due to thermal radiation from a hot   neutron
star   rather   than   synchrotron   radiation   from accelerated
particles.
     Despite these two examples of active pulsars associated
with   supernova remnants, as a class pulsars probably cannot
account for the number of cosmic rays in the galaxy.   There
is   a   conspicuous   absence   of pulsars in most of the SNRs.
The observational work on X-ray emission   from   pulsars   has
been   carried out by the Columbia University group.   Several
older pulsars are detected as X-ray   sources   (D.   Helfand,
Colloquium).   However, the X-ray emission could be explained
by thermal mechanisms involving a hot neutron   star   surface
which   is   not   directly   related   to   the   acceleration   of
particles.
     There are two uncommon examples of neutron stars   where
the   X-ray   images   do   provide evidence for the presence of
high energy particles.   These objects are unusual;   they are
in the centers of shell-like X-ray structures that look like
supernova remnants, and the neutron stars are members   of   a
binary   system.   The   high   energy   particle signatures are
X-ray jets emanating from a compact central   X-ray   emitting
object.   The   objects   are   SS433   and   G109.1-1.0.   The
extraordinary properties of   SS433   have   been   reviewed   by
Margon   (1982).   SS433 emits two opposing beams of unionized
gas at a quarter of the velocity of light.   Figure 6a is   an
X-ray   image   of SS433 obtained with the IPC of the Einstein
Observatory (Seward et al., 1980).   The   X-ray   emission   is
diffuse   extending   in   two   large   jets   to   bulges   on the
radio-shell of W50,   The brightness of   the   jets   increases
outward   from   SS433   forming   rather   elongated   lobes that
resemble radio lobes in radio galaxies.   This has   motivated
the   suggestion   that   SS433   is   a   miniature version of an
active galaxy.   Thus, if we   can   explain   the   acceleration
process   in   SS433,   perhaps we can extrapolate the model to
radio galaxies, Seyfert galaxies, and quasars.   One   problem

is making an analogy to active galaxies is the absence of radio emission from the SS433 jets. It is not clear whether or not this can be explained by the large difference in scale.

The unusual object Gl09.1-1.0 was discovered serendipitously by Gregory and Fahlman (1980) in an IPC field of the Einstein Observatory. In Figure 6b, we see a central object which is a compact binary system, and a shell-like morphology suggestive of a supernova remnant. A jet of X-ray emission emerges from the central and merges with the shell. Only half of the shell of Gl09.1-1.0 is seen probably because its interaction with an adjoining dense cloud region has resulted in premature cooling. With a diameter of 38 pc, the intrinsic size of Gl09.1-1.0 is about half of that of W50.

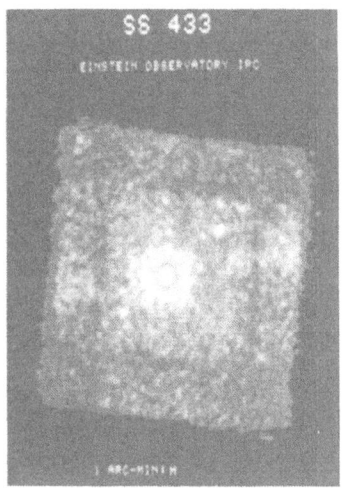

Figure 6a. X-ray image of SS433 (Seward et al., 1980). Faint jets are seen on both sides of the central object extending to the boundaries of the picture.

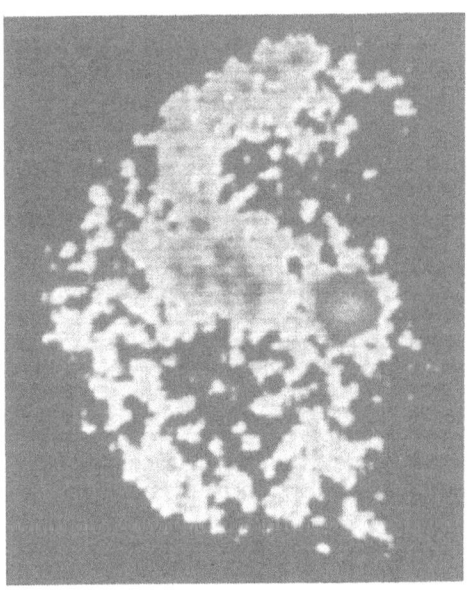

Figure 6b. X-ray image of Gl09.1-1.0 (Gregory & Fahlman, 1980).

There are two possible ways by which particle beams might produce the diffuse X-ray emission. First, a relatively low energy particle beam could produce X-rays thermally by generating shocks in the ambient medium. For an assumed X-ray spectral temperature of $10^7$ K, the derived electron density is ~$10^{-1}$ $cm^{-3}$ and the total thermal energy contained in the plasma is $1 \times 10^{49}$ erg. If the plasma has been built up over $10^3$ years, the required energy input for heating is $3 \times 10^{38}$ erg s$^{-1}$ , comparable in magnitude to the kinetic energy flow in the beams derived from the optical spectroscopic data. Only $3 \times 10^{-5}$ of this energy is required to account for the energy radiated as diffuse X-rays. We conclude that the beam kinetic energy might be dissipated in heating the plasma.

A second possibility is that the diffuse emission, which is detected at ~0.5-3 keV and may thus be relatively hard, is synchrotron radiation. The in situ production of relativistic particles by beams in regions distinct from a central compact source is clearly suggestive of the type of processes that have been invoked for double-lobed radio galaxies and QSO's. A self consistent synchrotron model for the X-ray emission would require a magnetic field $B \approx 10^{-5}$ gauss, $3 \times 10^{42}$ electrons with Lorentz factors $\gamma \approx 10^{8}$ , and energy-loss lifetimes of ~$10^{3}$ years. Such electrons would have a total energy content of ~$3 \times 10^{44}$ erg in order to account for the X-ray emission. Since the electron lifetime is comparable to its expected residence time in the beam (if $v \approx 0.258c$), each relativistic electron radiates a significant fraction of its energy during the period of transit to the edge of the SNR.

Since there is no clear radio counterpart to the X-ray beams discernable above the radio background of the SNR, W50, a lower limit of $\alpha \approx -0.8$ ($S \propto \nu^{\alpha}$) may be set for the spectral index of the beam emission between the X-ray and radio region. It is worth noting that if $\alpha \approx -0.6$, similar to that in the radio for SS433 itself, then the minimum energy in the form of relativistic electrons is $10^{47}$ erg, and could be as high as $3 \times 10^{48}$ erg if the (unobservable) protons are included.

### 2.4 Supernova Remnant Acceleration of Cosmic Rays

The acceleration processes that have received the greatest attention in recent years involve shock waves in supernova remnants. It is clear that the rate of energy release of the ejecta from a supernova explosion at $10^{51}$ ergs every $10^{9}$ sec is sufficient to maintain the energy density of cosmic ray particles. The key question is at what stage in the evolution of a supernova remnant does the transfer of energy take place. The Fermi acceleration process discussed above could, in fact, be powered by supernova remnants. The terminal stages in the evolution of a supernova remnant are cavities or bubbles of low density hot gas in the interstellar medium and moderate velocity neutral clouds. It is these clouds that would be participating in the Fermi acceleration process. Alternatively, cosmic ray acceleration may take place at a much earlier stage in the evolution of a supernova remnant. If the acceleration takes place during the relatively early phases of a supernova remnant; that is, the first $10^{4}$ years, the transfer must be through a shock wave acceleration process. The kinetic energy of particles in the supernova remnant is less than 1 MeV, so a more than $10^4$ factor of increase is needed to explain the typical cosmic ray proton. The radio emission from the shells of supernova remnants is an indication that at least high energy electrons are present. Radio emission is produced by synchrotron radiation in the magnetic field being carried by the ejecta. X-ray measurements reveal the structure of the shock waves because invariably hot gas is created in their vicinity.

A conclusion as to whether supernova remnants possess the attributes expected of shock waves that are presently accelerating particles is beyond the scope of this paper and probably requires more theoretical work to state what is expected in the appearance of the shock. A series of X-ray images obtained by the Einstein Observatory reveal that strong shock waves are indeed present in supernova remnants, at least up to the age of $10^4$ years. They also reveal a lack of neutron stars in the younger remnants, the ones which are expected to be the pulsars that rotate rapidly and accelerate particles.

### Young SNR's

The X-ray images of Cas A (Fig. 7) and Tycho (Fig. 8) are characterized by expanding shells with a clumpy sub-substructure that is very suggestive of turbulence. The turbulence is probably brought about by the deceleration of ejecta in the interstellar medium. It is possible that these clumps participate in a Fermi-type mechanism on a faster time scale and in a more localized environment than

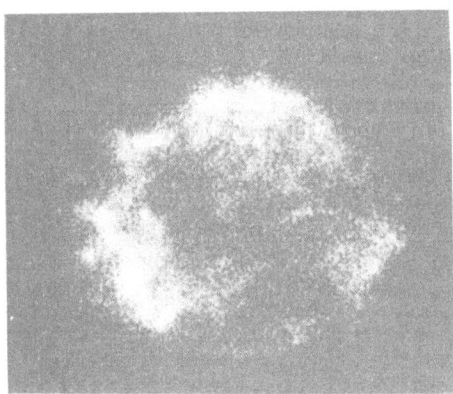 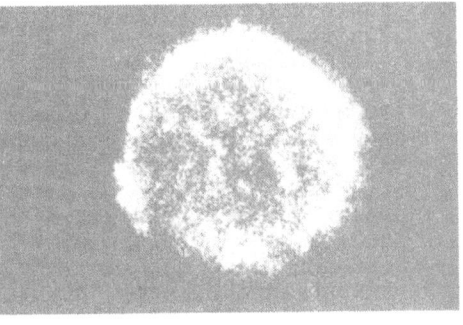

Figure 7. High resolution X-ray image of Cas A (Murray et al., 1979). Cas A and Tycho show complex shock wave structures.

Figure 8. High resolution X-ray image of Tycho (Gorenstein et al., in preparation).

as originally proposed by Fermi for low velocity
interstellar clouds and magnetic inhomogeneities. The work
of Axford, Jokipii, Blandford, Bell, Chevalier, McKee, and
others involving models of acceleration in turbulent media
are relevant in this regard. The generally good correlation
of radio emission with X-ray emission is evidence that high
energy electrons are physically present near the shocked
material. However, there is no proof of in situ
acceleration as the electrons may have been accelerated
earlier and carried along in a compressed magnetic field
that is being transported by the ejecta. Strong X line
emission in Tycho and Cas A, as observed by the Goddard
Space Flight Center group on the Einstein Observatory
(Becker et al., 1980), not only proves that the X-rays have
a thermal origin but that the SNR is highly enriched with
elements such as Si, S, A, and Ca. Our understanding of
thermal emission suggests that the strong lines give a true
indication of an abundance anomaly and are not merely the
result of a peculiarly preferential state of ionization. As
cosmic rays have a near cosmic elemental distribution, the
overabundance of these elements may be an embarrassment to
models in which shock waves in young SNR's are the primary
mechanism of cosmic ray acceleration. On the other hand,
SN1006 has a lack of line emission and a power law continuum
spectrum which implies that the X-ray have a synchrotron
origin. Electrons that produce X-rays by synchrotron
radiation are much more energetic than those that radiate
only at radiofrequencies. Since their lifetime for
synchrotron emission in the X-ray band is short, the
electrons in SN1006 are probably being accelerated now. It
is interesting that there is no pulsar or other indication
of a neutron star in any of these three young remnants where
it would be expected to be most intense. Thus, in contrast
to the Crab Nebula, there is no single center of particle
acceleration. This is an argument against pulsars being a
significant source of cosmic rays.

Shock waves are found in older supernova remnants. An
X-ray image of the Vela SNR was made from 37 fields of the
IPC of the Einstein Observatory (Fig. 9). Although the
angular resolution is much cruder than the Cas A or Tycho
HRI image because Vela is a factor 4-6 closer, it can be
said that clumps as large as those found in Tycho are not
present here. They would be about 1 to 2 arcminutes in size
and almost resolvable. Thus, the shock wave appears to be
less turbulent than in the younger remnants. Lines are
present in the X-ray spectrum of the Vela SNR and the Cygnus
Loop, which is indicative of thermal emission produced by a
shock wave. The O abundance is nearly cosmic. This is not
surprising as these older remnants are in the adiabatic
expansion phase where swept up interstellar material
dominates the mass of the ejecta. So if cosmic rays are
accelerated in this phase or beyond, their composition would
reflect the cosmic elemental abundance as observed. With

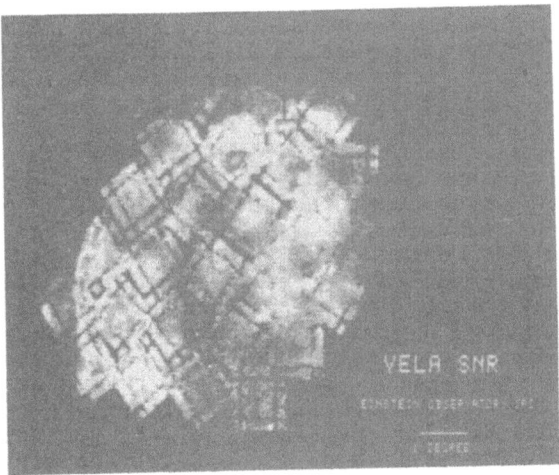

Figure 9. X-ray image of the supernova remnant in Vela which is believed to be $10^4$ years old. High temperature shock waves are present over region approximately 5° x 7° made up from 37 fields of the Einstein Observatory's Imaging Proportional Counter.

less turbulence and less clumping, the details of the acceleration process must necessarily be different for older SNR's than for younger ones, but the strong X-ray flux reveals that shock waves persist for at least $10^4$ years.

## 3. Extragalactic Sources

A substantial amount of the Einstein Observatory's time has been devoted to extragalactic sources. Much of this time has involved the study of quasars and other galaxies with active nuclei, and clusters of galaxies. There are two objects thus far where results have been obtained that are directly relevant to high energy particles. The two objects are Cen A (Feigelson et al., 1979) and M87 (Schreier et al.). In both cases, X-ray jets are detected that must surely be associated with high energy particles. The most interesting feature of Cen A is an X-ray jet that emanates from the nucleus of Cen A towards the northeast in the direction of an inner radio lobe. There are other features including X-ray lobes which are seen in IPC images. The X-ray jet consists of a series of collimated knots extending from < 8 arcsec to 4 arcmin ($\leq$ 0.2 to 5.5 kpc). The knots are about 150 pc in radius and each emits $\sim 10^{39}$ ergs/sec in the 0.5 - 4.5 keV band. Thermal and synchrotron emission processes are compatible with the X-ray data. Spectral data that could in principle distinguish between the two, but it

would require much more sensitive instruments. Thermal
models based upon the ejection of massive clouds of hot gas
from the nucleus at speeds of several thousand km/sec could
explain the cooler material which is seen as optical
filaments embedded within it.

The final object in my discussion is the giant
elliptical galaxy M87, also well known as the radio source
Virgo A. One noteworthy feature of M87 that is seen in
visible light and radio is a jet of knot-like structures
that emerges from the nucleus towards the northwest. The
light is polarized and free of lines indicating that its
origin is synchrotron radiation. Another result that has
been derived by optical astronomers from measurements of
stellar densities and stellar velocity dispersion is that
the center of M87 contains a compact object of
$5 \times 10^9 M_\odot$ (Young et al., 1978; Sargent, et al., 1978)
which could only be a black hole. However, there are
alternate interpretations of the optical measurements which
do not require a central massive black hole.

X-ray studies of M87 have resulted in several new
results. The dominant X-ray feature is the existence of a
hot gaseous corona which appears to be in hydrostatic
equilibrium. From simple arguments, it is possible to
demonstrate that M87 has a massive dark halo which extends
to 230 kpc or more from the center (Fabricant et al., 1980).
The mass of the dark halo is $3 \times 10^{13} M_\odot$. Further
investigations of the temperature distribution of the
gaseous corona corroborate this interpretation. Hot gas
around M87 accounts for 95% of the X-ray flux. The
remainder includes two features that are connected with
particles. Figure 10 is an HRI image of the central region
of M87. X-rays are seen from the center of M87 and from the

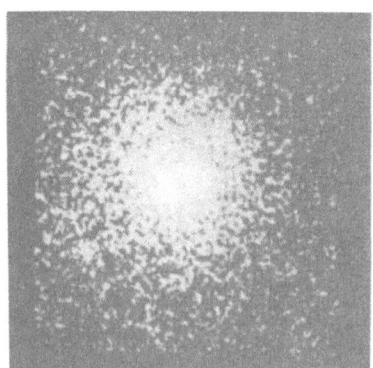

Figure 10. Central region of
M87 showing concentrations of
emission at the center and the
brightest knot in the jet
(Schreier et al.).

brightest knot in the jet. The X-ray flux is in good
agreement with an extrapolation of the optical synchrotron
spectrum. The two regions are separated by 14", which is
equal to 1100 kpc at the distance of M87. The regions are

extended, and there is some X-ray emission in between.. The detection of X-ray synchrotron radiation extends the maximum frequency by a factor of $10^3$. This increases the maximum electron energy required by a factor of 30 to $10^5$ GeV and decreases the synchrotron lifetime by the same factor. The decrease in the electron radiative lifetime is critical. It is now equal to about 500 years which is much less than the 3000 year light travel time between the two X-ray emitting regions. Thus, there must be local acceleration of electrons. The rate of electron acceleration at the bright knot is equal to that at the nucleus. Prior to the X-ray observations one could have made a viable model in which the entire jet could be explained by electrons ejected from the nucleus. Now, in analogy to SS433, we could say that relatively low energy protons are ejected from the nucleus. Interaction with gas clouds along their path leads to the formation of shocks. In SS433, only $10^{-5}$ of the initial particle energy appears as thermal X-rays. If the fraction of the ejected particles' energy that ultimately appears as $10^5$ GeV electrons is larger than $10^{-5}$, then synchrotron radiation will overwhelm the thermal radiation from the shock wave. Otherwise, thermal radiation from the heated clouds will dominate.

The second feature of M87 that we can associate with high energy electrons is shown in Figure 11. This image

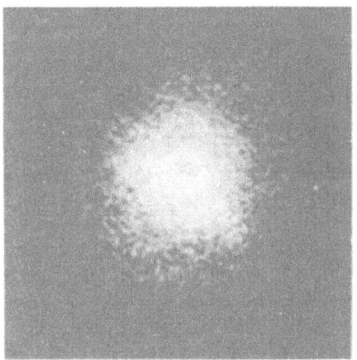

Figure 11. Larger region of M87 showing lobes to the southwest and the east (Schreier et al.). Comparisons to radio maps of this region indicate that these lobes are probably due to inverse ·Compton scattering of high energy electrons off the 2.7K microwave background.

covers a larger region than the previous one. Two lobes of X-ray emission emanate from the, center, one towards the east, the other to the southwest and are detectable out to 3.5' or 17 kpc. This feature is also seen in radio. The lobes are consistent only with inverse compton scattering of the 2.7 K microwave background by the electrons responsible for the radio. The X-ray flux is order of magnitude above the extrapolation of the radio spectrum; so a synchrotron radiation mechanism is not adequate. The enhancement is probably not of thermal origin. Additional heating of the gas by electrons is more likely to reduce rather than increase the X-ray flux in our 0.5-2 keV band, which is nearly optimum for detecting X-rays from the hot gas corona. The temperature of the corona is $3 \times 10^7$ K and it is most likely in hydrostatic equilibrium under the gravitational influence of the galaxy. Additional local heating would increase the temperature and <u>decrease</u> the density to maintain pressure equilibrium within the corona. Since X-ray emission is proportional to the square of the density, the X-ray flux would decrease contrary to the observations. Inverse compton scattering of the 2.7 K microwave background by the same electrons responsible for the radio emission is a self-consistent explanation. For ultrarelativistic electrons ($\gamma \gg 1$), inverse compton scattering increases the photon energy a factor of $\gamma^2$. From the microwave to X-ray region, the frequency increase factor is $10^7$ or $\gamma = 3000$. The characteristic energy of the electrons must be about 1.5 GeV. This is much less than the energy of the electrons responsible for synchrotron radiation in the jet as might be expected out at the distance of the lobes. The excess X-ray emission from the lobes is $1.2 \times 10^{40}$ ergs/sec.

## 5. Summary and Conclusions

Models of two-stage cosmic ray acceleration can find some encouragement in the widespread presence of X-ray flare activity in M stars implying that sources of particle injection are ample. However, the problem of accounting for heavier particles remains.

In addition to the Crab and Vela pulsars, there are relatively few other neutron star systems that are accelerating particles. Jets of X-ray emission imply that particle beams are ejected from neutron star binary systems in SS433 and G109.1-1.0. They survive out to distances of tens of parsecs without much deflection. Radio shells indicate that both are associated with supernova remnants. Otherwise, there is not much X-ray evidence that pulsars are accelerating particles. Supernova remnants are characterized by turbulent shock waves that could be sites of particle acceleration.

X-ray jets are found in Cen A and M87 that have a synchrotron radiation origin. Electron energies of $10^5$ GeV are required to account for them. Lobes in M87 can be explained as inverse Compton radiation of 1.5 GeV electrons off the universal 2.7°K microwave background. These electrons extend out to a distance of 17 kpc from the center of M87, at least a factor of 15 further than from the center of our own galaxy.

## References

Becker, R.H., et al., 1980, Ap.J. (Letters), 235, L5.

Eichler, D., 1979, Ap.J., 229, 419.

Fabricant, D., Lecar, M., and Gorenstein, P., 1980, Ap.J., 241, 552.

Feigelson, E., et al., 1979, Ap.J. (Letters), 234, L69.

Gorenstein, P., Tucker, W.H., and Seward, F., in preparation.

Gregory, P. and Fahlman, G., 1980, Nature, 287, 805.

Haisch, B.M., et al., 1980, Ap.J. (Letters), 242, L99.

Lanzerotti, L.O., 1977, article in The Solar Output and Its Variation, O.R. White (ed.), Colorado Associated University Press, Boulder, CO.

Margon, B., 1982, article to appear in Accretion Driven Stellar X-ray Sources, W.H.G. Lewin and E.P.J. van den Heuvel (eds.), Cambridge University Press.

Murray, S.S., et al., 1979, Ap.J. (Letters), 234, L69.

Rosner, R. and Vaiana, G., 1978, Ap.J., 222, 1104.

Sargent, W.L.W., et al., 1978, Ap.J., 221, 731.

Schreier, E., Feigelson, E., and Gorenstein, P., to be submitted to Ap.J.

Seward, F., Grindlay, J., Seaquist, E., and Gilmore, W., 1980, Nature, 287, 806.

Stern, R.A., Underwood, J.H. and Antiochos, S.K., 1981, I.A.U. Circular No. 3585.

Topka, K.P., 1980, Ph.D. thesis, Harvard University.

Van Speybroeck, et al., 1979, Ap.J. (Letters), 234, L45.

Young, P.J., et al., 1978, Ap.J., 221, 721.

SUPERNOVAE, YOUNG REMNANTS, AND NUCLEOSYNTHESIS

Robert P. Kirshner
Department of Astronomy
University of Michigan
Ann Arbor, Michigan 48109
U.S.A.

ABSTRACT

Supernovae and supernova remnants may be intimately linked to cosmic ray problems as the site of nucleosynthesis, the site of particle acceleration, or both. Although direct evidence for the origin of cosmic rays in supernovae remains elusive, observations of supernovae and their remnants by optical telescopes now provides good evidence that nuclear processing has taken place in these stars. The energetics and spectra of extragalactic supernovae immediately after the explosion indicate that Type I supernovae produce large amounts of iron peak elements. In a complementary way, spectroscopy of the 300 year old galactic remnant Cassiopeia A demonstrates that nuclear processing through oxygen burning took place in that object, which may have been a 15 to 25 solar mass star. The recent discovery of a handful of remnants with abundances like those in Cas A leads to the hope that a detailed correspondence between abundance patterns in these remnants and models for stellar interiors will provide new insight into the last stage of stellar evolution.

## 1. Introduction

One aim of cosmic ray research has been to infer the properties of the source. In its broadest conception, this would mean knowing the biography of each nucleus, from the day it emerged from the primordial cauldron of the Big Bang, through the nuclear cooking of stellar interiors, out to a dangerous passage across the interstellar medium to its ultimate reward of provoking a spark in somebody's chamber. As illustrated in Figure 1, supernovae have been popular candidates for significant events in nucleosynthesis and for the acceleration of cosmic rays for nearly 50 years.

Although supernovae are widely believed to have a central role to play in nucleosynthesis both as the endpoint of static stellar burning and as the locus of explosive processing, the actual evidence that supernovae do these wonderful things has accumulated only recently. In this contribution, I would like to emphasize new results on the chemical abundances in supernovae that give empirical evidence that nuclear burning beyond helium synthesis really takes place. The data come from two sources: extragalactic supernovae studied in the first year or two after discovery and from supernova remnants that are less than 1000 years old.

## Be Scientific with OL' DOC DABBLE.

Figure 1-- This 1934 scientific publication provides
a terse summary of Zwicky's model for cosmic ray origin
and neutron star formation as well as Ol' Doc Dabble's
empirical approach to science.

## 2.  Extragalactic Supernovae-- An Iron-clad Case?

World-wide monitoring programs detect new stars in other galaxies
about 10 time each year.  The reports are circulated by IAU telegrams
so that interested observers can begin work promptly.  Supernovae can
be classified by their spectra into two chief types imaginatively
labelled Type I (SN I) and Type II (SN II) (see Oke and Searle 1974).
The SN I have spectra with no hydrogen lines and are found in all types
og galaxies.  The SN II have strong hydrogen lines and are found near the
spiral arms of spiral galaxies: they are presumably massive and short-lived
stars which evolve so rapidly that they explode near the place where
they were formed.

The SN I's have been of particular interest in recent years because two lines of argument suggest that they are dominated by iron peak elements( for numerous references, see the recent conference proceedings edited by Meyerott and Gillespie (1980) and by Wheeler (1980)). The indirect argument is based on the energetics of the eruption and the evolution of luminosity with time. The direct agument hinges on the identification of $Fe^+$ and $Fe^{++}$ emission lines in the optical spectrum several months after maximum light.

Figure 2 shows the evolution of the luminosity of a SN I-- the light curve. During the peak of luminosity most of the energy comes from a continuum whose temperature is about 12 000 K at a radius of a few times $10^{14}$ cm (10 AU!) (Kirshner, Arp, and Dunlap 1975). The outer layers of the supernova expand at 12 000 km/sec and the photosphere colls down to 6000 K during the first four weeks after maximum light (Kirshner et. al. 1973a, 1973b). After this early peak, the spectrum shifts to a group of blended emission lines whose luminosity decreases very nearly exponentially for at least 700 days(Kirshner and Oke 1975) with a half-life of about 60 days. This long exponential tail has led to several suggestions that the energy source for SN I's might be radioactive decay.

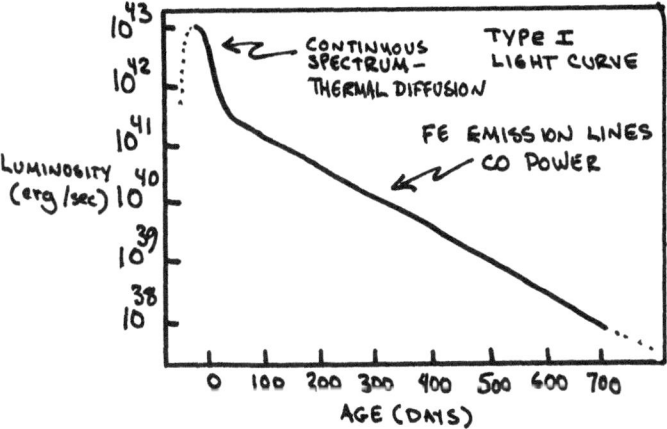

Figure 2-- Light curve for a SN I (after Kirshner and Oke 1975).

The modern incarnation of these models stems from the suggestion by Colgate and McKee(1969) that the decay chain $^{56}Ni-^{56}Co-^{56}Fe$ might be significant. In recent models (Arnett 1979, Colgate, Petschek, and Kriese 1980, Chevalier 1981) this radioactive decay supplies the energy both for the peak of luminosity, which is dominated by considerations of energy balance and thermal diffusion in the expanding star, and for the exponential tail where the energy goes into heating an electron gas that collisionally excites optical emission lines.

These models start with the plausible suggestion that the star which explodes to become a SN I might be a carbon/oxygen white dwarf. The

thermonuclear detonation of a compact star near the Chandrasekhar mass
should lead to the production of several tenths of a solar mass of the
doubly magic nucleus $^{56}$Ni.  Nickel decays with a six day half-life to $^{56}$Co
by emitting positrons and gamma rays.  In the models (and presumably in
supernovae) the positrons annihilate and the gammas are absorbed at first,
but as it expands, the supernova envelope becomes progressively more
transparents to the gammas.  This initial radioactive energy source heats
the supernova envelope and produces the thermal emission seen near
maximum light.  As the star expands in the first few weeks up to $10^{15}$cm
(100 AU!) it begins to turn transparent to optical photons and the
diffusion picture breaks down.

   At about this point, the subsequent decay of $^{56}$Co to $^{56}$Fe becomes
significant.  Calculations of the decay energy deposited as heat amid
the elctrons in the expanding nebula by this decay (77 day half-life)
provide a very good match to the energy observed from SN I in the
exponential part of the light curve.

   This indirect argument for substantial quantities of iron peak
elements in SN I's is based on plausible nuclear physics and provides a
good quantitative match to the energetics.  Its impact is greatly increased
by the direct observation of iron in the spectra of SN I's at late times.
Once the thermal continuum fades, the spectra of SN I's consist of broad
emission bands.  Kirshner and Oke (1975) showed that the superposition
of 216 [Fe II] lines provided a good match to the observed spectrum.
This idea has been developed by Meyerott (1980) and by Axelrod (1980)
who has calculated the ionization and excitation of iron peak elements
to produce a synthetic spectrum as shown in Figure 3.  The match between
the observations and this model is good, providing more evidence that
SN I's are rich in iron peak elements.

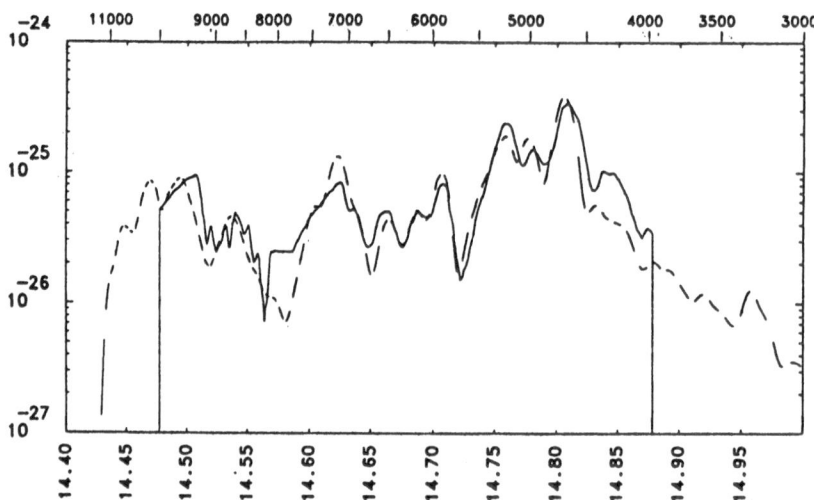

Figure 3--  Comparison between Kirshner and Oke's (1975) observations of
the SN I in NGC5253 (solid line) and Axelrod's (1980) calculation (dashed).

This model for SN I energetics and spectra is subject to additonal
observational tests.  If cobalt is decaying into iron, the spectra should
show a change in the cobalt to iron ratio.  The available data appear to
show this effect, but more observations are needed.  A second observational
check would be to show that young galactic remnants of Type I supernovae,
such as Kepler's and Tycho's have high iron abundance.  So far, the evidence
from X-ray spectra of these objects does not confirm an unusually high
iron abundance.  A third observation would be to observe the gamma rays
that escape from the supernova during the Ni-Co-Fe chain.  This would
provide a direct measure of the mass of nickel and allow the energetics
of the supernova to be investigated thoroughly.  The proposed Gamma Ray
Observatory will be able to detect gamma rays from supernovae in the
Virgo cluster and could provide conclusiove evidence on the nucleosynthesis of iron peak elements in Type I supernovae.


3.  Supernova Remnants: Some Bubbles of Oxygen?

It is convenient to think of supernova remnants in two classes:
young and old.  In young remnants, we see the actual material ejected by
the supernova explosion before it mixes with the surrounding interstellar
gas.  In old remnants, the swept-up matter dominates the ejecta and the
old remnants (such as the Cygnus Loop) retain no information about the
supernova explosion except the initial energy input.  The interstellar
shocks of old remnants may play an important role in cosmic ray acceleration and propagation, and they surely are significant in determining
the thermodynamic state of the interstellar gas, but they are not much
help in determining the sites of nucleosynthesis except as probes of the
interstellar abundances ( Blair, Kirshner, and Chevalier 1981 ).
For typical interstellar densities and supernova expansion velocities
the transition from young remnants to old ones (remnant adolescence)
takes place after 1000 years, so the young galactic remnants sometimes
have associated historical observations (see the delightful book by
Clark and Stephenson  1978 ).
One young remnant has provided particularly clear evidence for
nucleosynthesis: Cassiopeia A.  Cas A is a powerful non-thermal radio
source, a strong thermal X-ray emitter, and appears optically as a fragmentary ring of nebulosity.  Detailed work by Kamper and van den Bergh (1976)
shows that the optical filaments are of .two types: the "quasi-stationary
flocculi" which are slowly expanding, and the fast moving knots which has
space velocities in the range 4000-8000 km/sec.  The fast moving knots
have been traveling without deceleration since about 1667 A.D.
Spectra of the fast moving knots show clear evidence of very unusual
abundances: unlike other nebulae they show no emission lines of hydrogen,
and they have no lines of helium or of nitrogen (Kirshner and Chevalier
1977; Chevalier and Kirshner 1978).  Oxygen lines are present that arise
from neutral, once ionized, and doubly ionized atoms, so it seems very
unlikely that some bizarre ionization situation can be responsible for
the missing hydrogen lines. ı In additon to oxygen lines, the fast moving
knots in Cas A have prominent line from sulfur, argon, and calcium.  Based
on a shock model, the line strenghts can be interpreted as abundances.
The result is that about half the mass in a typical fast moving knot is

oxygen, with the rest made up of sulfur, argon, and calcium. There can
be no doubt that this material, which must have been nearly all hydrogen
and helium when the star formed, has undergone advanced stages of nuclear
burning.

Individual knots show varying amounts of oxygen compared to sulfur,
argon, and calcium, as though the ejecta were not thoroughly mixed
(Chevalier and Kirshner 1979). As an extreme example, Figure 4 shows a
knot which has only oxygen lines.

The pattern of abundances observed in Cas A corresponds well to the
abundances expected in the inner layers of a massive star. The computation
by Weaver, Zimmerman, and Woosley (1978) illustrated in Figure 5 shows
that a massive star at the brink of collapse has substantial zones which
are rich in oxygen and zones which have the products of burning oxygen,
chiefly silicon, sulfur, argon, and calcium. Since silicon does not have
emission lines that we would expect to observe, it seems fair to
characterize the abundances in Cas A as oxygen plus varying· amounts of
oxygen-burning products. It is plausible to identify the fast moving
knots with poorly mixed debris from the inside of a massive star.

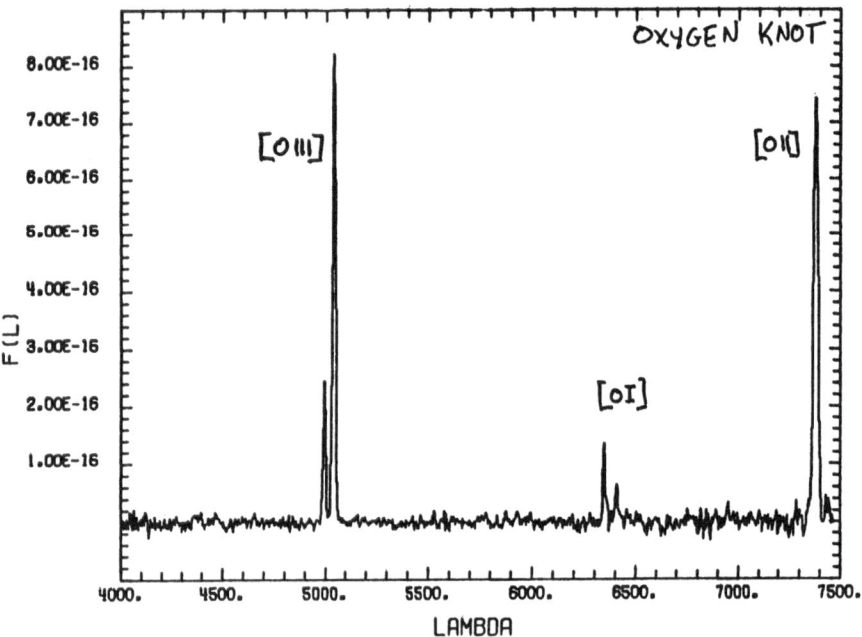

Figure 4-- The spectrum of a fast moving knot in Cas A. To the limit of
these observations, oxygen is the only element present in this knot.

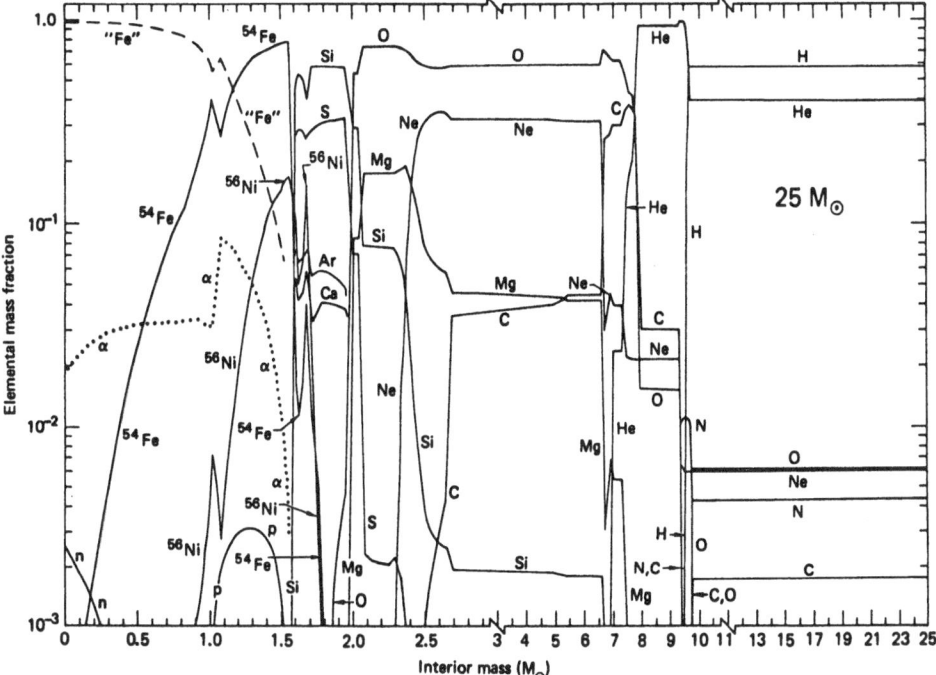

Figure 5 -- The structure of a 25 solar mass star at the onset of collapse (Weaver, Zimmerman, and Woosley 1978). Note the large zones of oxygen and of the oxygen burning products sulfur, argon, and calcium.

The idea that these elements were produced in a massive star has independent confirmation based on the X-ray data presented by Fabian et.al (1980). They show that about 15 solar masses of hot plasma are required to produce the observed X-rays.

A few other young remnants show silimar abundance patterns in fast-moving, undiluted debris. Knots with no hydrogen, but strong oxygen lines have been identified in the galactic remnant G292.0+1.8 (Goss et.al. 1980), in two remnants in the Large Magellanic Cloud N132D (Lasker 1978, 1980) and 0540-69.3 (Mathewson et.al. 1980),and in the extraordinary remnant in the distant galaxy NGC 4449(Balick and Heckman 1978, Blair and Kirshner 1980).

The NGC 4449 object, about 5 Megaparsecs away, has about 25 times the radio luminosity of Cas A (Seaquist and Bignall 1978), shows powerful lines of neutral, once ionized and doubly ionized oxygen but no hydrogen lines, and has an expansion velocity of 3500 km/sec. Recent observations with the Einstein X-ray Observatory show that the remnant in NGC 4449 is also a luminous X-ray source (Blair, Kirshner, and Winkler 1982).

If each of these objects is the result of the explosion of a massive star, then detailed observations and detailed comparison with models of stellar interiors may provide new insight into the structure of stars on the day they die.

## ACKNOWLEDGEMENTS

I am grateful to the organizers of this meeting for giving me the opportunity to talk to so many cosmic ray physicists and for their patience. This work was supported by an Alfred P. Sloan Research Fellowship, the U.S. National Science Foundation grants AST 76-17600 and AST 80-5050 and NASA grant NAG-8341.

## REFERENCES

Arnett,W.D. 1979, Ap.J. Letters 230,L37.
Axelrod,T.S. 1980, Ph.D. Thesis, Univ. of Calif- Santa Cruz.
Balick,B. and Heckman.T. 1978, Ap.J. Letters 226,L7.
Blair, W.P. and Kirshner, R.P. 1980, Ap.J. 236,135.
Blair, W.P., Kirshner, R.P., and Chevalier, R.A. 1981, Ap.J. 247,879.
Blair, W.P., Kirshner. R.P., and Winkler, P.F. 1982. Ap.J. inpress.
Chevalier, R.A. 1981, Ap.J. 246,267.
Chevalier, R.A. and Kirshner, R.P. 1978, Ap.J. 219,931.
Chevalier, R.A. and Kirshner, R.P. 1979, Ap.J. 233,154.
Clark, D.H. and Stephenson, F.R. 1977, The Historical Supernovae, Pergamon
     Press: Oxford.
Colgate, S.A. and McKee, C. 1969, Ap.J. 157,623.
Colgate, S.A., Petschek, A.G., and Kriese, J.T. 1980, Ap.J. Letters 237,L81.
Fabian, A.C. et.al. 1980, M.N.R.A.S. 193,175.
Goss, W.M. et.al. 1980, M.N.R.A.S. 193,901.
Kamper, K. and van den Bergh, S. 1976, Ap.J. Suppl. 32,351.
Kirshner, R.P. et.al.  1973a, Ap.J. Letters 180,L97.
Kirshner, R.P. et.al.  1973b, Ap.J. 185,303.
Kirshner, R.P., Arp, H.C., and Dunlap, J.R. 1975, Ap.J. 207,44.
Kirshner, R.P. and Chevalier, R.A.  1977, Ap.J. 218,142.
Kirshner, R.P. and Oke, J.B. 1975 Ap.J. 200,574.
Lasker, B. 1978, Ap.J. 223,109.
Lasker, B. 1980, Ap.J. 237,765.
Mathewson, D.S. et.al. 1980, Ap.J. Letters 242,L73.
Meyerott, R.E. 1978, Ap.J. 221,975.
Meyerott, R.E. 1980, Ap.J. 239,257.
Meyerott, R.E. and Gillespie, G.H. (eds.) 1980, Supernovae Spectra, Ameri-
     can Institute of Physics: New York.
Oke, J.B. and Searle, L. 1974, Ann. Rev. Astron. Astrophys. 12, 315.
Seaquist, E.R. and Bignell, R.C. 1978, Ap.J. Letters 226, L5.
Weaver, T.A., Zimmerman, G.B. and Woosley, S.E. 1978, Ap.J. 225,1021.
Wheeler, J.C. (ed.) 1980, Texas Workshop on Type I Supernovae, U of Texas
     Press: Austin.

NUCLEOSYNTHESIS, MASS LOSS AND STELLAR EVOLUTION

C. de Loore
Astrophysical Institute
Vrije Universiteit Brussel
Pleinlaan 2, B-1050 Brussels
Belgium

## 1. Introduction

Stars are formed from material of the interstellar medium with a given composition : X (the fraction of hydrogen), Y (the fraction of helium), Z (the fraction of the sum of the heavy elements, which can be splitted in $Z_1$, $Z_2$, $Z_3$,... representing these heavy elements).

The produced stars evolve, i.e. as a function of time the mass, radius, luminosity, temperature and density of the different stellar layers, from center to surface, change; due to nuclear reactions also the chemical composition of the different layers, varies.

A part of the stellar matter is, during the evolution, restituted to the interstellar medium, whether or not altered by the nuclear processes. In material of deep layers, the composition is changed by nuclear reactions, hence if such material is restituted, it has a chemical composition different from the initial one, hence X', Y', Z' ($= Z_1'$, $Z_2'$, $Z_3'$,...). This can occur in the following ways :

1. mass loss by stellar wind during core and shell hydrogen burning
2. mass loss during the red giant stage
3. mass loss due to Roche lobe overflow in the case of close binaries (mass exchange and mass loss)
4. mass loss of stars during the Wolf-Rayet stage
5. the ejection of outer layers at the stage of planetary nebulae
6. the supernova explosion of a single star, leaving a compact object as remnant, or leading to complete disruption
7. the supernova explosion of one or two components of a massive close binary.

Low mass stars pass through a planetary nebula stage, where a small fraction of the stellar matter is transferred to shells, which resolve afterwards, so that this matter returns to the interstellar medium. The stars evolve later into white dwarfs. Massive stars pass through a supernova stage, where most of the stellar matter is restituted to the interstellar medium.

## 2. General characteristics of the stellar evolution

### A'. Single stars

1) $\boxed{M < 1.2\ M_\odot}$

Stars with masses below 1.2 $M_\odot$ do not develop a convective core, and hydrogen burning proceeds from layer to layer, without mixing of the original hydrogen and the produced helium. Hence the core is in radiative equilibrium, while in the outer parts convective layers are present. Nuclear reactions proceed by the proton-proton cycle and the energy production ε depends moderately on the temperature, according to

$$\varepsilon \sim T^3$$

The evolutionary time scale is very long (of the order of $10^9$ years). Only a small fraction of the stellar matter is returning to the interstellar medium.

The stars pass through a planetary nebula stage; they lose a part of their mass, which returns to the interstellar medium. The remnants end their life as white dwarfs (He, C-O dwarfs).

2) $\boxed{1.2\ M_\odot < M < 10\ M_\odot}$

Stars with masses exceeding 1.2 $M_\odot$ develop a convective core; at the end of core hydrogen burning their evolutionary tracks show a special feature, absent in the tracks of lower mass stars. The tracks turn backwards to the blue part of the Hertzsprung-Russell diagram, as a consequence of the contraction of the star. Later on hydrogen is ignited in shells, and the stellar radius increases again, which results in a redward motion in the HRD. Stars with masses between 1.2 and 10 $M_\odot$ show similar evolutionary tracks, reflecting the overall features of the star, i.e. a contraction followed by rapid expansion. The energy is produced by the CNO-cycle, with a stronger dependence on the temperature than in the case of the proton-proton cycle :

$$\varepsilon \sim T^{12}$$

The stars evolve faster, in time scales of the order $10^7$-$10^9$ years. The final phase is the white dwarf stage.

3) $\boxed{M > 10\ M_\odot}$

Stars with masses larger than 10 $M_\odot$ show a composition discontinuity in the layers around the convective core caused by the expansion of the core, as a consequence of the helium enrichment. The opacities are determined by bound-free transitions and electron scattering. The importance of scattering grows with the mass and becomes dominant for stars more massive than 10 $M_\odot$. This has an important repercussion on the evolution of stars with masses in this range.

For stars with masses between 1.2 and 10 $M_\odot$ the convective core shrinks during hydrogen fusion. The outer layers of the core are in radiative equilibrium, hence in the regions left behind by the shrinking core the composition does not change and the hydrogen abundance from limb to centre increases monotonously.

In stars with larger masses the composition changes suddenly and a convectively neutral domain is formed. This situation is called semiconvection. In these regions the energy transfer occurs radiatively; a partial mixing takes place, just to satisfy the equilibrium criterion of convective neutrality. Semiconvection is not very important during core hydrogen burning, but the importance increases during hydrogen shell burning and helium fusion. For more details see de Loore, 1980, and Dellaporta, 1971. The evolutionary time scale is short : $10^6$-$10^7$ years. The stars finish their life by a supernova explosion where most or all the mass of the star is restituted to the interstellar medium.

Stars more massive than 15 $M_\odot$ lose a fraction of their initial mass

(10 to 20%) during core hydrogen burning or shell hydrogen burning, by stellar wind mass loss. After the supernova explosion a neutron star (or a black hole ) remains, or the star is completely disrupted so that no remnant is left.

## B. Binaries

The evolution of binaries is determined by three parameters:

i) the masses $M_1$ and $M_2$ of primary and secondary respectively;

ii) the mass ratio $q = \dfrac{M_2}{M_1}$ ;

iii) the orbital period of the system.

If the primary mass is lower than 15 $M_O$ the star passes through a stage of mass transfer and mass loss. If the primary mass is larger than 10 $M_O$ the mass ratio distribution shows a double peak, one around 1, and another at a lower q value (0.3 to 0.4). The primary has first a stage of mass loss by stellar wind. Afterwards the radius of the primary can exceed the Roche radius, and mass exchange and mass loss occur.

## 3. Mass loss in luminous stars

### a) Observations, general considerations

Observations of the ultraviolet spectrum of early type stars reveal that these stars are losing mass at rates of $10^{-8}$-$10^{-5}$ $M_Oyr^{-1}$; the material is accelerated to velocities of the order of some $10^3$ km s$^{-1}$. Although the mechanism for the production of the mass loss is not yet known, it is generally accepted that the radiation pressure in the ultraviolet resonance lines accelerates the matter to these large velocities. All luminous stars with log $L/L_O$ > 4.3 ($M_{bol}$ < -6) have observable mass loss rates; less luminous stars show mass loss effects only when their rotational velocities are sufficiently large, i.e. v sin i > 200 kms$^{-1}$ (Snow and Marlborough, 1976; Lamers and Snow, 1978). The position of mass losing stars is shown in Figure 1. As can be seen the mass loss limit mentioned before applies to all stars with $T_{eff}$ between 7500 K and 40000 K, for $M_{bol}$ < -6. Evolution of stars during core hydrogen burning occurs more or less at constant luminosity,

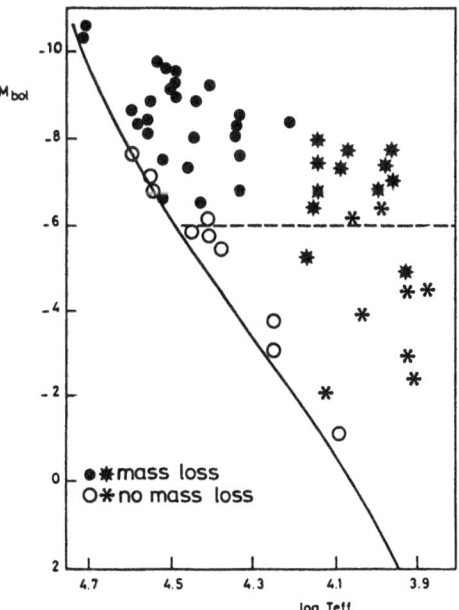

circles: Snow and Morton, 1976
asterisks: Lamers et al., 1980a

*Fig.1. The position of stars with observable mass loss in the HR diagram. Only stars brighter than $M_{bol}$=-6 show mass loss effects, unless they have large rotational velocities.*

and this means that stars not losing mass at the zero age main sequence
(ZAMS) will not show large mass loss effects during later evolutionary
stages, while stars revealing mass loss effects at the ZAMS will continue to lose mass during later evolutionary phases.

b) Mass loss rates

     Estimates of mass loss rates can be derived from P Cygni profiles
in ultraviolet spectra (Morton, 1967; Lamers and Morton, 1976; Conti and
Garmany, 1980; Gathier et al., 1981) from optical spectra, mainly Hα
(Hutchings, 1976; Klein and Castor, 1978), infrared photometry (Barlow
and Cohen, 1977), from radio observations with the VLA at λ 6 cm (Abbott
et al., 1981). From the infrared excess measured by Barlow and Cohen
(1977) a dependence of the mass loss rate on the luminosity of $\dot{M} \sim L^{1.1}$
may be derived, compatible with the predictions of the radiation pressure
driven wind model of Castor et al. (1975)(if the radiation pressure parameter $\alpha = 0.90$).
     From the radio observations of Abbott et al. (1981) a dependence of
$\dot{M} \sim L^{1.56}$ can be derived. Lamers et al.(1980b) and Conti and Garmany
(1980) examined the correlation between the mass loss rates of O type
stars and the luminosity class or gravity on the basis of UV spectra of
O stars. They found a strong dependence on luminosity class, and that
the mass loss rates of Of stars are much larger (a factor$\geq$4) than those
of other O stars.
     The mass loss rates derived from the UV, IR or optical spectra in-
clude large uncertainties, since for the determination of the rates a
number of parameters that cannot be observed have to be introduced, the
velocity law $v(r)$, the ionization fraction and a photospheric radiation
field has to be assumed. Mass loss rates derived from radio observations
are much more reliable. They are based on observable parameters : the
stellar distance, the radio flux, the terminal stellar wind velocity.
The mass loss rates derived from the UV, IR, optical and radio are shown
in Figure 2. As may be seen in the figure the mass loss rates from Of
stars are systematically larger than for normal O stars.
     For all O stars the following relation between $\dot{M}$ and luminosity
can be derived :

$$\log \dot{M} = +1.5 \log L - 14$$

and this can be compared with the Abbott et al. relation

$$\log \dot{M} = +1.56 \log L - 14.4$$

In Of stars and type I O stars this is

$$\log \dot{M} = +1.18 \log L - 12.65$$

while for normal type V-III O stars the expression is

$$\log \dot{M} = +1.68 \log L - 16.81$$

which leads to a mass ratio of 5-10 between Of and O stars.

     Andriesse (1979) derived a mass loss rate equation from the fluctua-
tion theory, where $\dot{M}$ depends on M, R and L as

$$\log \dot{M} = 1.5 \log L - 13.23 + 2.25 \log R - 2.25 \log M$$

Lamers (1981) has used a number of early type stars from O3 to B9, including O and Of type stars, of spectral class I to V, to examine the dependence of the mass loss rate on the stellar parameters. He finds an expression

$$\log \dot{M} = 1.42 \log L - 15.35 + 0.61 \log R - 0.99 \log M$$

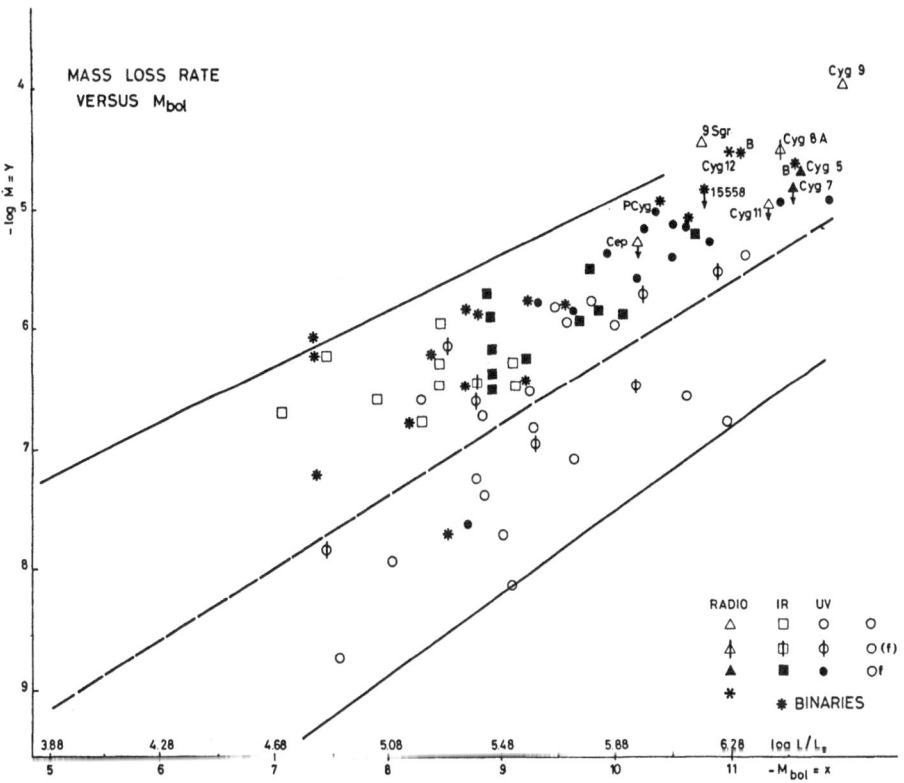

Fig.2. *Mass loss rates as a function of the luminosity*

Mass loss rates of the order of $10^{-4}$ $M_\odot yr^{-1}$ for Wolf-Rayet stars have been determined from infrared observations by Barlow and Cohen (1977). Willis et al. (1979) estimate the mass loss of γ Vel at $1.1$ $10^{-4}$ $M_\odot yr^{-1}$. Abbott et al. (1981) have determined mass loss rates for 10 Wolf-Rayet stars with the NRAO-VLA at λ 6 cm and found values ranging from 1 to $5$ $10^{-5}$ $M_\odot yr^{-1}$. As was already known before, the mass loss rates of Wolf-Rayet stars are not strongly correlated with the stellar luminosity or the spectral type. The mass loss rates as well as the luminosities are depending on the distance which is not accurately known. If the distance is larger the mass loss rates decrease, but also the luminosities. In that case the position of the stars in the diagram shifts to the left and to the bottom, so that they still agree with the general trend.

An overwiew of the mass loss rates for stars defined by their position in the HRD is depicted in Figure 3.

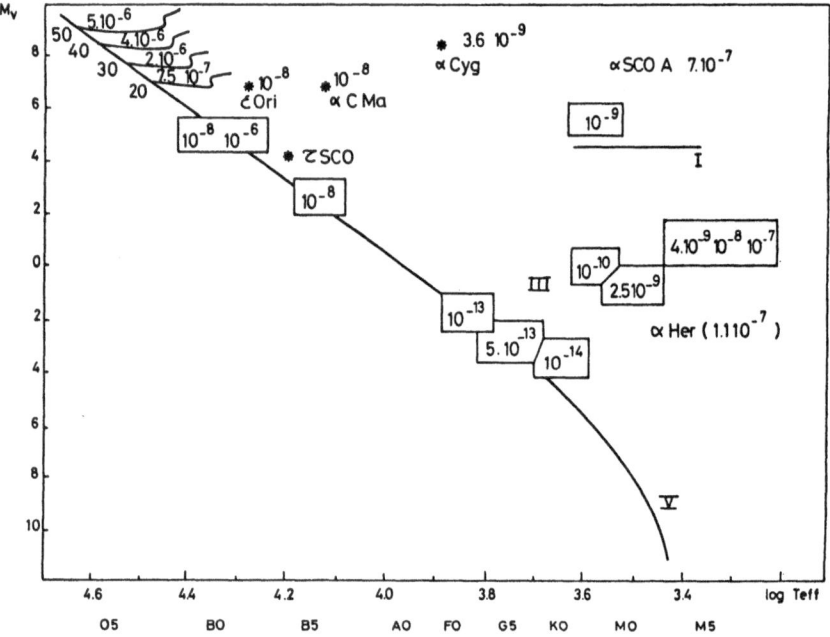

Fig.3.  Mass loss rates for stars of various spectral types

## 4. Nuclear burning phases

### a) Hydrogen and helium burning;changes of the abundances

Massive stars contain, when they arrive at the zero age main sequence (ZAMS) after contraction, convective cores with temperatures of the order of $3-4 \cdot 10^7$ K and densities of the order of $1.5-4$ gcm$^{-3}$ where hydrogen burning occurs by the CNO-tricycle. The extent of these convective cores for stars with masses ranging from 20 to 140 $M_O$ is shown in Table 1. An example of the hydrogen and helium abundance changes in the centre of the core of a 100 $M_O$ star is shown in Figure 4. The convective core shrinks and disappears when hydrogen in the core is exhausted. The star contracts and some time later ignites hydrogen in a shell.

Also the abundances of heavy elements (C,N,O) change during the CNO-cycle. The change of these abundances by weight as a function of time is shown in Figure 5; the upper time scale (in $10^6$ yrs) shows the change of the abundances at the stellar surface, the lower time scale (in 100 yrs) shows the abundance changes in the core. As may be seen after 400 years these elements have reached their equilibrium abundances.

For temperatures of the order of $10^8$ K and densities of the order of $10-10^3$ He ignition in the core is started. Helium is depleted, and

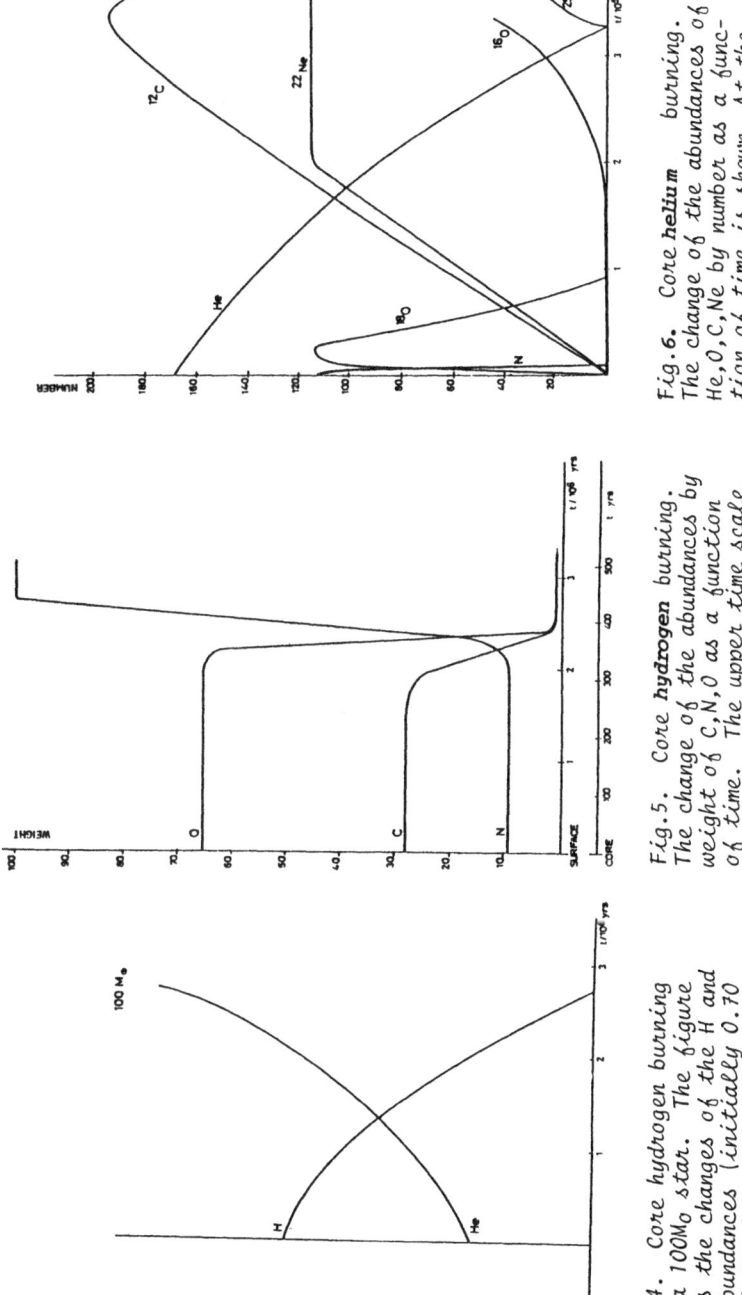

Fig.4. Core hydrogen burning for a 100M⊙ star. The figure shows the changes of the H and He abundances (initially 0.70 and 0.27) by weight as a function of the evolutionary time.

Fig.5. Core hydrogen burning. The change of the abundances by weight of C,N,O as a function of time. The upper time scale depicts the situation in the core; the lower time scale corresponds with the surface.

Fig.6. Core helium burning. The change of the abundances of He,O,C,Ne by number as a function of time is shown. At the extreme right the change of the C-abundance and the production of 25Mg is indicated.

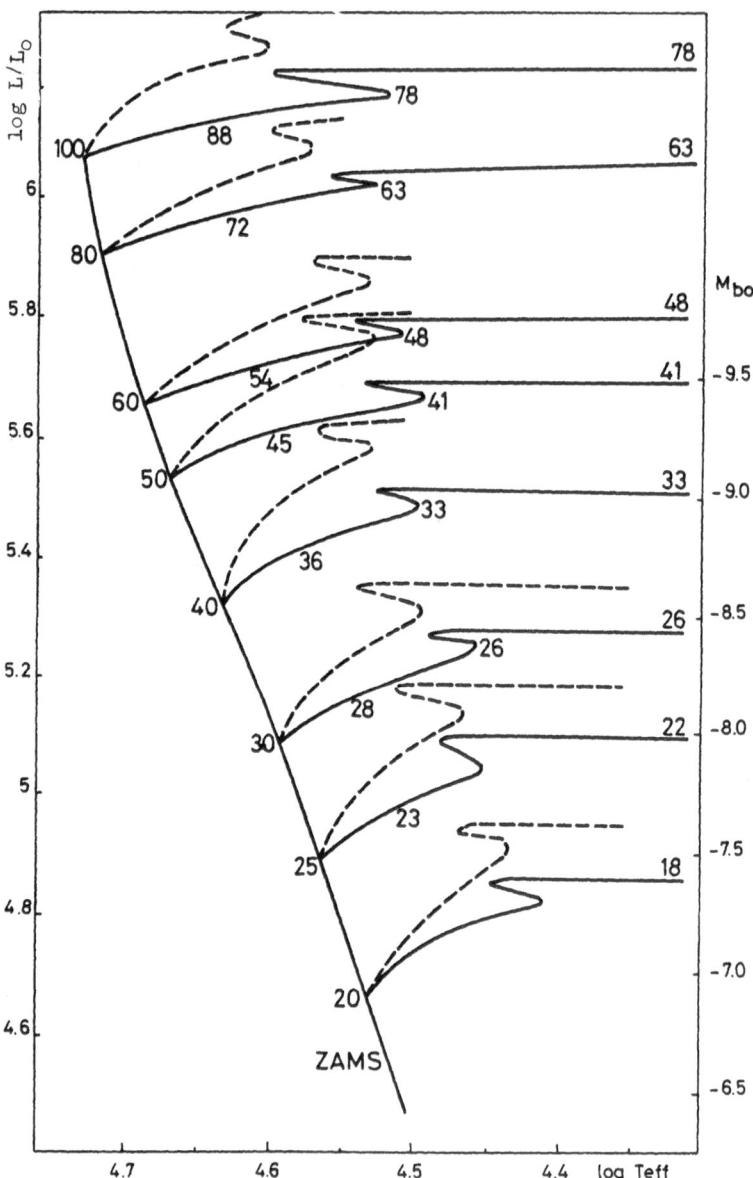

Fig. 7.  Evolutionary tracks for massive stars.
Full lines: tracks with mass loss (N=100)
Dashed lines: conservative tracks

*Table 1.* *The mass of the convective core for initial masses from 20 to*
*100 $M_O$, expressed in $M_O$ and the remnants of the primary after*
*the Roche lobe overflow (RLOF) for binaries*

| Initial mass ($M_O$) | Initial convective core ($M_O$) | Remnant after RLOF ($M_O$) |
|---|---|---|
| 20 | 9 | 5.4 |
| 30 | 16 | 9.7 |
| 40 | 24 | 14.3 |
| 60 | 40 | 26 |
| 80 | 58 | 36 |
| 100 | 78 | 48 |
| 120 | 95 | 58 |
| 140 | 110 | 70 |

$^{18}O$, $^{22}Ne$, $^{12}C$ are formed; rapidly $^{18}O$ is destructed and $^{16}O$ is formed.
The change of the abundances of these elements as a function of time is
depicted in Figure 6.

### b) Hydrogen and helium burning - evolutionary tracks

As a consequence of the nuclear reactions and the subsequent energy
production the stellar structure changes : temperature, radius, density
and luminosity at different depths in the star are adapted to the new
situation, i.e. the star evolves. A practical way to follow the evolution
of the star is to describe the evolutionary tracks in a Hertzsprung-
Russell diagram, i.e. the change of luminosity and effective temperature
as a function of time.

Evolutionary tracks for stars with initial masses between 20 and
100 $M_O$ are shown in Figure 7 for the case when mass loss is taken into
account (full lines) and for the case of constant mass (dashed lines).
The tracks are computed by de Loore, De Grève and Vanbeveren (1978) and
by de Loore, De Grève and Lamers (1977). A moderate average mass loss,
roughly in agreement with the observations was taken into account, according to

$$\dot{M} = 100 \ L/c^2 \quad (\dot{M} \text{ in } gs^{-1}, L \text{ in erg } s^{-1})$$

where L denotes the stellar luminosity, and c the velocity of light. This
reduces to

$$\dot{M} = 6.8 \ 10^{-12} \times L \quad (\dot{M} \text{ in } M_O yr^{-1}, L \text{ in solar units})$$

Andriesse (1979,1980) starts from the idea that mass losses are a consequence of large stochastic velocities finding their origin in the instability of mass flows proceeding outwards in subphotospheric layers. He
derives a mass loss equation from the classical thermodynamic fluctuation
theory. In this context the mass loss rate M can be expressed as

$$\dot{M} \sim L^{3/2} \ (R/M)^{9/4}/G^{7/4}$$

Evolutionary tracks with this mass loss equation were calculated by
Andriesse et al. (1981) for 100, 60 and 40 $M_O$. The tracks are shown in
Figure 8.

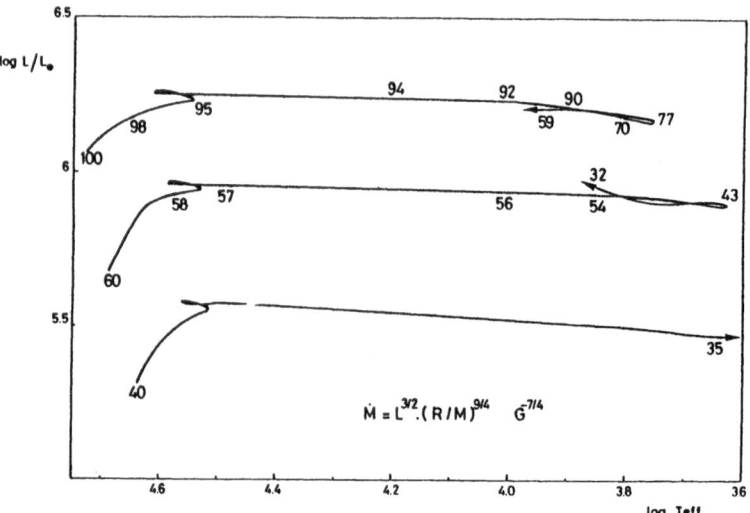

.Fig.8. Evolutionary tracks with mass loss according to
        Andriesse, Packet, de Loore, 1981

## c) The effect of overshooting in the evolution

The transport of the energy produced by nuclear reactions in the
stellar interior occurs by radiation or by convection. Stars more massive than 1.2 $M_0$ have convective cores, where mixing is very efficient
and where a substantial fraction of the energy is transported outwards
by turbulent convection. The classical theory used for the treatment of
these convective cores is the mixing length theory (Böhm-Vitense, 1958).
In this theory it is assumed that individual elements of matter travel
over a distance l, the mixing length, of the order of a pressure scale
height

$$l = \alpha \, H_p = \alpha \, \frac{RT}{\mu g}$$

($H_p$ pressure scale height, $\alpha$ = 0.8-1.5, R gas constant, T temperature,
$\mu$ molecular weight, g gravitational acceleration) conserving entropy,
and then lose their identity, i.e. dissolve while they transfer their
energy to the surrounding medium, where again elements are formed,
rising over a distance l and so on. Within the framework of the mixing
length theory, mixing of stellar material stops there where the acceleration of the moving elements vanishes. According to Schwarzschild and
Härm (1958) convection stops when

$$\nabla_{rad} \left(= \frac{d \ln T}{d \ln P_g}\right)_{rad} < \nabla_{ad} \left(= \frac{d \ln T}{d \ln P_g}\right)_{ad}$$

with T temperature and $P_g$ gas pressure).
According to Ledoux (1947) a molecular gradient term should be included.
Applying the Schwarzschild or the Ledoux criterion means that one
assumes that the mixing ends where the acceleration of the elements is

zero. However these cells can penetrate the radiative layers outside
the convective core since their velocity at the boundary determined with
the Schwarzschild-Ledoux definition is not zero. The real boundary of
the convective core is that shell where all velocities vanish. This is
called overshooting, and has the effect that the convective cores become
larger, hence the mixing is more efficient, and the hydrogen burning
lifetime will be longer. A comparison of the mass of the convective
core defined in the classical way or, with overshooting, calculated by
**Bressan et al.** (1981) is shown in Table 2.

Table 2. Comparison of the mass of the convective core (in $M_\odot$) calculated with the Schwarzschild criterion and with overshooting
(Bressan et al. (1981)

| stellar<br>mass | mass of the core<br>Schwarzschild crit. | mass of the core<br>with overschooting |
|---|---|---|
| 100 | 78 | 85 |
| 60 | 40 | -47 |
| 20 | 9 | 11.2 |

Evolutionary tracks for masses of 2, 4, 6 and 9 $M_\odot$ with
overshooting calculated by Bressan et al. (1981) are shown in Figure 9.

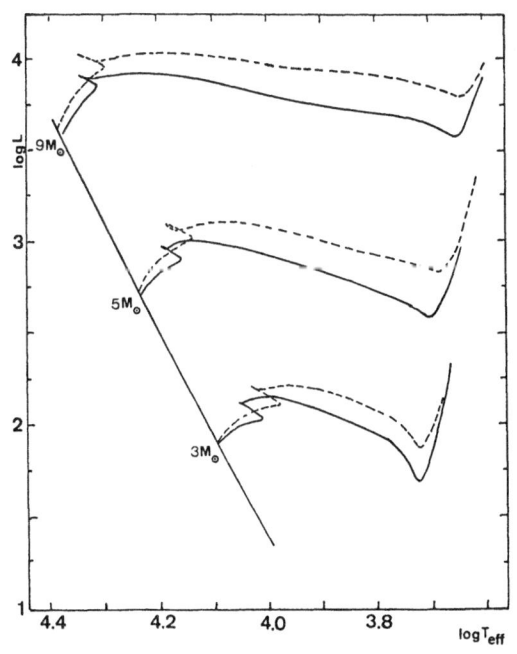

Fig.9

Evolutionary tracks
with non local treat-
ment of convection in
the core.

Full line:Schwarzschild
criterion

Dashed line:overschoo-
ting ($\ell/H_p = 1$)

(Maeder & Mermilliod,
1981)

In larger cores occurs
a better mixing.
The magnitudes are in-
creased by $0^m\!.6$ to $0^m\!.8$.

d) Change of the chemical composition of the outer layers by mass loss

Mass loss can expose layers where N has been enhanced by the CNO tricycle (Dearborn and Eggleton, 1977). Significant differences will result depending on whether the mass was removed during stages near the ZAMS or during phases near H-depletion. Dearborn and Blake (1979) calculated evolutionary tracks for initial masses of 15, 30 and 60 $M_O$ with mass loss, up to a presupernova configuration, for different mass loss rates. The results for a 30 $M_O$ star are shown in Figure 10.

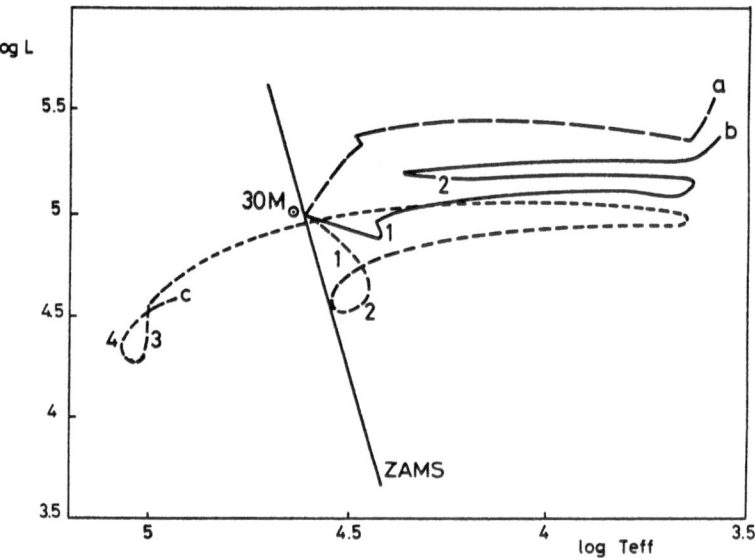

Fig.10. *Evolutionary tracks for a 30 $M_O$ star computed by Dearborn and Blake (1979) for different mass loss rates.*

| | $\dot{M}_{in}$ | $\dot{M}_{final}$ | $\bar{\dot{M}}$ | $M_c$ | final state |
|---|---|---|---|---|---|
| a | 0 | 0 | 0 | 11.5 | Red Giant |
| b | $1.7 \ 10^{-6}$ | $4.3 \ 10^{-6}$ | $2.5 \ 10^{-6}$ | 8.2 | Red Giant |
| c | $2.5 \ 10^{-6}$ | $2.8 \ 10^{-6}$ | $2.6 \ 10^{-6}$ | 4.5 | Wolf Rayet |

1 Point where $^{14}N = {}^{12}C$
2 Point where $^{14}N = {}^{16}O$
3 $^{14}N$ enhanced He core
4 $^{12}C$ enhanced He core

A **star** of 100 $M_O$ has at the ZAMS a convective core of 77.5 $M_O$; the remnant at the end of core hydrogen burning calculated with N=100 ($M = 100 \ L/c^2$, $\dot{M} \sim 8.6 \times 10^{-6} \ M_O yr^{-1}$) of 77.42 $M_O$ shows the outer layers of this original core, where convective mixing and nuclear burning changed the original X=0.7, Y-0.27 composition. The final atmospheric value for hydrogen is X=0.67. If we convert the mass loss rate found by Abbott et al. (1981)($\dot{M}$ = 3.8 $10^{-5}$ $M_O yr^{-1}$) into an N-value, we find N=400. With this value even deeper layers will show up at the surface, and the hydrogen abundance will still be lower. For an initial 20 $M_O$ model the convective core is 8.55 $M_O$; even for a mass loss rate with N=500 a remnant of 11 $M_O$ is left, hence the atmospheric layers still show the original composition. In this case not enough material is removed to show at the surface material changed by nuclear burning.

For a primary of 20 $M_O$ in a binary system, where a remnant of 5.6 $M_O$ is left after mass exchange and mass loss, the situation is different, and here material contaminated by nuclear processes appears at the surface. The hydrogen abundance in this case has dropped to X ~ 0.2.

The variation of the chemical composition of the outer layers as a consequence of stellar wind mass losses is shown in Table 3 for N=300 and in Table 4 for N=500 (de Loore, De Grève, Vanbeveren, 1979).

*Table 3. The evolutionary time (in $10^6$ yrs) since the ZAMS, necessary to show the hydrogen abundance of column 1 in the stellar atmosphere (N=300)*

| $X_{at}$ | 100 $M_O$ | 80 $M_O$ | 60 $M_O$ |
|------|------|------|------|
| 0.7 | 0 | 0 | 0 |
| 0.6 | 1.80 | 2.25 | 2.95 |
| 0.5 | 2.43 | 2.86 | 3.66 |
| 0.4 | 2.88 | 3.23 | |
| 0.36 | 2.99 | | |

*Table 4. The evolutionary time (in $10^6$ yrs) since the ZAMS, necessary to show the hydrogen abundance of column 1 in the stellar atmosphere (N=500)*

| $X_{at}$ | 100 $M_O$ | 80 $M_O$ | 60 $M_O$ |
|------|------|------|------|
| 0.7 | 0 | 0 | 0 |
| 0.6 | 1.58 | 2.00 | 2.62 |
| 0.5 | 2.24 | 2.62 | 3.48 |
| 0.4 | 2.73 | 3.17 | 4.15 |
| 0.3 | 3.16 | 3.65 | |
| 0.25 | 3.36 | 3.85 | |
| 0.21 | 3.50 | | |

As may be seen from Figure 5 due to stellar wind mass loss also the atmospheric nitrogen increases and the carbon and oxygen abundances decrease. It is not yet clear if it is observationally possible to confirm this N-overabundance, although there are indications for overabundances of N in some O stars (Bisiacchi et al., 1981) derived from the N III λ 4514 line.

## 5. The evolution of binaries

During core hydrogen burning the star expands. When hydrogen in the core is exhausted the star contracts, but rapidly expands again during shell hydrogen burning and core helium burning. This is depicted in Figure 11 for a 60 $M_O$ star.

In the case of a single star the expansion of the star to values of 100 times the initial radius occurs unhampered. In the case of a binary however, the companion interferes and limits the rate of expansion of the more massive star. The evolution of close binary systems is deter-

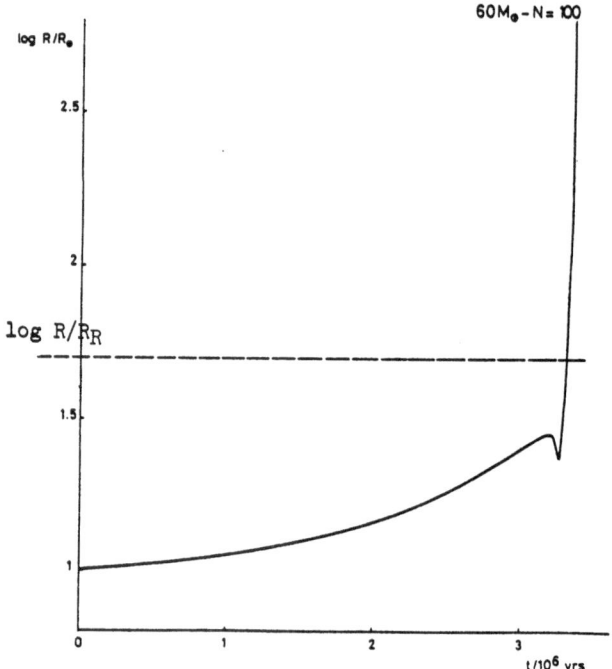

Fig.11. The variation of the radius of a star of 60 $M_O$ (evolution calculated with mass loss, N=100).

mined by considerations of Roche lobe overflow or by tidal interaction.
The two components can be considered as point masses surrounded by equipotential surfaces. The common equipotential surface in this special
case of point masses is called the Roche surface, determined by the
Roche radius $R_R$, - the radius of a sphere with volume equal to the Roche
volume -, a function of the distance A of the two components and their
mass ratio q (= $M_1/M_2$, with $M_1$ the mass of the primary and $M_2$ the mass
of the secondary). The Roche radius can be calculated as

$$R_R = A \ (0.38 + 0.2 \log M_1/M_2)$$

for q values between 0.3 and 20 (Paczynski, 1966).
When the radius of the evolving star exceeds $R_R$ this star has to lose
mass, in order to keep its radius within the Roche radius. This situation is shown in Figure 11 by the dashed line.
    A part of the matter expelled by the primary is accreted by the
secondary and the remaining part leaves the system (= non-conservative
evolution). The parameters of the system (semi-major axis, period)
change. In this way the initial mass ratio is inverted. Vanbeveren et
al. (1979) have calculated non-conservative close binary evolution. As
an example the evolution of a 40 $M_O$ + 20 $M_O$ close binary system is shown
in Figure 12. The two components are first losing mass due to stellar
wind during 4.6 $10^6$ years and then, during the shell hydrogen burning,

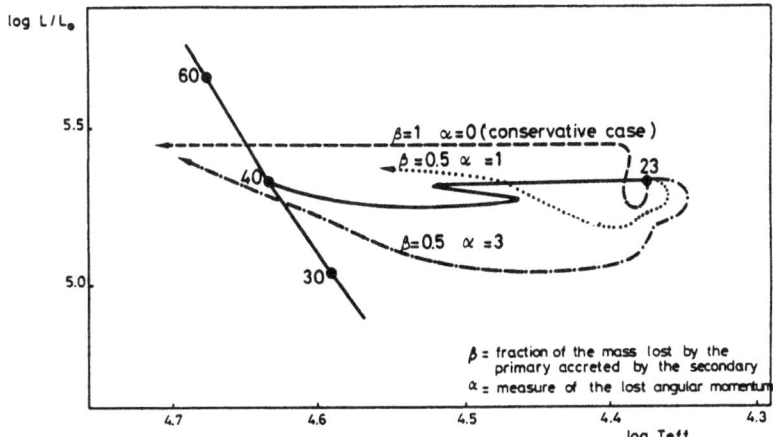

**Fig.12.** *Non-conservative evolution for a 40 $M_O$ + 20 $M_O$ system for various values of the parameters α and β. In full lines the evolutionary track with stellar wind mass loss, N=300.*

the 23 $M_O$ remnant of the primary starts mass loss due to Roche lobe over-flow, leaving at the end after ~12000 years a remnant of 11 $M_O$.

In the conservative case, β=1, i.e. all the mass lost by the primary is accreted by the secondary, the secondary mass is increased to 29 $M_O$. With the assumption that all the mass leaves the system (β=0) the final secondary mass is ~17 $M_O$, and for the intermediate, probably the most realistic case, β=0.5, which means 50% of the mass lost by the primary is accreted by the secondary, the other 50% leaving the system, the final secondary mass is 23 $M_O$.

Several assumptions for angular momentum losses can be considered: α=0 means conservation of angular momentum; larger values, α=1 or α=3 correspond to various angular momentum losses, α=3 corresponding with the case where ~50% of the angular momentum is carried away by the matter leaving the system. An overview of the variation of the system parameters for this case is given in Table 5.

**Table 5.** *Non-conservative evolution of a 40 $M_O$ + 20 $M_O$ system.*
   *1. Stellar wind mass loss: N=300 : phase 1*
      *End phase 1: $M_1$=23.1 $M_O$; $M_2$=16.9 $M_O$; P=10.2 d*
   *2. Different cases of mass transfer determined by different*
      *β- and α-values. $t_1$ is the start of the overflow phase,*
      *$t_2$ the end.*

|  |  | $M_1$ | $M_2$ | P(d) | $t_2-t_1$ (yrs) |
|---|---|---|---|---|---|
| β=0 (conservative) | α=0 | 11 | 29 | 18.8 | 12200 |
| β=0.5 | α=1 | 10.7 | 23.1 | 6.8 | 11150 |
|  | α=3 | 10.3 | 23.2 | 2.9 | 7700 |
| β=0 | α=0 | 11.2 | 16.9 | 62.5 | 11300 |
|  | α=1 | 11.0 | 16.9 | 22.3 | 13300 |
|  | α=3 | 10.2 | 16.9 | 2.4 | 14100 |

The computations lead to the following conclusions :
  a) as a consequence of the first phase (stellar wind mass loss) the
mass ratio of the two components changes as well as the orbital period;
  b) mass and momentum losses have practically no influence on the mass
and structure of the primary at the end of RLOF. The core of the star
governs the evolution regardless of the situation of the outer layers;
  c) cancelling the conservative assumptions during the RLOF phase
affects principally the period; this is mainly determined by α.
  A consequence of stellar wind mass losses and non-conservative evolution is a considerable enrichment of the interstellar medium of material processed by nuclear reactions. This is illustrated in Table 6.

Table 6.  Core and stellar masses at various evolutionary stages and the
          expelled matter.

| PRIMARY | | SECONDARY | | PRIMARY | | SECONDARY | | MATTER RESTI-TUTED TO ISM | |
|---|---|---|---|---|---|---|---|---|---|
| Initial mass | Mass of initial convective core | Initial mass | Mass of initial convective core | Remnant after stellar wind | Remnant after Roche lobe overflow | Remnant after stellar wind | Mass after accretion (β = 0.5) | Original composition | Changed composition |
| 100 | 77 | 80 | 58 | 77 | 43 | 63 | 80 | 40 | 17 |
| 60 | 40 | 40 | 24 | 48 | 23 | 33 | 45.5 | 23 | 8.5 |
| 40 | 24 | 30 | 16 | 24 | 9 | 20 | 27.5 | 26 | 7.5 |
| 30 | 16 | 20 | 9 | 26 | 9 | 18 | 26.5 | 13 | 1.5 |

In the case of the most massive binary of this table, 100 $M_O$ + 80 $M_O$, an
amount of 57 $M_O$ is restituted to the interstellar medium during the wind
phase and the mass exchange phase, composed of 40 $M_O$ of matter of the
initial composition and 17 $M_O$ of processed material (He- and N-enhanced).
  Binary evolutionary computations with mass and/or angular momentum
loss for low mass stars were carried out a.o. by Yungelson (1973) and
Plavec et al. (1973). Yungelson (1973) calculated mass transfer according to case B for a system consisting of a 1.5 $M_O$ primary and a 1.3 $M_O$
secondary, assuming that a quarter of the matter expelled by the primary,
carrying with it a given amount of angular momentum, leaves the system.
The result for the computed case is that the final period is smaller
than in the conservative case. Plavec et al. (1973) calculated the mass
transfer for a 7 $M_O$ star with a convective enevelope, assuming that a
fraction f of the matter lost by the mass loser leaves the system, carrying with it a fraction g of the specific angular momentum. The value of
the conservative evolution and the influence of the various parameters
on the final state was examined by De Grève et al. (1978). They concluded that using conservative evolution the observed X-ray binaries can
only be explained if extreme mass ratios for the initial ZAMS systems
are adopted together with extremely small periods. Taking into account
mass loss from the system, and allowing angular momentum losses these
restrictions can be removed.

## 6. The production of Wolf-Rayet stars

The 6th Catalogue of Galactic Wolf-Rayet stars (van der Hucht et al., 1981) contains 158 galactic Wolf-Rayet stars and 18 among them are known binaries with solutions for the orbit. A study of these systems by Massey (1981) reveals that the average mass of a WR star is ~20 $M_O$, ranging from 10 to 50 $M_O$; the period range of these systems is similar to that of their O type progenitors. From a comparison of the projected orbital separations and eccentricities of O type binaries and WR binaries Massey concludes that only the most massive O systems evolve into WR systems, which agrees with the conclusion of Vanbeveren and de Loore (1980) that early WN and WC systems result from ZAMS masses exceeding 50 $M_O$.

A scenario where O type stars are converted into Of stars, becoming Wolf-Rayet stars in a more advanced stage, has been proposed by Conti (1976). Comparison of observations of Wolf-Rayet stars with evolutionary sequences (WR of population I or close binaries) reveals that,in order to convert a massive star into a WR star, a significant part of its matter has to leave. In this way matter which was initially present in the convective core appears at the surface, matter containing elements produced by hydrogen fusion, by the CNO-cycle (see section 4). A mechanism to take away a sufficiently large amount of matter is the mass transfer process in close binaries, by Roche lobe overflow. WR population I stars could be produced by a huge mass loss process at very short time scales, after the red giant phase. Computations for such stars have been carried out by Maeder (1981a,b). With this scenario WN stars can be produced, starting from initial masses exceeding 20 $M_O$.

WN stars contain hydrogen in their atmospheres : $N(H)/N(He)$ ~ 1 to 2; for normal stars this ratio is ~10. Willis and Wilson (1978) and Nussbaumer et al. (1979) have determined C, N and He-abundances for WN5 and WN6 stars and for two WC stars. The abundances of hot WN stars reveal that their atmospheres contain essentially helium, hence on an evolutionary viewpoint these stars are in a stage following hydrogen core and shell burning, since all hydrogen has to be converted into helium in order to have a helium atmosphere.

For WC stars we observe $3\alpha$-products at the surface which means that these stars are more evolved, and deeper layers appear at the surface. In conclusion one can say that the WN phase occurs between the stages where $X_{atm} \simeq 0.2$ and $C_{atm} \neq 0$. The WC phase occurs after this phase.

Different possibilities for the production of Wolf-Rayet stars, single stars and binaries have been examined by Maeder (1981c). His conclusions are the following :

a) The scenario of Conti (1976): It is assumed that O type stars are converted into Of stars, later on in transition WN7-9 stars, transformed into classical WN stars and then into WC stars. Computations have been performed by de Loore, De Grève and Vanbeveren (1978) and by Chiosi et al. (1978). Evolutionary models for masses in the range 40 to 100 $M_O$, with mass loss rates as observed for Of stars (4 to 7 $10^{-6}$ $M_O yr^{-1}$) were computed by Nöels et al. (1981) in order to follow the surface composition of H, He, C, N and O. The result of these computations is that at the surface equilibrium CNO-products appear, i.e. enhanced N and less O, during the core hydrogen phase. In this way late WN stars can be produced.

b) Internal mixing and diffusion (Schatzman and Maeder, 1981): In massive stars (M >50 $M_O$) the radiative viscosity becomes very high, and the associated diffusion coefficients are sufficiently large to produce a substantial mixing during their main sequence lifetimes. This turbulent mixing will, for massive stars, lead to the production of Wolf-Rayet stars (Maeder, 1981c).

c) The Wolf-Rayet phase as post red supergiant phase for stars with masses > 20 $M_O$: The different phases are then: O, red supergiant, WN, WC.

d) Binary evolution with mass transfer: Computations were carried out by Vanbeveren, De Grève, van Dessel and de Loore. The different phases are then: O+O, WR+O, later on compact object + WR.

## 7. Nucleosynthesis in massive stars with mass loss

It is clear that the study of the evolution of massive stars with mass loss is of great importance for nucleosynthesis and the chemical evolution of galaxies.

Yields of heavy elements from massive stars have been calculated by Arnett (1978). However in the case of mass loss some assumptions, valid for the case of stars with constant mass, have to be revised, when helium synthesis is concerned: indeed in that case the deep convection zones present in red supergiants will mix some processed material (He and $^{14}N$) from the core into the envelopes. In models with mass loss, one has to take into account the produced and ejected elements such as He, and heavy elements in WC stars. Maeder (1981d) has calculated the stellar yields in He and heavy elements for stars with masses ranging from 9 to 170 $M_O$, for a constant mass, and for "intermediate" and strong mass loss rates. The results are shown in Figure 13. The conclusions of Maeder are:

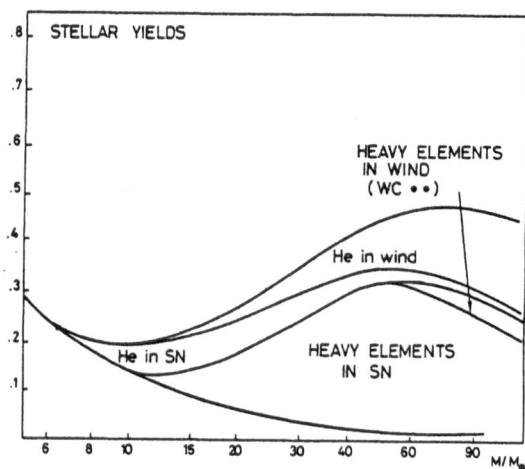

Fig.13. *The yields in helium and heavy elements for stars of different masses (Maeder, 1981).*

Masses < 60 $M_O$
The yields in He and metals do not depend strongly on M, since the core mass, hence the internal evolution is not altered and there is also some degree of compensation between the matter ejected in stellar wind mass loss models and the matter mixed into the envelope at constant mass.

Masses > 60 $M_O$
The yields in metals are reduced by high mass loss, and He preserved in this way and ejected in the wind increases the stellar yield. The median star contributing to the synthesis of heavy elements is an initial 32 $M_O$ star. For the helium production this is 18 $M_O$.

## 8. Final phases of single stars

Models for the late evolution of massive stars show that their cores can collapse and that the bounce of the core may give rise to the SN explosion (Bruenn, Arnett and Schramm, 1977). The supernova event leaves a neutron star, which can be a pulsar. The rate of type II SN in the galaxy is ~2-3 per century (Tammann, 1977). This rate is smaller than the estimated birthrate for pulsars (15-16 per century)(Taylor and Maucherat, 1977). The processes leading to the final phases of single stars and binaries are not completely known. Probably a wide variety of possibilities exists, depending a.o. on the initial mass.

According to Nomoto (1981) two types of SN occurs, i.e. supernovae with carbon deflagration for stars with masses between (6±2) and (8±1) $M_O$, and the electron capture supernovae for masses between 8 and 10 $M_O$. Carbon deflagration destroys the star completely and no remnant is left. Stars with masses between 8 and 15 $M_O$ are probably responsible for classical SN of type II. The core of O, Ne, Mg implodes and a remnant in the form of a neutron star is left (Barkat et al., 1974 ; Wheeler, 1978). The exact process is not known. It is believed that SN of type II produce pulsars and contribute very little to nucleosynthesis. Stars with large masses develop Fe cores during the dynamical collapse and produce also neutron stars. These stars contribute much to nucleosynthesis.

A mechanism for the explosion of type I SN is given by Chevalier (1981) as a nuclear runaway in an accreting white dwarf and an explosion (Mazurek and Wheeler, 1980; Sugimoto and Nomoto, 1980).

The basic explosion mechanisms are:
- carbon detonation (a supersonic burning front)
- carbon deflagration (a subsonic burning front involving energy transfer, convection, conduction)
- helium detonation

Carbon is expected to ignite in a degenerate CO core when $\rho_{cen} > 2 \ 10^9$ g/cm3, corresponding to a white dwarf mass of 1.4 $M_O$. The detonation wave propagating through the star, brings the material to $^{56}Ni$. This occurs for all the matter, except the outer 0.006 $M_O$ where Si-group elements may be present. The final energy in the supernova explosion is ~1.6 $10^{51}$ erg, and the white dwarf is completely disrupted.

Also possible is that C deflagrates. Since the propagation velocity of a deflagration front is not known all possibilities are open. Nomoto, Sugimoto and Neo (1976) have computed models. An idea about the ejecta:
Fe peak : 1.02 $M_O$
Ne Si group : 0.28 $M_O$
CO : 0.10 $M_O$

It is also possible that a He white dwarf occurs in a binary system after mass transfer and mass loss. Nomoto and Sugimoto have performed computations and conclude that a central detonation starts when the white dwarf mass is ~0.7 $M_O$. The explosion ejects this 0.7 $M_O$ of Fe peak nuclei with a net energy of 2 $10^{51}$ erg and no remnant is left.

Colgate et al. (1980) have calculated models, and tried to fit those to observations. They succeeded in deriving the ejected mass, and the expansion velocities. They concluded that the ejected masses are ~0.5 $M_O$ and they found that if the exploding star is a white dwarf of 1.4 $M_O$, presumable a neutron star is left. This could be a clue for models of galactic bulge sources, that are probably neutron stars + red dwarf companions. A number of these objects are found in globular clusters, and then it is generally thought that they were formed by a capture process (computations of Sutantyo, 1975; Fabian, Pringle, Rees, 1975). An average of 3.3% of all stars in cluster cores collide in $10^{10}$ years (Hills and Day, 1976). Hills and Day estimate that for an average of ~3.3% of all globular cluster stars a collision occurred during the lifetime of the galaxy. Adopting then as average lifetime for the X-ray source $5 \cdot 10^8$ years, an average age for globular clusters of $1.2 \cdot 10^{10}$ yrs, and a steady state situation of 8 globular cluster sources (as observed) we find a formation rate of 1 per $8.5 \cdot 10^9$ years. With the adopted collision rate this leads to the fact that per cluster some 40 neutron stars are required to provide the adopted formation rate. From the initial mass function one knows that a few thousand stars more massive than 8-10 $M_O$, lower limit mass necessary for the production of a neutron star, were present. Only 1% of this number is necessary to explain the observations (van den Heuvel, 1980).

Weaver et al. (1978) have computed the evolution of a 15 and a 25 $M_O$ population I star from ZAMS to core collapse. Since during advanced evolutionary stages the mixing time for convective layers becomes comparable to the nuclear burning time scale, convection and also semi-convection had to be treated time dependent (Eggleton, 1971; Scalo and Ulrich, 1973; Arnett, 1974 , 1977). Convective overshooting is taken into account by slow mixing in a time of the order of a radiation diffusion time of non-convective regions adjacent to convective ones. Convective heat transport is assumed to occur in regions where the material satisfies the Ledoux criterion. Material is considered as dynamically unstable to convection if the Ledoux criterion is satisfied, and only as secularly unstable if the Schwarzschild, but not the Ledoux criterion is satisfied.

The initial composition of the stars was X=0.7, Y=0.28 and the initial configuration was a cloud with a mean density of 0.5 $gcm^{-3}$ and a temperature profile adjusted to give hydrostatic equilibrium. The stars rapidly relax to the usual ZAMS structures.

The general structure of the 15 and 25 $M_O$ presupernova stars is that of a red supergiant star with a high density mantle ($3 \cdot 10^{10}$ cm in radius), a nearly constant density envelope ($\rho = 10^{-8}$ $gcm^{-3}$) with a temperature of ~$10^5$ K, extending out to a photospheric radius of $3.9 \cdot 10^{13}$ cm or $6.7 \cdot 10^{13}$ cm for a 15 or 25 $M_O$ star respectively. The stars have evolved into an onionskin structure with H, He, C, Ne, O and Si-burning convective shells, separated by density gradients, sufficiently steep to prevent mixing. An example of such a structure is given in Figure 14, for a 16 $M_O$ during silicon burning, calculated by Arnett (1973).

The time of hydrostatic silicon burning is sufficiently long to allow neutralization of the resulting iron-peak elements by electron capture during the Si-burning. The masses of the resulting iron cores are 1.56 and 1.61 $M_O$ respectively. During the early collapse phase photodesintegration of the iron peak elements to $\alpha$-particles is the main source of energy loss; neutrino losses due to neutronization are a factor 30, and due to thermal plasma processes a factor 1000 less important. The elements between oxygen and calcium show in the 25 $M_O$ star an enhancement of ~ 35, relative to the solar abundances; for the lighter elements the enhancement is systematically larger. The ratios of neighbouring elements agree pretty well with the corresponding solar ratios. The larger quantity of $^{54}Fe$ and rare highly neutronized iron peak species makes it

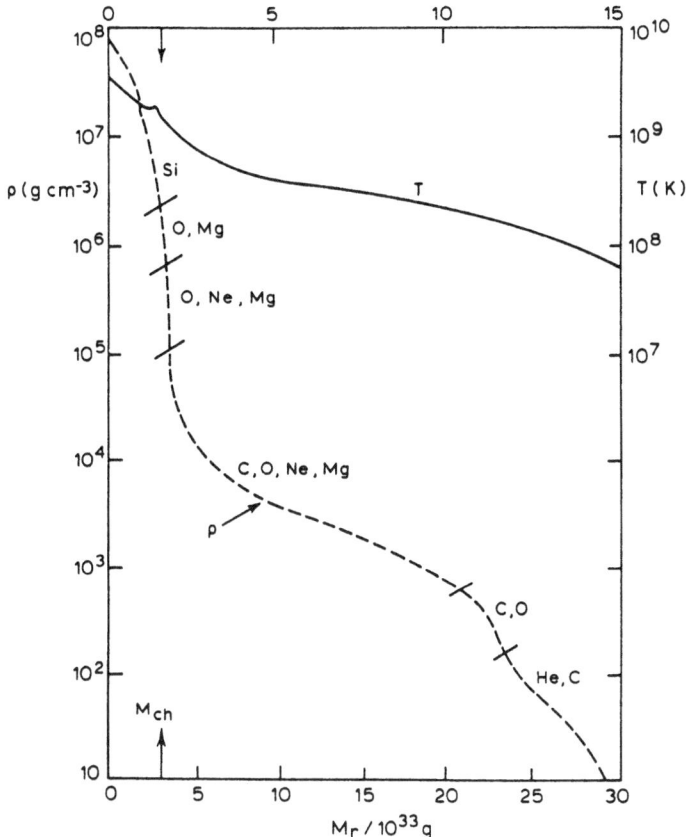

Fig.14.    The structure of a 16 M_O star during Si-burning
          (Arnett, 1977).

necessary that not very much matter inside the core boundary can escape
to the interstellar medium (<0.03 M_O), in order to prevent an overproduc-
tion of these isotopes.  Hence nearly the complete core has to collapse
to form a neutron star.  The 25 M_O star can eject 6.2 M_O of heavy elements; the 15 M_O star can eject ≤ 1.1 M_O.  This has the consequence that,
since, according to the Salpeter initial stellar mass function, 15 M_O
stars are born a factor $(15/25)^{-1.35}$ more frequently than 25 M_O stars,
the ratio of the production of heavy elements by 25 M_O or by 15 M_O stars
is about a factor of 3.  The abundances of some elements are given in
Figure 15.

     For carbon and nitrogen an underproduction is found; for the pro-
duction of these elements in lower mass stars, convective mixing and
mass loss in planetary nebulae might be invoked, or perhaps unprocessed
carbon, ejected in lower mass supernovae.

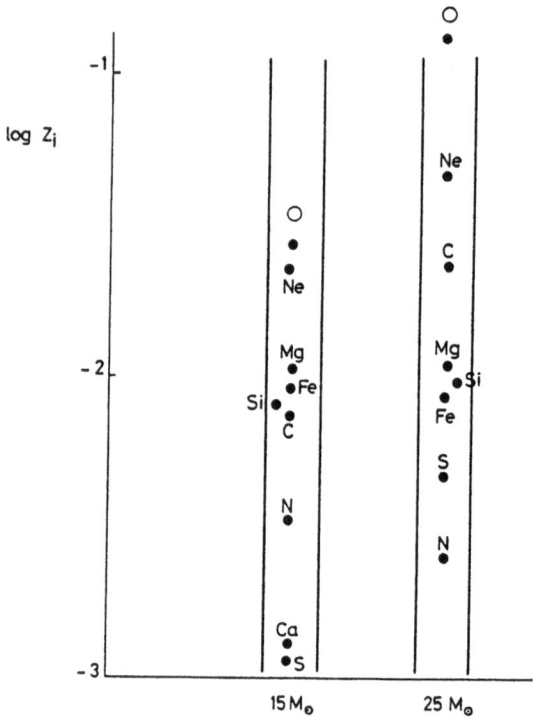

*Fig.15.* The abundances of various elements in massive stars (15 $M_O$ and 25 $M_O$) at the presupernova stage.

## 9. Late phases of massive close binary evolution

Massive close binaries will evolve after a phase of mass transfer and mass loss into a system containing a He-primary and a massive companion (Wolf-Rayet phase). The helium star will evolve through further successive nuclear burning stages into a compact object (neutron star or black hole) after a supernova explosion (van den Heuvel and Heise, 1972), observable after a certain time as X-ray binary. Evolutionary scenarios for X-ray binaries were calculated by Tutukov and Yungelson (1973), de Loore et al. (1974,1975) and de Loore and De Grève (1975a,b, 1976). The lifetime of the X-ray stage is of the order of $10^5$ years for X-ray binaries with an optical companion of ~20 $M_O$ (Savonije, 1979). At a given moment this optical component will fill its Roche lobe. Since the accretion rate of the compact companion is limited by its Eddington limit (~$10^{-8}$ $M_O yr^{-1}$), and the mass loss rate of the mass losing companion is larger, a common envelope will be formed. Common envelope stages were proposed by Paczynski and computations were performed by Taam et al. (1978), Tutukov and Yungelson (1979).

For the further evolution two possible scenarios exist :
  a) the neutron star is engulfed by its companion in the common envelope.
In this way a red supergiant with a compact core is formed (Thorne and
Zytkow, 1977)
  b) the larger part of the matter expelled by the non compact companion
leaves the system carrying away a large amount of angular momentum, leading to a very small orbit (van den Heuvel and de Loore, 1973). The supernova explosion is probably not symmetric; according to computations of
De Cuyper (1981), de Loore et al. (1974), with the assumption that an
extra kick of 100 $kms^{-1}$ reflects this asymmetry, a shell expansion velocity of ~10000 $kms^{-1}$ and a post supernova remnant of 1.5 $M_O$, the probabilities for the system to remain bound are of the order of 70-80% for
initial primary masses between 40 and 160 $M_O$ for all mass ratios. For
lower masses of the primary (20-40 $M_O$) the systems are never disrupted.
However, probably larger kick velocities exist, especially for systems
with large mass primaries, and in this case the disruption probabilities
are much larger. Vanbeveren and de Loore (1982) examined the kick velocities and their influence on the final status (bound or disrupted) of
massive close binaries. A further investigation of the kick velocities
and the connected disruption probabilities is very important for the
occurrence of OB runaways.

As well in the case of spiralling in as in the case of Roche lobe overflow with reverse mass transfer and mass loss the outer layers with the
original composition leave the primary, and helium rich layers appear at
the surface. The O star will evolve into a helium star, hence will show
more and more Wolf-Rayet characteristics. The expelled matter will be
stored in a shell or a cloud around the system. This evolutionary stage
is a second Wolf-Rayet stage; since the center of gravity of the systems
has a large velocity due to the supernova-explosion, these Wolf-Rayet
stars are called run-away Wolf Rayet stars (HD 50896, HD 93131, HD 96548,
HD 192163, HD 197406, HD 164270).

  The final state of the primary of a massive close binary system is
a neutron star (or black hole) if the mass is sufficiently large, or a
white dwarf. The critical mass has a value of 10-15 $M_O$ (Massevitch et
al., 1976). De Loore and De Grève (1976) performed computations for
binary systems with primaries between 10 and 15 $M_O$, and they could fix
a critical mass of ~14 $M_O$. Primaries of lower mass undergo two successive stages of mass transfer, leading finally to a white dwarf. As an
example the evolution of a 10 $M_O$ + 8 $M_O$ system, calculated by De Grève
and de Loore (1976) is shown in Figure 16.
  The lifetime of the secondary in a binary system after the mass
transfer phase can be estimated from computations by Doom and De Grève
(1981) adopting the following conditions :
  a) mass transfer starts after the hydrogen phase, before He-exhaustion
  b) 50% of the matter leaves the system during the mass transfer phase.
  Comparing the lifetimes of the secondaries with the remnant lifetime
of the primaries reveal that only for the case where the initial mass
ratio is sufficiently small (q < 0.8 for stars with masses exceeding
60 $M_O$, and q < 0.9 for lower mass stars) the succession of the various
evolutionary phases is: main sequence phase, Wolf-Rayet phase, explosion,
OB-runaway, X-ray stage, reverse mass transfer phase, Wolf-Rayet runaway,

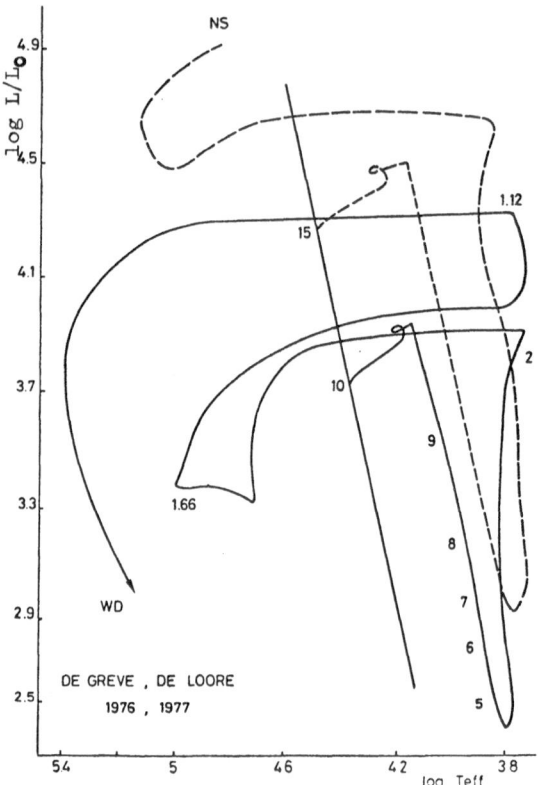

Fig.16. *The evolution of the primary of a* $10M_O$ + $8M_O$ *system, evolving into a white dwarf after two mass exchange phases in full line. The decreasing masses of the primary are indicated. For comparison, the evolution of a primary of* $15M_O$, *evolving into a neutron star is depicted as a dashed curve.*

explosion of the secondary . If on the contrary the masses are nearly equal (q ~ 1) the scenario is modified, in the sense that the OB runaway phase and the X-ray stage are skipped: the secondary fills its Roche lobe nearly immediately after the mass transfer phase of the primary.

The secondary has already evolved into a helium star before the explosion of its companion, hence before the neutron star is produced. In this case the normal X-ray stage is absent in the evolutionary scenario. Also Wolf-Rayet binaries consisting of two helium stars can be produced. The possibility that the system will be disrupted cannot be ruled out, and this could represent a possibility for the formation of single Wolf-Rayet stars.

REFERENCES

Abbott, D.C., Bieging, J.H., Churchwell, F. 1981, Astrophys.J., in press.
Andriesse, C.D. 1979, Astrophys.Space Sci. 61, 205.
Andriesse, C.D. 1980, Astrophys.Space Sci. 67, 461.
Andriesse, C.D., Packet, W., de Loore, C. 1981, Astron.Astrophys.95, 202.
Arnett, W.D. 1978, Astrophys.J. 219, 1008.
Arnett, W.D. 1974, Astrophys.J. 194, 373.
Arnett, W.D. 1977, Astrophys.J.Suppl. 35, 145.
Arnett, W.D. 1973, in D.N. Schramm and W.D. Arnett (eds.) "Explosive
    Nucleosynthesis", p.236.
Barkat, Z., Reiss, Y., Rakavy, G. 1974, Astrophys.J.Letters 193, L21.
Barlow, M.J., Cohen, M. 1977, Astrophys.J. 213, 737.
Bisiacchi, G.F., López, J.A., Firmani, C. 1981, preprint.
Böhm-Vitense, E. 1958, Z. Astrophys. 46, 108.
Bressan, A.G., Bertelli, G., Chiosi, C. 1981, Astron.Astrophys., submitted.
Bruenn, S.W., Arnett, W.D., Schramm, D.N. 1977, Astrophys.J. 213, 213.
Castor, J.I., Abbott, D.C., Klein, R.I. 1975, Astrophys.J. 195, 157.
Chevalier, R.A. 1981, Astrophys.J. 246, 267.
Chiosi, C., Nasi, E., Sreenivasan, S.R. 1978, Astron.Astrophys. 63, 103.
Colgate, S.A., Petschek, A.G., Kriese, J.T. 1980, Astrophys.J.Letters
    237, L81.
Conti, P.S. 1976, Mem.Soc.Roy.Sci.Liège, 6e Série IX, 193.
Conti, P.S., Garmany, C.D. 1980, Astrophys.J. 238, 190.
Dallaporta, N. 1971, in M. Hack (ed.) Colloquium on Supergiant Stars,
    p.250.
Dearborn, D., Blake, J.B. 1979, in "Mass Loss and Evolution of O-type
    Stars", eds. P.S. Conti and C. de Loore.
Dearborn, D., Eggleton, P. 1977, Astrophys.J. 213, 448.
De Grève, J.P., de Loore, C., van Dessel, E.L. 1978, Astrophys.Space Sci.
    53, 105.
Doom, C., De Grève, J.P. 1981, preprint.
Eggleton, P. 1971, Monthly Notices Roy.Astron.Soc. 151, 351.
Fabian, A.C., Pringle, T.E., Rees, M. 1975, Monthly Notices Roy.Astron.
    Soc. 172, 15.
Gathier, R., Lamers, H.J.G.L.M., Snow, T.P. 1981, Astrophys.J., in press.
van den Heuvel, E.P.J., Heise, J. 1972, Nature Phys.Sci. 239, 67.
van den Heuvel, E.P.J. 1980, in "Extragalactic X-ray Astronomy", Proc.
    Erice Advanced Study Institute, eds. G. Setti and R. Giacconi.
van den Heuvel, E.P.J., de Loore, C. 1973, Astron.Astrophys. 25, 387.
Hills, J.G., Day, C.A. 1976, Astrophys.J.Letters 17, L87.
van der Hucht, K., Conti, P.S., Lundström, I., Stenholm, B. 1981, Space
    Sci.Rev. 28, 3.
Hutchings, J.B. 1976, Astrophys.J. 203, 431.
Klein, R.I., Castor, J.I. 1978, Astrophys.J. 220, 902.
Lamers, H.J.G.L.M., Snow, T.P. 1978, Astrophys.J. 219, 504.
Lamers, H.J.G.L.M., Morton, D.C. 1976, Astrophys.J.Suppl. 32, 715.
Lamers, H.J.G.L.M., de Jager, C., Macchetto, F.M., Praderie, F. 1980a,
    preprint.
Lamers, H.J.G.L.M., Paerels, F., de Loore, C. 1980b, Astron.Astrophys.
    87, 68.
Lamers, H.J.G.L.M. 1981, Astrophys.J., in press.
Ledoux, P. 1947, Astrophys.J. 105, 305.

de Loore, C. 1980, Space Sci.Rev. 26, 113.

de Loore, C., De Grève, J.P., Vanbeveren, D. 1978, Astron.Astrophys. 67, 373.

de Loore, C., De Grève, J.P., Vanbeveren, D. 1979, 22nd Coll. Liège.

de Loore, C., De Grève, J.P., Lamers, H.J.G.L.M. 1977, Astron.Astrophys. 61, 251.

de Loore, C., De Grève, J.P., van den Heuvel, E.P.J., De Cuyper, J.P. 1974, Proc. 2nd IAU Reg. Meeting, Trieste.

de Loore, C., De Grève, J.P., De Cuyper, J.P. 1975, Astrophys.Space Sci. 36, 219.

de Loore, C., De Grève, J.P. 1975a, Mém.Soc.Roy.Sci.Liège, 6e Série, Tome VIII, P.399.

de Loore, C., De Grève, J.P. 1975b, Astrophys.Space Sci. 35, 241.

de Loore, C., De Grève, J.P. 1976, IAU Symp. No 73 "Structure and Evolution of Close Binaries", eds. P. Eggleton, J. Whelan, S.M. Mitton.

Maeder, A. 1981a, Astron.Astrophys., in press.

Maeder, A. 1981b, Astron.Astrophys., in press.

Maeder, A. 1981c, Astron.Astrophys., in press.

Maeder, A. 1981d, Astron.Astrophys., in press.

Maeder, A., Mermilliod, J.C. 1981, Astron.Astrophys. 93, 136.

Massey, P. 1981, Astrophys.J. 246, 153.

Mazurek, T.J., Wheeler, J.C. 1980, Fund.Cosm.Phys. 5, 193.

Morton, D.C. 1967, Astrophys.J. 147, 1017.

Noels, A., Conti, P.S., Gabriel, M., Vreux, J.M. 1981, Astron.Astrophys., in press.

Nomoto, K. 1981, Proc. IAU Symp. No 93 "Fundamental Problems in the Theory of Stellar Evolution", eds. D. Lamb, D. Schramm, D. Sugimoto.

Nomoto, K., Sugimoto, D., Neo, S. 1976, Astrophys.Space Sci. 31, L37.

Nussbaumer, H., Schmutz, W., Smith, L.J., Willis, A.J., Wilson, R. 1979, in "The First Year of IUE", ed. Willis A.J.

Paczynski, B. 1966, Acta Astron. 16, 231.

Plavec, M., Ulrich, R.K., Polidan, R.S. 1973, Publ.Astron.Soc.Pacific 85, 769.

Roxburgh, I. 1978, Astron.Astrophys. 65, 281.

Savonije, G. 1979, Astron.Astrophys. 71, 352.

Scalo, J.M., Ulrich, R.K. 1973, Astrophys.J. 183, 151.

Schatzman, E., Maeder, A. 1981, Astron.Astrophys. 96, 1.

Schwarzschild, M., Härm, R. 1958, Astrophys.J. 128, 348.

Snow, T.P., Marlborough, J.M. 1976, Astrophys.J.Letter 203, L87.

Snow, T.P., Morton, D.C. 1976, Astrophys.J.Suppl. 32, 429.

Sugimoto, D., Nomoto, K. 1980, Space Sci.Rev. 25, 155.

Sutantyo, W. 1975, Astron.Astrophys. 44, 227.

Taam, R.F., Bodenheimer, P., Ostriker, J.P. 1978, Astrophys.J. 222, 269.

Tammann, G.A. 1977, in "Supernovae", ed. D.N. Schramm, p.95.

Taylor, J., Manchester,R.1977, Astrophys.J. 215, 885.

Thorne, K.S., Zytkow, A. 1977, Astrophys.J. 212, 832.

Tutukov, A.V., Yungelson, L.R. 1973, Nauch.Inform. 27, 58.

Tutukov, A.V., Yungelson, L.R. 1979, Acta Astron. 29, 665.

Vanbeveren, D., de Loore, C. 1982, in "Wolf-Rayet Stars : Observations, Physics, Evolution", IAU Symp. No 99, eds. C. de Loore and A.J. Willis.

Vanbeveren, D., De Grève, J.P., van Dessel, E.L., de Loore, C. 1979, Astron.Astrophys. 73, 19.

Vanbeveren, D., de Loore, C. 1980, Astron.Astrophys. 86, 21.
Weaver, T.A., Zimmerman, G.B., Woosley, S.E. 1978, Astrophys.J. 225,1021.
Wheeler, J.C. 1978, Mem.della Soc.Astron.Italiana 49, p.349.
Willis, A.J., Wilson, R., Macchetto, F., Beeckmans, F., van der Hucht, K.,
    Stickland, D.J. 1979, "The First Year of IUE", London.
Willis, A.J., Wilson, R. 1978, Monthly Notices Roy.Astron.Soc. 182, 559.
Yungelson, L.R. 1973, Sov.Astron.A.J. 16, 864.

ACCELERATION OF COSMIC RAYS BY SHOCK WAVES

W.I. Axford
Max-Planck-Institut für Aeronomie
D-3411 Katlenburg-Lindau 3
Germany

1.  Introduction

The problem of the origin of galactic cosmic rays is a particularly difficult one despite the fact that rather detailed measurements of the properties of cosmic rays can be made, at least in the vicinity of the Sun. The current situation has been well reviewed by Lingenfelter (1) who points out that there are several linked problems to be solved, namely the question of sources and acceleration mechanisms, propagation within the galaxy, escape from the galaxy and of course solar modulation which affects the interpretation of the observations, especially below $\sim$ 1 GeV/nuc. It is usually supposed that these problems can be treated separately, so that the sources (perhaps supernovae, pulsars, black holes, flare stars, etc.) simply provide cosmic rays with given elemental and isotopic abundances and given spectra, which then propagate independently by diffusion through the interstellar medium, producing secondaries and perhaps losing energy as they do so until they eventually leave the galaxy by some means, which is usually described in terms of a "free escape" boundary condition to the diffusion equations.

In recent years there has been renewed interest in the possibility that the acceleration of cosmic rays should occur, not in discrete sources, but in the diffuse interstellar medium, as a consequence of shock waves associated with supernova remnants (2-5). Since the supernova remnants concerned are rather large and indeed tend to dominate the whole interstellar medium (6) it is becoming clear that the problems of acceleration and propagation of cosmic rays cannot be so easily separated. A further difficulty is concerned with the escape of cosmic rays from the galaxy which may be associated with a galactic wind (7-9) which is partly driven by cosmic ray pressure and therefore not an independent process. These complexities give added interest and significance to the role of cosmic rays in the dynamics of the interstellar medium but of course also make the traditional problems of cosmic ray physics much more difficult to treat.

We will attempt here to review the current status of investigations into various aspects of the problem of shock acceleration of cosmic rays. Due to space limitations and the uneven nature of our progress, not every aspect can be treated in detail and many of the problems associated with non-linear effects (section 10) are left to a later review. The reader is referred to other review articles for further information (10-14) and also to related work on the acceleration of particles at the Earth's bow shock (15 and references therein).

The idea that collision-free shocks exist was first put forward by Gold in 1953 as an explanation for the short duration of the sudden commencements of geomagnetic storms (16). Disucssions of the processes

by which such shocks can accelerate energetic particles have until recently been rather fragmentary with numerous authors proceeding along
roughly the same paths quite independently of each other.

The first observations of such an effect occurring in interplanetary space was made in 1959 by Dorman et al. (17-19) who noted that
small ($\sim$ 1%) increases of cosmic ray intensity sometimes occur before
the sudden commencement of a geomagnetic storm which is usually followed
by a much larger ($\sim$ 10%) "Forbush" decrease of intensity. Large ($\sim$ 100%)
enhancements of solar energetic particle fluxes prior to geomagnetic
storm sudden commencements were discovered by Reid and Axford (20,21) in
1962, using high latitude riometer observations. Spacecraft observations
of such "pre-SC energetic particle enhancements" were reported by Bryant
et al. (22) in 1962. Since that time very detailed observations have
become available of intensity increases associated with solar flare
shock waves (23-25), corotating interaction regions (26-28) and the magnetospheric bow shocks of the Earth (15,29-31) and Jupiter (32,33). Two
classes of event can be distinguished, namely (a) smooth, quasi-exponential increases occurring in front of a shock wave, lasting many hours in
the case of interplanetary shocks, and (b) short-lived intensity bursts
("shock spikes") lasting typically tens of minutes and exhibiting rather
high field-aligned anisotropies (34-36). In addition, there is clear
evidence for strong non-thermal heating of the shocked plasma (37,38)
and for escaping particle beams upstream from shocks which may represent
ions and electrons which have been directly accelerated from the plasma
itself (15,39,40). It is an advantage of the shock acceleration theory
for cosmic rays that, in contrast to many other theories, the mechanisms
involved can be observed directly and in great detail in interplanetary
space.

The first attempts to deal with theoretical aspects of shock acceleration were made by Dorman and Freidman (41) and Shabansky (42) on
the one hand, Parker (43), Wentzel (44,45) and Hudson (46) on the other,
all considering scatter-free acceleration as reviewed in section 4.
Hoyle (47) used the Rankine-Hugoniot conditions for a shock, including
the cosmic ray pressure and enthalpy and assuming that the background
gas behaves isotropically to obtain a very interesting result which is
a special case of the non-linear theory reviewed in section 10. The significance of scattering in the medium on either side of the shock was
noted independently by several authors, in particular Schatzman (48),
Jokipii and Davis (49,50) and Van Allen and Ness (51). Little progress
could be made in this direction, however, until suitable transport equations became available and the transition conditions for the cosmic ray
distribution function at the shock were understood (52-56). Numerous
solutions of these equations with the appropriate transition conditions
(56,11) were obtained by Fisk (57-59) but with the assumption that the
energetic particle spectrum has a specific power law form and overlook-
ing a particular solution of the equations. A numerical solution of the
transport equations simulating a solar flare event with a propagating interplanetary shock wave was given by Morfill and Scholer, taking into account shock reflection and second order Fermi acceleration (59). Independently, and more or less simultaneously, Krimsky (2), Axford, Leer
and Skadron (3), Bell (4) and Blandford and Ostriker (5) found the solution of the steady, one-dimensional problem (see section 5) and noted
its obvious applications to the cosmic ray acceleration problem, es-

pecially in terms of the efficiency of the process, the possibility of achieving power law spectra and the prevalence of shock waves in the interstellar medium.

A summary of the observed properties of galactic cosmic rays and the inferences which can be drawn from them is given in section 2. It should be remembered that some of these inferences are based on the most simple-minded view of cosmic ray propagation in the galaxy, ignoring the complications mentioned above, and hence may eventually have to be revised. Furthermore, the observations are not always entirely consistent with each other and it has been necessary to choose particular values for various parameters which may also require revision. These remarks also apply to section 3 where we review the basic properties of the interstellar medium with the aim of providing representative values of various important parameters as a basis for subsequent discussion. It is already emphasized at this stage that cosmic rays must play an important role in the dynamics of the interstellar medium since, if the medium is sufficiently strongly coupled to the cosmic rays to account for their acceleration, then the reverse must be true and the cosmic rays therefore affect the dynamics of the medium itself. This has important implications for the behaviour of supernova remnants, the nature of supersonic turbulence in the interstellar medium and for the formation of a galactic wind.

The basic elements of scatter-free acceleration of energetic particles by shock waves are outlined in section 4. We emphasize that although some interesting and significant phenomena can result from simple reflection and transmission of particles at shock waves without additional scattering, this does not provide a satisfactory basis for an explanation of galactic cosmic ray acceleration. The essential features of the more important mechanism in which scattering is included are outlined in section 5. It is shown that for a steady, plane shock with given scattering properties upstream and downstream (i.e. given cosmic ray diffusion coefficients) cosmic rays are efficiently and irreversibly accelerated in such a manner that power law energy/momentum spectra are a natural consequence. The important modifications associated with time dependence and losses due to interaction with the background medium are described in section 6 and the effects of non-planar geometry in section 7. These additional effects, which at some stage must be considered in any attempt to model the real situation, lead in general to spectra which are not power-law in form.

The possibility that galactic cosmic rays are accelerated by supernova shocks is discussed in section 8. It is evident that there are many attractive and plausible features of such a scheme. However, there are some conceptual difficulties notably with regard to the role played by cosmic rays in the dynamics of the supernova remnants and the means by which the cosmic rays and the remnants eventually merge with the interstellar medium and flow into the galactic halo from which they eventually escape. Nevertheless, with some reasonable but perhaps not totally justifiable assumptions, Blandford and Ostriker (60,61) have been able to construct a model which accounts for the observed spectra of cosmic ray primaries rather well. It should be noted however, that this is again based on the assumption that the cosmic rays do not affect the dynamics of the interstellar medium so that the model goes only one step beyond the familiar "leaky box" approach by including a consistent source mechanism.

One of the difficulties encountered by the shock acceleration theory is that it requires that there be sufficient scattering in the interstellar medium by magnetic field irregularities for the acceleration to be efficient up to energies exceeding certainly $10^3$ GeV/nuc and hopefully up to $10^5$-$10^6$ GeV/nuc where there is some evidence that changes in the spectrum occur (62-64) and the anisotropy begins to increase (65,66). This problem, which has been emphasized in particular by Ginzburg and Ptuskin (67) and Cesarsky and Lagage (68) and is discussed in section 9, leads one rather quickly to conclude that a completely non-linear treatment is necessary. This must take into account the generation, amplification and damping of hydromagnetic waves in a medium with strong supersonic turbulence and streaming cosmic ray fluxes (4,69-73). A brief outline of the present state of this aspect of the shock problem is given in section 10.

## 2. Galactic Cosmic Rays

The differential spectrum of primary galactic cosmic ray protons has the form $j(T) \propto T^{-\mu_1}$, $\mu_1 \sim 2.65$, in the kinetic energy range $10 \leq T \leq 10^6$ GeV (Figure 1). At lower energies ($1 \leq T \leq 10$ GeV) the spectrum is somewhat flatter and is noticeably affected by solar modulation. The spectrum becomes slightly steeper in the region above $10^6$ GeV and there are possibly further changes of slope at higher energies (62-64). The break in the spectrum at $\sim 10^6$ GeV, together with the observation that the anisotropy increases from very low values ($10^{-2}$-$10^{-1}$%) to high values (1-100%) in the range $10^6$-$10^{12}$ GeV suggests that extragalactic particles become progressively more prominent at higher energies (65,66). We know essentially nothing about the spectrum at energies $T \ll 1$ GeV due to the overwhelming effects of solar modulation, although it is often assumed for convenience (but without real justification) that it has the form of a power law in total energy, namely $(T + T_0)^{-\mu_1}$, where $T_0$ is the rest mass energy (see Figure 2).

The spectra of other primary species appear to be roughly the same as that of protons when expressed in terms of energy per nucleon (64,75, 76), but again nothing can be said about the forms of the unmodulated spectra at low energies. The relative abundances of the primaries correspond approximately to the solar and local galactic abundances deduced by Cameron and others as far as the heavier elements are concerned (77). However, the lighter elements appear to be less efficiently accelerated so that H, He and C, N, O, for example, are suppressed by factors of order 30, 20 and 2-5, respectively, with respect to the relative abundances of Si, Ca, Fe, etc. (78-81). There is also a tendency for the neutron-rich isotopes of Ne, Mg, etc. to be somewhat overabundant in the cosmic radiation in comparison with solar system abundances (80,82) which may well reflect a genuine difference between the compositions of the Sun (representing the interstellar medium $\sim 4 \times 10^9$ years ago) and the cosmic ray source material (possibly representing the interstellar medium during the past $10^{7-8}$ years) (83,84). It is interesting and possibly significant that the relative abundances of elements in the galactic cosmic ray source are rather similar to those of solar energetic particles (80,85).

If we relate the differential number density $U_i(T)$ of a species i

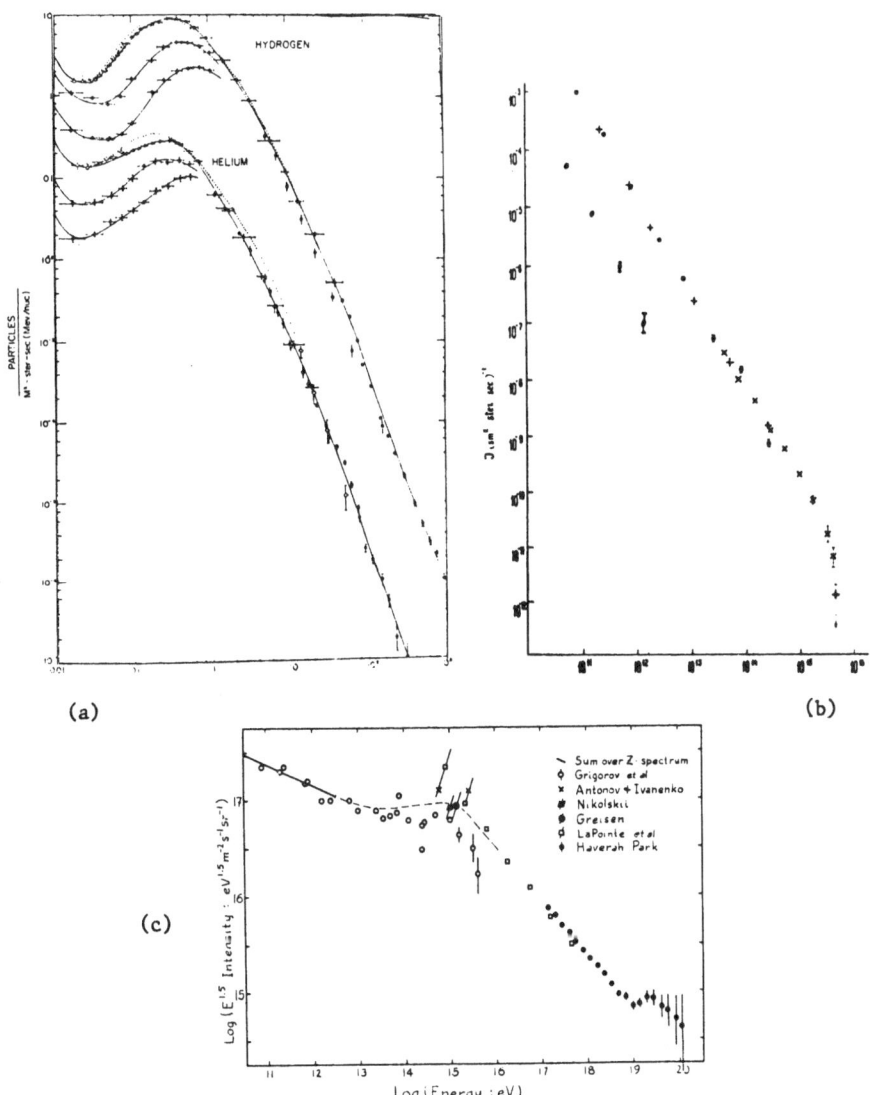

(a)

(b)

(c)

Fig. 1. (a) Differential energy/nucleon spectra of protons and alpha-particles for various phases of the solar cycle (compiled by Webber (215)). (b) Satellite measurements of the integral energy/nucleon spectra of protons and alpha-particles (Grigorov et al. (216, 217)). (c) Integral total energy spectra obtained by a variety of techniques in the range $10^{12}$–$10^{20}$ eV (62).

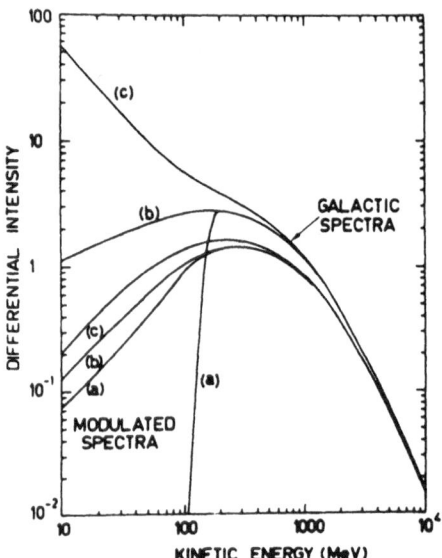

**Fig. 2.** Model calculations of the modulated spectra of galactic cosmic ray protons observed at the Earth, corresponding to three different assumed forms of the unmodulated spectrum (74). The modulation parameter is φ = 0.140 GV. Note the insensitivity of the unmodulated spectrum below ∿ 200 MeV to the form of the modulated spectrum, indicating the difficulty in interpreting observations in this range and of determining the unmodulated spectrum at low energies.

to its differential intensity by means of $j_i(T) = vU_i(T)/4\pi$ where v is the particle speed corresponding to a kinetic energy per nucleon of T, then the total (internal) energy density and pressure of the cosmic radiation, $W_c$ and $p_c$ respectively, can be expressed as

$$W_c = \sum_i \int_0^\infty A_i T\, U_i(T)dT \,, \qquad p_c = \sum_i \int_0^\infty \frac{1}{3}\, \alpha\, A_i T\, U_i(T)dT \qquad (2.1)$$

where $A_i$ is the atomic mass of the species i and $\alpha = (T + 2T_0)/(T + T_0)$. From observations made at solar minimum and with $\alpha \sim 1$, it is deduced that $W_c \sim 1.6 \times 10^{-12} erg/cm^3$ and $p_c \sim 0.6 \times 10^{-12} dyn/cm^2$ (86). These estimates may well be too low by a small factor as result of a neglect of residual modulation at solar minimum and the possible existence of substantial undetected fluxes in the range $T \lesssim 1$ GeV/nucleon (see Figure 2). The total energy density (enthalpy) of the observed cosmic rays is $H_c = W_c + p_c \sim 2.2 \times 10^{-12} erg/cm^3$.

The cosmic ray electrons are evidently primary particles but have a somewhat softer spectrum than that of protons, presumably as a result of inverse Compton and synchrotron losses. The bending of the measured electron spectrum at high energies is consistent with an age of the order of $2 \times 10^7$ years (87-90). Assuming diffusive propagation, it can be shown that the cosmic ray electrons we observe directly must originate within $\sim 1$ kpc of the Sun (89). At a given kinetic energy the electron flux is $\sim 100$ times smaller than that of protons (90), but the integral fluxes may not differ so much if it should be the case that there are many more electrons in the range below 1 GeV.

Secondary particles produced by nuclear interactions with the interstellar medium usually have a differential spectrum of the form $j(T) \propto T^{-\mu_2}$, $\mu_2 \sim 3$, in the range $3 \lesssim T \lesssim 10^2$ GeV/nucleon (76,91-93). Since the secondary production spectrum should be approximately the same as that of the primaries, the difference between the spectral indices $\mu_1$ and $\mu_2$ suggests that the residence time in the galaxy ($\tau$) depends on energy, with the more energetic particles escaping more easily:

$$\tau = \tau_0 (T/T_1)^{-0.35} \,, \qquad T > T_1 \sim 3 \text{ GeV/nuc} \qquad (2.2)$$

where $\tau_0$ is the residence time for $T = T_1$ (94,95). The fact that $\mu_2 > \mu_1$ is significant in that it implies that cosmic rays cannot be the result of a general and continuous acceleration process operating throughout the interstellar medium since this would produce a secondary spectrum flatter than that of the primaries, i.e. $\mu_2 < \mu_1$ (96-99). Furthermore, it suggests that the primary source spectrum has the form $j(T) \sim T^{-2.3}$ for $10 \le T \le 10^6$ GeV.

The observed fluxes of secondaries indicate that for primaries with $T < T_1$, $\bar{\rho} v \tau \sim 7$ v/c gm/cm$^2$, where $\bar{\rho}$ is the mean density of the interstellar medium in the volume occupied by cosmic rays while they are retained by the galaxy (94,95,100-102). Measurements of the flux of the radioactive isotope Be$^{10}$ suggest that $\bar{\rho}$ is equivalent to $\sim 0.22$ H atom/cm$^3$ and $\tau_0 \sim 2 \times 10^7$ years (103-105). Since the mean density of matter within $\pm$ 100 pc of the central plane of the galaxy is $\sim 0.4$ H atom/cm$^3$

(106), the cosmic ray storage volume in the above-mentioned sense must have a half thickness of $\sim$ 200 pc and a total volume $V_g \sim 8 \times 10^{66}$ cm$^3$ (assuming a disc of 15 kpc radius). The mean power of the cosmic ray source must therefore be $V_g(W_c+p_c)/\tau_0 \sim 4 \times 10^{40}$ erg/sec. It should be noted that these conclusions are based on the assumption that no significant post-acceleration of secondary particles occurs; this would have the effect of reducing the primary path length.

It is evident from the inhomogeneous distributions of the non-thermal radio emission associated with cosmic ray electrons (107) and the diffuse gamma-ray emission due to interactions of cosmic ray nuclei with the interstellar medium (108), that the distribution of cosmic rays in the galaxy is not uniform. This has implications for the cosmic ray sources since they should presumably be in some way associated with the regions in which the cosmic ray intensity is higher than average. Accordingly supernova remnants and the vicinities of OB associations are possible source candidates. There is evidence also that the cosmic ray intensity is noticeably higher towards the centre of the galaxy (109, 110). This may indicate that a peculiar source exists at the centre, or that there is a higher concentration of more normal sources, or simply that conditions are such that a higher cosmic ray pressure can be maintained in this region (109,111). It seems unlikely, however, that the bulk of the cosmic rays we observe can come from such a distant source since the electrons we observe directly probably originate from within $\sim$ 1 kpc of the Sun and there are no strong reasons for believing that the nuclear component behaves differently.

A consequence of the inhomogeneity of the distribution of cosmic rays is that as a result of the peculiar motions of the Sun and the sources and the time variations of the latter, the Sun must at times find itself in regions of higher than average cosmic ray intensity. The fact that the observed intensity does not seem to have fluctuated by more than $\pm$ 30% during the past $10^5$-$10^6$ years and has remained roughly constant (to within a factor $\sim$ 2) over the last $10^8$ years therefore places constraints on the sizes and number of the source regions (112, 113).

On the basis of the above considerations it appears that only supernovae and their remnants can provide enough power to account for the galactic cosmic rays (86,12). The overall efficiency of the acceleration process is of the order of 10% but since energy exchange is not a one-way process (i.e. cosmic rays can perform work on the medium in which they find themselves) the acceleration mechanism(s) must be capable of working at close to 100% efficiency at least temporarily. Moreover, we must rely on sources existing within $\sim$ 1 kpc of the Sun to provide the local cosmic rays and the main acceleration must occur within a relatively small fraction ($\ll$ 10%) of space in order to leave the spectra of secondaries largely unaffected and to avoid a high probability for the observation of large temporal variations at the Sun. The observed small anisotropy for $T \lesssim 10^6$ GeV suggests that many sources may be involved and/or that there is a considerable amount of scattering of cosmic rays in the interstellar medium.

In order to achieve a single power law in $10 \leq T \leq 10^6$ GeV, it seems that a dominant acceleration process must exist. Since the source composition is rather normal apart from the suppression of lighter elements such as H, He, CNO, etc., relative to Si, Fe, etc., it is neces-

sary that the particles accelerated should be extracted from a plasma
of normal composition, having in particular no significant dust component which would tend to reduce the supply of non-volatile elements.
Finally, the acceleration must be relatively prompt in the sense that it
takes particles from thermal (plasma) to cosmic ray energies in a time
which is short compared with $\tau$ and with the minimum time scale of any
loss mechanism that can occur at intermediate energies. Otherwise the
abundance distribution would be distorted in favour of H, He, etc. relative to Si, Fe, etc. in disagreement with the observations (114).

From many points of view it appears that shock acceleration in the
hot component of the interstellar medium should be considered a promising source for galactic cosmic rays because (a) the mechanism is
efficient, (b) supernova shocks can provide the necessary power, (c) the
hot interstellar medium combines normal composition with sufficiently
low densities to minimize any effects of energy losses, (d) power law
energy spectra can be achieved under suitable circumstances, and (e) the
shock waves which determine the spectrum occupy only a small fraction
(<< 10%) of interstellar space.

## 3. The Interstellar Medium

It is sufficient for our purposes here to assume a rather simple
model for the interstellar medium and its extension into the galactic
halo. We assume that the interstellar medium can be divided into three
phases: hot, intermediate and cold, of which the hot component occupies
most of space and the cold component contains most of the mass (6,115).

The hot interstellar medium (HISM) is the result of supernova explosions and stellar winds and may be regarded as being in a state of
supersonic turbulence (116). It should contain little dust, so that it
is a plasma of essentially "normal" composition which is suitable as a
source region for galactic cosmic rays. For the purposes of further discussion we assume representative values of the number density (n), tem-
perature ($T_g$) and magnetic field (B) to be n = $3\times10^{-3}$ cm$^3$, $T_g$ = $7\times10^5$ K
and B $\leq$ 1-3 $\mu$Gauss. The HISM occupies most of the cosmic ray containment volume $V_g$ and hence involves a mass of the order of $2\times10^7$ $M_\bullet$.

The cold component has densities greater than 1/cm$^3$, temperatures
less than $10^4$ K and magnetic field strengths of the order of 3 $\mu$Gauss.
The gas contains much dust and hence does not have normal composition,
being deficient in non-volatile elements in particular. The volume of
space occupied by the cold component does not exceed 0.2 $V_g$ and is probably substantially smaller. The total mass of the cold component is of
the order of $2\times10^9$ $M_\bullet$.

It is difficult to estimate the amount of material involved in
the intermediate phase which comprises the denser HII regions as well
as the transition zones between the hot and cold components where energy
transfer by heat conduction and radiation plays a role. The importance
of this component in controlling the dynamics of the HISM depends somewhat critically on whether heat conduction is impeded by magnetic fields
and microscopic plasma instabilities (117,118).

The supernova rate in the outer parts of the galaxy near the Sun
has been estimated to be of the order of 1/10-1/30 per year (119). Assuming that each explosion involves the release of $10^{51}$ ergs, the
average supernova power is $\sim$ $10^{42}$-$3\times10^{42}$ erg/sec which is sufficient to

account for the cosmic ray source power requirement. Other possibly significant sources, which like supernovae tend to be associated with young
hot stars, are stellar winds ($\lesssim 10^{41}$ erg/sec) and HII regions ($\lesssim 10^{40}$
erg/sec) (12,120). The mass injection rate from supernovae and stellar
winds is $\sim 2M_\odot$/year, most of which is likely to go directly into the
HISM. In any case, the turn-over time for the interstellar medium appears to be at most of the order of $10^9$ years, which is less than the
age of the solar system and therefore allows some change in average composition to occur. It is conceivable that the turnover time for the HISM
could be much smaller ($\sim 10^7$ years) if the gas is injected directly from
supernovae and stellar winds and leaves the galaxy in the form of a
galactic wind.

The pressures and internal energies associated with the plasma,
cosmic rays and magnetic fields in the HISM are as follows:

$$P_g \sim 1.1\times10^{-12}, \quad P_c \sim 0.6\times10^{-12}, \quad P_M \sim 0.04\text{--}0.4\times10^{-12} \text{ dyn/cm}^2 ;$$

$$W_g = 1.6\times10^{-12}, \quad W_c = 1.8\times10^{-12}, \quad W_M = 0.04.\text{--}0.4\times10^{-12} \text{ ergs/cm}^3.$$

The speed of sound ($c_s$) in the HISM, which is the minimum shock speed,
is estimated as follows:

$$c_s^2 = \gamma p_g/\rho + \gamma_c p_c/\rho + B^2/4\pi\rho = c_g^2 + c_c^2 + c_A^2 , \tag{3.1}$$

$$c_g \sim 165 , \quad c_c \sim 110 , \quad c_A \sim 35\text{--}110 , \quad c_s \sim 200\text{--}230 \text{ km/sec}.$$

Here $\gamma = 5/3$ and $\gamma_c = 1 + \overline{\alpha}/3$ are the specific heat ratios for the
plasma and cosmic ray gas, respectively.

Note that the ß of the plasma is large, especially if the contri-
bution of cosmic rays is included: $\beta_g = p_g/p_M \sim 3\text{--}30$; $\beta = (p_c + p_g)/p_M \sim$
4–40. The total energy density in the HISM is $H = \Sigma p + \Sigma W \sim 6\times10^{-12} \text{erg/cm}^3$.
Hence the total energy per unit mass is $10^{15}$ergs/gm, corresponding to a
speed of $\sim 450$ km/sec, which exceeds the speed of galactic rotation ($v_r$
$\sim 250$ km/sec) and the escape speed ($v_{esc} \sim 340$ km/sec) at the Sun's location ($\sim 8$ kpc from the centre of the galaxy). It is clear then that
the HISM does not have to partake in Keplerian motion around the galaxy
as do the stars and cold component of the interstellar medium. Furthermore, unless held back in some manner, it will flow away from the disc
of the galaxy forming a galactic wind (7,8).

Assuming that the lower energy cosmic rays and plasma move together,
an upper limit to the mass flux involved in the galactic wind can be
estimated from the total mass of the HISM and the residence time for
cosmic rays, namely $2M_\odot$/year. This is an acceptable level of mass loss
for the galaxy and implies an outflow speed $V_z \sim 15$ km/sec at the outer
edge of the cosmic ray containment volume. An indication of the expected
nature of the galactic wind is given in Figure 3. Note that the halo is
likely to be roughly spherical since the scale height of the medium, including the cosmic rays, is of the order of 10 kpc. The halo, as well as
the whole cosmic ray containment volume, should be maintained in a state

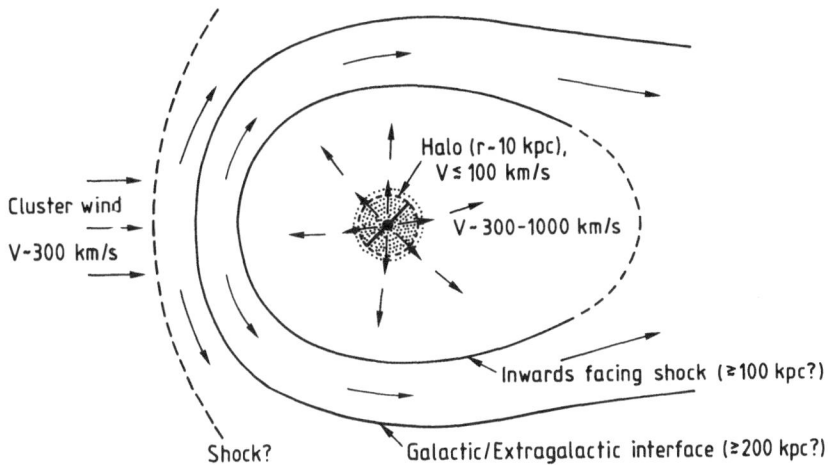

**Fig. 3.** Possible configuration of the galactic wind and its interaction with the local cluster medium. The numbers shown are mere guesses intended to give some idea of what might be expected. Since the plasma density decreases rapidly away from the galaxy (first exponentially and further out as the inverse square of the distance), supernova shock waves should become stronger as they propagate through the halo and out into the galactic wind region. Particle acceleration can occur at such shock waves and also at the terminal shock. Extragalactic cosmic rays should be modulated at low energies in a manner similar to the solar modulation of galactic cosmic rays.

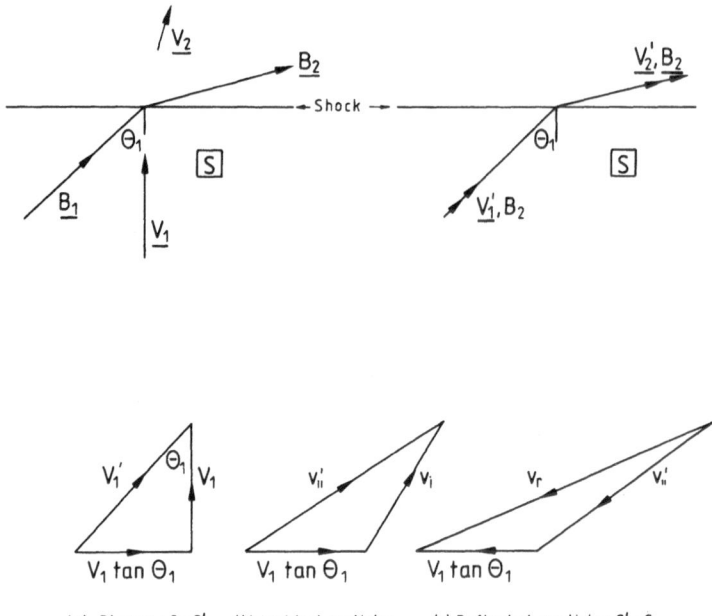

(a) Plasma : S→S'   (b) Incident particle:   (c) Reflected particle: S'→S
                        S→S'

**Fig. 4.** Above: The flow configuration in the shock frame S (the upstream plasma flow is normal to the shock plane) and in the frame S' in which the electric field vanishes (the plasma flow is parallel to the magnetic field everywhere).
Below: (a) the transformation of the plasma flow S → S';
(b) the transformation of the velocity of an incident particle with velocity $v_i$ in S, to $v_{\shortparallel}'$ in S';
(c) the transformation of the velocity of a reflected particle, $v_{\shortparallel}'$ in S' to $v_r$ in S. Note that $v_r \gg v_i$.

of strong supersonic turbulence by interacting supernova shock waves. In
particular, shock waves which run out into the halo should tend to speed
up and strengthen so that the gas must be continually heated in a manner
similar to that envisaged in early theories for the heating of the solar
corona (121).

If it is assumed that the escape of cosmic rays from the containment volume is diffusive it is possible to estimate a value for the diffusion coefficient normal to the plane of the galaxy as:

$$\kappa_z = L^2/3\tau \sim 2 \times 10^{26} (T/T_1)^{0.35} cm^2/sec , \qquad (3.2)$$

where $L \sim 200$ parsecs is the distance to the escape boundary. The corresponding mean free path is $\lambda_z \sim 0.06 (T/T_1)^{0.35} pc$, which is larger than
the gyroradius $r_g \sim 10^{-6} (T/T_0) pc$, provided $T \lesssim 10^6$ GeV. Large-scale
convection by the galactic wind in the containment volume could have im-
portant effects at low energies since $V_z L/\kappa_z \sim 3$ for $T = T_1$, if $V_z = 10$ km/sec.

## 4. Scatter-free Shock Acceleration

In a perpendicular shock (i.e. propagating perpendicular to the
magnetic field) it has been shown that, if scattering can be neglected,
particles interacting with the shock conserve their magnetic moments
(122-126,11). Thus,

$$p_{\perp 1}^2/B_1 = p_{\perp 2}^2/B_2 \qquad (4.1)$$

where subscripts 1/2 refer to upstream/downstream conditions respectively and p is the particle momentum. Subsequent expansion of the
medium back to the original magnetic field strength will return the particle energy to its original value unless pitch angle scattering occurs
during the intermediate phase. We cannot rely only on this mechanism to
account for the acceleration of galactic cosmic rays, however it demonstrates that shocks have an important effect on the existing cosmic
rays which cannot avoid having their energy density enhanced by $\sim 4$ and
(for electrons) their synchrotron volume emissivity enhanced by a factor
$\sim 100$ for strong shocks.

Oblique shocks are more interesting in that incident particles can
be reflected back upstream with a considerable increase of energy in
favourable circumstances (43-46,11,14,123,127-132). In order to understand this mechanism it is advisable to make a transformation from the
"shock frame" (S) in which the shock is stationary and the incident flow
normal to it, to the frame (S') in which the ambient electric field
vanishes (Figure 4). The transformation involves a velocity change
$(- V_1 \tan\theta_1,0,0)$ which makes the bulk flow velocity parallel to the magnetic field on both sides of the shock. In the simplest situation where
magnetic reflection occurs, the particle energy is conserved in S' and
it can be shown that for $\theta_1$ sufficiently large, the particle magnetic
moment is also approximately conserved (12,14,130). Using primes to de-

note quantities in S', the condition for reflection is $T_\perp' > (B_1/B_2)T'$ (i.e. the particle must have a sufficiently large pitch angle in S'). On transforming back to S, it is found that for reflected particles:

$$\Delta T_{ref} = 2mv_{\shortparallel}' \, V_1 \, \sin\theta_1 \, \tan\theta_1 \qquad\qquad (4.2)$$

and $T_{\perp 2} = (B_2/B_1) \, T_{\perp 1}$ for non-relativistic transmitted particles. If the incident particle is relativistic, $\Delta T/T \sim 2(V_1/c)\sin\theta_1\tan\theta_1$, which can be large if $\theta_1 \to \pi/2$. For low energy particles $\Delta T/T$ can exceed unity but the effect is not interesting unless the initial particle energy is already of order $(mV_1^2\tan^2\theta_1/2)$. Thus, for example, an incident particle having a guiding centre moving with the upstream medium cannot be reflected unles its initial energy exceeds $1/2 \, mV_1^2(B_2+B_1\tan^2\theta_1)/(B_2-B_1)$, which is already large if $\tan^2\theta_1$ is large enough for the energy change on reflection to be substantial ($\Delta T = 2mV_1^2\tan^2\theta_1$).

This mechanism cannot be a basis for the acceleration of galactic cosmic rays (133) since it depends on the prior existence of a "seed" population which is more energetic than the background plasma and therefore subject to severe energy losses between shock encounters which would suppress the heavier particles, contrary to observation (114). Nevertheless, the mechanism exists and must be taken into account, for example in consideration of the shock transition conditions in analyses based on cosmic ray transport equations in scattering media (see section 5). Furthermore, in interplanetary space where suprathermal "seed" particles are quite prevalent, the mechanism can produce the "shock spike" phenomenon as a result of slow variations of the magnetic field direction which temporarily make $\theta_1 \to \pi/2$.

A variant on the mechanism has been proposed by Sonnerup (134) who suggests that reflection processes other than magnetostatic may be effective and that as a consequence a small fraction of the incident thermal plasma might be reflected and accelerated. Without specifying the reflection mechanism, it is easily shown that the energy of reflected ions would be $mV_1^2(1+4\tan^2\theta_1)/2$ if the incident Mach number is large and magnetic moments are conserved. Observations indicate that this process may be operating on solar wind particles at the Earth's bow shock (135) which suggests that it might be able to provide an injection mechanism for galactic cosmic rays in supernova shocks. It is of course necessary to be sure that the effect is different from magnetostatic reflection of suprathermal particles and/or leakage of shock-heated ions upstream, both of which would produce beams with energies $\propto \tan^2\theta_1$ as a result of the velocity "filter" arising from the interaction of the solar wind with a finite shock.

It is interesting to speculate on possible reflection mechanisms other than the simple magnetotstatic one usually assumed. One possibility is an electric field normal to the shock surface which tends to reflect ions and holds electrons behind the shock. A simple analysis shows that for a cold (high Mach number) incident plasma, such an electric field would have to have a potential drop of at least $mV_1^2/2e$ ($\sim 1$ kV in the solar wind) to have a significant effect, whereas one would expect potential drops of about $kT_e/e$ ($\sim 30$–$50$ volts) if the shock is not perpendicular, so that the effect would be of secondary importance at best.

A second possibility is reflection from the magnetic field "overshoot" that occurs at the front of quasi-perpendicular shocks, where $B_2$ may be locally a factor $\sim$ 2 greater than predicted by the Rankine-Hugoniot conditions (136); however, for a high Mach number incident plasma the magnetostatic reflection requirement is difficult to satisfy so that even with the assistance of an electric potential drop it seems unlikely that this could explain the observations. A third possibility is reflection by strong turbulence which could scatter the particles back upstream. This cannot be Alfvén wave turbulence, however, since in the frame S', Alfvén waves are convected away from the shock at a speed exceeding $V_1\tan\theta_1$ and for $\tan\theta_1 \gg 1$ the reflection would involve an energy loss which is barely compensated in transforming back to the shock frame. To be effective, the waves must be stationary or even move towards the shock in S'; large amplitude whistlers, which are prevalent in oblique shocks, could satisfy this requirement although it is not immediately clear whether they could efficiently scatter ions.

5. Acceleration by Shocks in Scattering Media

In contrast to the scatter-free mechanism described in section 4, in which particles make a single interaction with the shock before being reflected or transmitted downstream, the presence of scattering permits multiple interactions to occur and hence leads to much more efficient acceleration. It is possible (4,137-140) to deduce the simplest result for acceleration by a plane steady shock by random walk arguments as discussed by Chandrasekhar (141), but in general it is best to use transport equations which describe the behaviour of the particle distribution in terms of convection and diffusion in the scattering medium, including energy changes due to various causes. In terms of the differential current $S(x,T)dT$ and number density $U(x,T)dT$ in $(T, T+dT)$, the one-dimensional forms of these equations are (56):

$$\frac{\partial U}{\partial t} + \frac{\partial S}{\partial x} = -\frac{1}{3} V \frac{\partial^2}{\partial x \partial T} (\alpha TU) + Q , \qquad (5.1)$$

$$S = CVU - \kappa \frac{\partial U}{\partial x} = V(U - \frac{1}{3} \frac{\partial}{\partial T} (\alpha TU)) - \kappa \frac{\partial U}{\partial x} , \qquad (5.2)$$

which can be combined to yield a Fokker-Planck equation for U (52-56):

$$\frac{\partial U}{\partial t} + \frac{\partial}{\partial x} (VU) = \frac{\partial}{\partial x} (\kappa \frac{\partial U}{\partial x}) + \frac{1}{3} \frac{\partial V}{\partial x} \frac{\partial}{\partial T} (\alpha TU) + Q . \qquad (5.3)$$

Here $V(x)$ is the speed of the scattering medium, Q represents particle sources and sinks and non-adiabatic energy changes, $\kappa(x,T)$ is a diffusion coefficient related to the spectrum of magnetic irregularities in the medium (142) and C is the Compton-Getting coefficient (143,144). In deriving these equations, which are correct to a factor $(1 + O(V/v)^2)$, it is assumed that the distribution function is nearly isotropic (S $\ll$

vU/3), the Compton–Getting transformation is linear and particle inertia
effects are negligible ($\partial S/\partial t \ll v^2 S/3\kappa$). In terms of U, S the differential intensity is $j = vU/4\pi$ and the anisotropy is $\xi = 3S/vU$.

Consider the situation in which a shock wave is situated at $x = 0$
facing in the negative x-direction so that $V = V_1$ in $x < 0$, $V = V_2$ in
$x > 0$ with $V_1 > V_2$. We assume that $U \to U_1(T)$ as $x \to -\infty$, U remains finite
as $x \to \infty$ ($U \to U_2(T)$) and there is a source $Q = Q_0(T)\delta(x)$ corresponding
to particle injection at the shock. It is easily shown that if the above
transport equations remain valid throughout, the transition conditions
appropriate to a sudden change in $V(x)$, $\kappa(x,T)$ etc., are:

$$U_1 = U_2 \; , \quad S_1 + Q_0 = S_2 \; , \tag{5.4}$$

where the subscripts refer to conditions upstream and downstream of the
transition respectively (56,11). [If the diffusion coefficient is anisotropic and the shock oblique, the condition on the current is $S_{n1} + Q_0 =
S_{n2}$.] In fact, the transport equations are not valid on such a small
scale (transition thickness $\ll$ mean free path) but a more detailed treatment shows that the conditions (5.4) are correct provided the distribution function does not become very anisotropic at the shock (11,145).
For low energy particles, shock reflection as described in section 4 may
introduce a correction $(1 + O(V/v))$ to these conditions which would not
invalidate the use of the transport equations elsewhere (145).

Provided $\int_0^x dx'/\kappa(x',T) \to 0$ as $x \to -\infty$, the solution of equations
(5.1), (5.2) in this situation is

$$U(x,T) = U_1(T) + \{U_2(T) - U_1(T)\} \exp \int_0^x \{V_1/\kappa(x',T)\}dx' \; , \quad x < 0 \; ,$$

$$= U_2(T) \; , \quad x > 0 \; , \tag{5.5}$$

where $U_2(T)$ is determined from the condition on $S(T)$ at $x = 0$:

$$\frac{1}{3}(V_1 - V_2)\frac{\partial}{\partial T}(\alpha T U_2) + V_2 U_2 = V_1 U_1 + Q_0 \; . \tag{5.6}$$

This equation can be integrated to yield

$$\alpha T U(T)\{T(T + 2T_0)\}^{\lambda_0/2} =$$

$$(V_1 \lambda_0/V_2)\int_0^T \{T'(T'+2T_0)\}^{\lambda_0/2}\{U_1(T')+Q_0(T')/V_1\}dT' \; , \tag{5.7}$$

where $\lambda_0 = 3V_2/(V_1 - V_2)$ (2–5). In particular if either $U_1(T)$ or $Q_0(T)/V_1$
have the form $\alpha_1 T_1 U_1 \delta(T - T_1)$, then

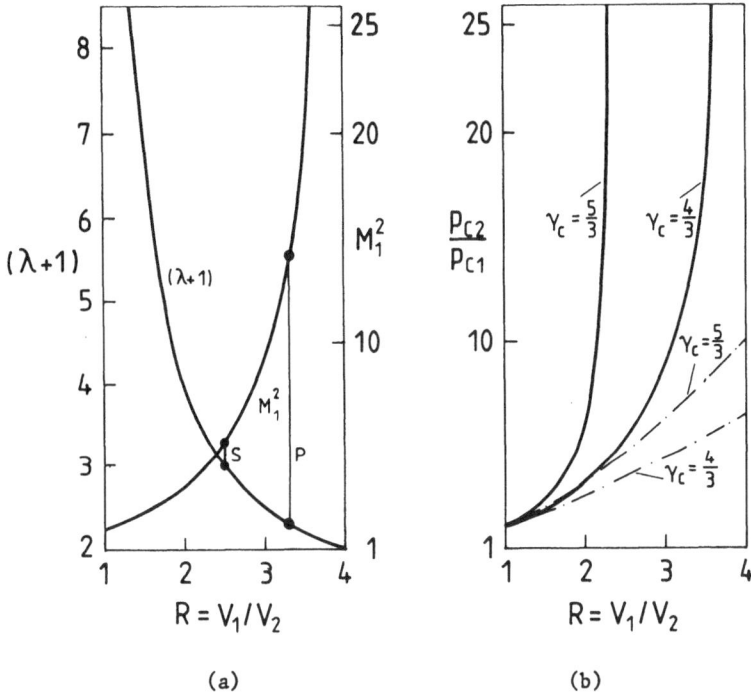

**Fig. 5.** (a) Spectral index ($\lambda$+1) as a function of compression ratio ($R = V_1/V_2$) and upstream Mach number ($M_1$) for a plane steady shock with monoenergetic injection (equation 5.9). The shocks responsible for the primary production spectrum correspond to the points labelled p and those which first affect the secondary spectrum to the points labelled s.
(b) Cosmic ray pressure ratio across a plane, steady shock ($p_{c2}/p_{c1}$) as a function of compression ratio for $\gamma_c = 5/3$, 4/3 from equation (5.11) (solid lines). The pressure ratio obtained from adiabatic compression with the same compression ratio is substantially less (dashed lines) except for weak shocks.

$$\alpha TU_2(T) = (V_1\lambda_0/V_2)\alpha_1 T_1 U_1\{T_1(T_1+2T_0)/T(T+2T_0\}^{\lambda_0/2} \quad , \quad T \geq T_1 \quad .(5.8)$$

Alternatively, if we treat $\alpha = \bar{\alpha}$ as constant, the equivalent solution is

$$U_2(T) = (V_1\lambda/V_2) \, U_1(T/T_1)^{-(\lambda+1)} \quad , \quad T \geq T_1 \quad , \tag{5.9}$$

where $\lambda = \lambda_0/\bar{\alpha}$. Thus the shock converts a monoenergetic distribution into a power law distribution above the initial energy, with a spectral index depending on the compression ratio $R = V_1/V_2 = 4M_1^2/(3+M_1^2)$ where $M_1$ is the shock Mach number ($\gamma = 5/3$) (see Figure 5a). For strong shocks, $M_1^2 \to \infty$, $R \to 4$ and hence $\lambda_0 \to 1$ and $(\lambda+1) \to 3/2$, for non-relativistic and $\to 2$ for relativistic particles, respectively. The pre-shock density increase is exponential (if $\kappa$ remains finite as $x \to -\infty$) and the spectrum in this region depends on the form of the diffusion coefficient (Figure 6a). Note that the power law spectrum is achieved only under the conditions we have assumed; if a 'free-escape' boundary ($U = 0$) is assumed at $x = -L$ for example, exponential spectra are obtained (146).

Equations (5.7) and (5.8) have a neater form if expressed in terms of the particle momentum (p) and the omni-directional component of the distribution function $F_0(p) = \alpha TU(T)/4\pi p^3$:

$$F_{02}(p) = (3/(V_1-V_2))Q_0(p/p_1)^{-\lambda_1} \quad , \tag{5.10}$$

where $\lambda_1 = (\lambda_0+3) = 3R/(R-1) \to 4$ for strong shocks.

It is instructive to consider the behaviour of the cosmic ray pressure according to these results. From (5.9) we obtain

$$P_{c2}/P_{c1} = V_1\lambda/V_2(\lambda-1) = R/\{1 - \bar{\alpha}(R-1)/3\} \tag{5.11}$$

and find that $P_{c2}/P_{c1} \to \infty$ for $R = 1+3/\bar{\alpha}$ (i.e. $R = 2.5$ for non-relativistic and $R = 4$ for relativistic particles). The divergence in the downstream pressure is a result of the flatness of the spectrum and indicates that shocks tend to bring particles to high energies. Furthermore, on comparing the cosmic ray pressure jump to that which would result from an equivalent adiabatic compression, one sees that shocks give more energy to the cosmic rays than an adiabatic change and the process is therefore irreversible, as would be expected from (5.9), (5.10) (see Figure 5b).

The effect of shock acceleration in the case in which the upstream cosmic ray spectrum is a power law is of particular interest in considering the post-acceleration of secondary particles. Using (5.7) with $Q_0 = 0$, $U_1(T) = U_1(T/T_1)^{-\mu}$ in $T \geq T_1$ and assuming $T \gg T_0$, one finds that if $(\lambda+1) \neq \mu$

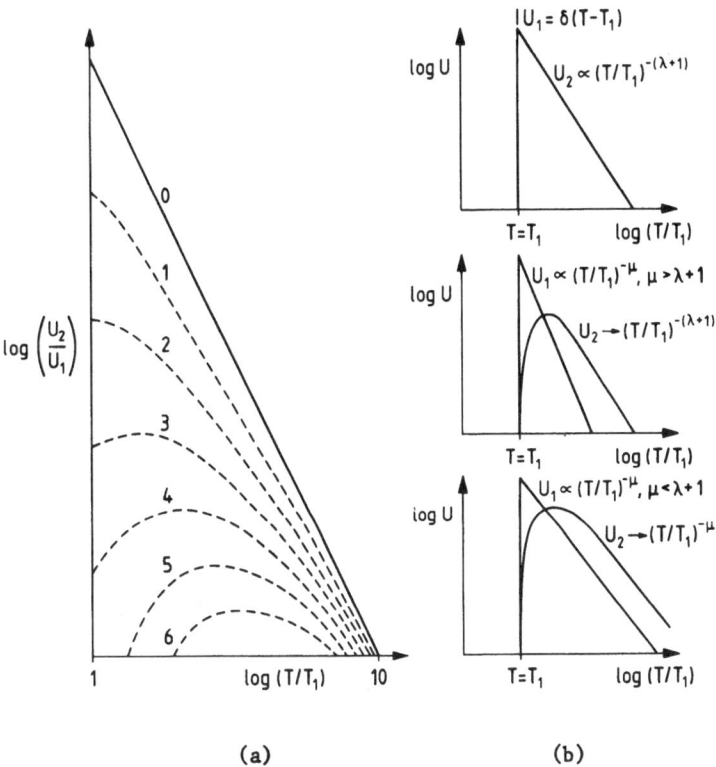

(a)  (b)

**Fig. 6.** (a) Spectra of particles accelerated by a strong $(V_1/V_2 = 4)$
plane, steady shock with monoenergetic injection at $T = T_1$ (equations
(5.5), (5.9)), at various distances $(V_1 x/\kappa_1)$ upstream and with $\kappa = \kappa_1(T/T_1)$. Note that the spectrum is a power law only at the shock and
downstream; ahead of the shock it tends to be peaked because high energy
particles can diffuse further upstream than low energy particles.
(b) The effect of a shock wave on an incident or injected spectrum which
has a power law form above a certain energy. If the initial spectrum is
softer than the shock would produce for monoenergetic injection (5.9,
5.10) the downstream spectrum is altered to the "shock" spectrum at high
energies. On the other hand, if the initial spectrum is flatter than the
"shock" spectrum the initial power law is preserved at high energies but
the spectrum is shifted upwards.

$$U_2(T) = \beta U_1 \{(T/T_1)^{-\mu} - (T/T_1)^{-(\lambda+1)}\} , \quad T > T_1 , \tag{5.12}$$

where $\beta = 1/(1-C_1(V_1-V_2)/V_1) = \lambda V_1/V_2(\lambda+1-\mu)$ and $C_1 = 1+\bar{\alpha}(\mu-1)/3$ is the upstream Compton-Getting coefficient. For $T \gg T_1$ we see that $U_2(T) \propto (T/T_1)^{-\mu}$ if $\mu < (\lambda+1)$ and $U_2(T) \propto (T/T_1)^{-(\lambda+1)}$ if $\mu > (\lambda+1)$ as shown in Figure 6b. Thus shock waves which are capable of generating the observed cosmic ray spectrum at high energies (i.e. $\lambda+1 = 2.3$) will change the spectrum of secondaries, but shocks with $(\lambda+1) \geq \mu_2 \sim 3$ will not; the corresponding compression ratios and shock Mach numbers are $R = 3.3$, $M_1^2 = 14.3$ and $R \leq 2.5$, $M_1^2 \leq 5$, respectively.

The solution given here for a plane, steady shock can be generalized to the case in which the diffusion coefficient is anisotropic. If the y-axis is taken to lie in the plane of the shock and the magnetic field to lie in the (x,z) plane, and if $\underline{V} = (V_1,0,0)$ in $x < 0$, then we must put $\kappa_1 = \kappa_{\shortparallel}\cos^2\theta_1 + \kappa_\perp\sin^2\theta_1$. The components of the current vector parallel to the shock surface are:

$$S_y = -\kappa_T\sin\theta_1 \frac{\partial U}{\partial x} , \quad S_z = -(\kappa_\perp - \kappa_{\shortparallel})\sin\theta_1\cos\theta_1 \frac{\partial U}{\partial x} , \tag{5.13}$$

where $\kappa_T = pv/3eB$ and $\kappa_{\shortparallel}$, $\kappa_\perp$ are the components of the diffusion coefficient parallel and perpendicular to the magnetic field direction, respectively. In the limit $\kappa_\perp \to 0$,

$$\frac{S_x - CV_1U}{S_z} = \frac{\kappa_{\shortparallel}\sin^2\theta_1 + \kappa_\perp\cos^2\theta_1}{(\kappa_{\shortparallel}-\kappa_\perp)\sin\theta_1\cos\theta_1} \to \tan\theta_1 , \tag{5.14}$$

thus for $\theta_1 \to \pi/2$ the particle distribution upstream of the shock is highly anisotropic and field-aligned much as in the case of scatter-free acceleration.

## 6. Plane Shocks With Energy Losses and Time Dependence

Solutions of the plane, one-dimensional problem discussed in section 5 allowing for time dependence with various initial conditions have been given by Fisk (147), Vasil'yev et al. (148), Toptygin (11) and Forman and Morfill (149). The effects of energy losses due to collisions, synchrotron emission and Compton scattering and adiabatic expansion, all expressed in terms of a loss time $\tau(T)$, have been considered by Bulanov and Dogiel (150,151) and Völk et al. (152-154).

If particles are injected at the shock, beginning at $t = 0$, with a monoenergetic spectrum (i.e. $U_1(T) = 0$, $Q_0(T,t) = Q_0\delta(T-T_1)H(t)$) and there are distributed losses of the form $Q(T) = -U/\tau$, equations (5.1) and (5.2) can be written

$$\frac{\partial \bar{S}}{\partial x} = -\frac{1}{3} V \frac{\partial^2}{\partial x \partial T} (\alpha T \bar{U}) - \bar{U}(s + \frac{1}{\tau}) , \tag{6.1}$$

$$\overline{S} = V(\overline{U} - \frac{1}{3} \frac{\partial}{\partial T} (\alpha T \overline{U})) - \kappa \frac{\partial \overline{U}}{\partial x} , \qquad (6.2)$$

where bars represent a Laplace transformation with respect to time $(e^{-st})$. Assuming $\kappa$, $V$ to be independent of $x$ both upstream and down-stream of the shock, solutions of these equations exist such that $\overline{U}$, $\overline{S}$ $\propto \exp \beta(s)x$, where

$$\beta = \beta_1(s) = \left[ 1 + \sqrt{\{1 + 4\gamma_1\kappa_1/V_1^2\}} \right] (V_1/2\kappa_1) \quad \text{in } x < 0 , \qquad (6.3a)$$

$$\beta = \beta_2(s) = \left[ 1 - \sqrt{\{1 + 4\gamma_2\kappa_2/V_2^2\}} \right] (V_2/2\kappa_2) \quad \text{in } x > 0 , \qquad (6.3b)$$

and $\gamma_i = s + 1/\tau_i$. Applying the continuity conditions (5.4) in the form

$$\overline{U}_1 = \overline{U}_2 = \overline{U}_o(T) , \quad \overline{S}_1 + Q_o\delta(T-T_1)/s = \overline{S}_2 , \quad \text{at } x = 0 , \qquad (6.4)$$

we eventually obtain for $\overline{U}_o(T)$ (the density of particles at the shock):

$$\overline{U}_o(s,T)/U_2(T) = \frac{1}{s} \exp \int_{T_1}^T - \frac{3}{2} \frac{(V_1A_1+V_2A_2)}{(V_1-V_2)} \frac{dT'}{\alpha T'} , \qquad (6.5)$$

where $U_2(T)$ is the steady state solution given by (5.8) with $Q_o = \alpha_1T_1U_1$ and $A_i = \sqrt{\{1+4\gamma_i\kappa_i/V_i^2\}}-1$. The effects of time dependence and energy losses are contained entirely in the exponential term on the right of (6.5) and they are not important if $A_1$, $A_2 \to 0$ (i.e. $4\gamma_i\kappa_i/V_i^2 \ll 1$).

Consider first the case of time dependence without energy losses ($\tau_i \to \infty$). Although the inversion of (6.5) is difficult in general, simple approximations can be found for large and small values of the time and an exact solution is available in a special case in which $A_1 = A_2$ and $\kappa_1$, $\kappa_2$ do not depend on energy. For small $s$ we can take $V_iA_i \sim 2s\kappa_i/V_i$ and hence obtain a rough representation of the solution for large time:

$$U_o(T,t) \sim U_2(T) H(t-\tau_a) , \qquad (6.6)$$

where

$$\tau_a = \frac{3}{(V_1-V_2)} \int_{T_1}^T \left[ \frac{\kappa_1}{V_1} + \frac{\kappa_2}{V_2} \right] \frac{dT'}{\alpha T'} . \qquad (6.7)$$

That is, at a given time $t$, no particles are seen at energies greater than a value $T$ such that $\tau_a(T) = t$, while at lower energies the spectrum approximates that of the steady state solution (5.8).

For the case in which $4\kappa_1/V_1^2 = 4\kappa_2/V_2^2 = t_o$ (independent of $T$) equation (6.5) can be inverted to yield

$$U_o(T,t)/U_2(T) = \{(T+2T_o)T/(T_1+2T_o)T_1\}^\nu \log\{(T_o+2T_o)T/(T_1+2T_o)T_1\}^\nu$$

$$\times \int_0^{t/t_o} \exp\left[-\frac{\nu^2}{4t'} (\log\{(T+2T_o)T/(T_1+2T_o)T_1\})^2 - t'\right] \frac{dt'}{\sqrt{(4\pi t'^3)}} , \quad (6.8)$$

where $\nu = 3(V_1+V_2)/4(V_1-V_2)$. Alternatively, in terms of particle momentum:

$$F_o(p,t,x=0)/F_{o2}(p) =$$

$$2\nu(p/p_1)^{2\nu}\log(p/p_1)\int_0^{t/t_o} \exp\left[-\frac{\nu^2}{t'} (\log p/p_1)^2 - t'\right]\frac{dt'}{\sqrt{(4\pi t'^3)}} , \quad (6.9)$$

where $F_{o2}(p)$ is the steady state solution given by (5.10). Using Laplace's method to estimate the integrals in these expressions for sufficiently large $t/t_o$ we find that they behave like (somewhat spread out) step functions with the "step" occurring at $t = \nu t_o \log(p/p_1) = \tau_a$ as defined in (6.7) (see Figure 7). For $t \gg \tau_a$ one finds that the right hand sides of (6.8) and (6.9) tend to unity as expected. [Note that these results are slightly different from those given by Toptygin (11) who effectively takes $Q_o(T,t) = Q_o\delta(T-T_1)\delta(t)$.]

The effects of energy losses on the steady state spectrum can be found by taking the limit $s \to 0$ in (6.5) and inverting:

$$U_o(T)/U_2(T) = \exp\int_{T_1}^T -\frac{3}{2} \frac{(V_1A_1+V_2A_2)}{(V_1-V_2)} \frac{dT'}{\alpha T'} , \quad (6.10)$$

where $A_i = \sqrt{\{1+4\kappa_i/V_i^2\tau_i\}}-1$. In the case in which $A_1$ and $A_2$ are independent of energy this integral is easily evaluated:

$$U_o(T)/U_2(T) = \left[(T+2T_o)T/(T_1+2T_o)T_1\right]^{-\eta} , \quad (6.11a)$$

or alternatively

$$F_o(p,x=0)/F_{o2}(p) = (p/p_1)^{-2\dot\eta} , \quad (6.11b)$$

where $\eta = 3(V_1A_1+V_2A_2)/4(V_1-V_2)$. The spectrum with losses is considerably steeper than without if $\eta = O(1)$. In general, we require that $\tau_1,\tau_2 \gg \tau_a$ if losses are not to effectively quench the shock acceleration mechanism. An example showing the spatial variations of various values of $A_1 = A_2 = \sqrt{(1+X)}-1$ is shown in Figure 8.

It should be noted that the "loss time" approximation to the energy loss term in (6.1) is not strictly valid except in the case of catastrophic losses due to nuclear interactions, where $1/\tau = vn\sigma$ and $\sigma \sim \pi(1.26x 10^{-13} A^{1/3})^2$ cm$^2$, with $n$ being the number density in the background

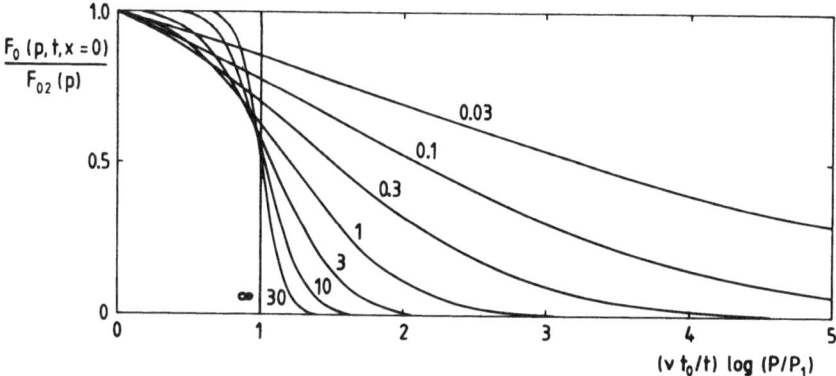

**Fig. 7.** The time dependent development of the spectrum at the shock for energy independent diffusion coefficients and monoenergetic injection (equation 6.9). We show here the ratio of the spectrum at a given time t to the equilibrium spectrum as a function of momentum.

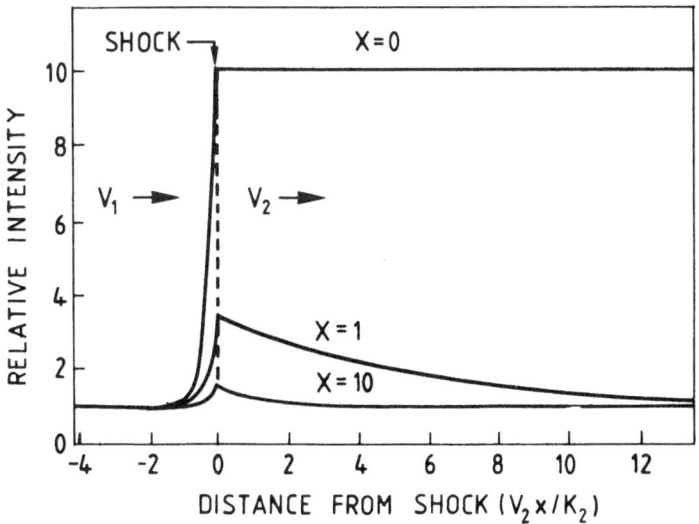

**Fig. 8.** The effect of losses ($X = 4\kappa_2/V_2^2\tau_2$) on the efficiency of shock acceleration (Völk et al. (152-154)).

medium and A the atomic number of the energetic particles. For continuous energy losses, the approximation is relatively crude, since in fact:

$$Q = - \frac{\partial}{\partial T} \left[ U \sum_i \left(\frac{dT}{dt}\right)_i \right] , \qquad (6.12)$$

where, in units of MeV/sec and with T expressed in MeV/nuc

$$\frac{dT}{dt} = 2.5 \times 10^{-9} \, B^2 T^2 \quad \text{for synchrotron losses } (e^+, e^-) ,$$

$$= 8 \times 10^{-16} \, nT \qquad \text{for Coulomb collisions },$$

$$= 4.8 \times 10^{-8} \, W_r T^2 \quad \text{for inverse Compton losses } (e^+, e^-),$$

$$= 4.5 \times 10^{-13} \, nZ^2 (c/v) \quad \text{for ionization losses },$$

$$= (2V/3R_s)\alpha T \qquad \text{for adiabatic losses },$$

with $W_r$ the energy density of photons, $Z$ the charge state of the particle and div $\underline{V} \sim 2V/R_s$ (155). In these cases, it is possible to use the "loss time" approximation as above, but it is obviously better to make a more exact analysis.

It is possible in principle to solve problems involving continuous energy losses analytically; such solutions have been given for cases in which the losses are adiabatic (156) and due to synchrotron and/or inverse Compton emission (157).

A somewhat related problem concerns the effect of second order Fermi acceleration occurring behind the shock (59,158-160,161). Here, following Webb (161), we take

$$Q = \frac{1}{p^2} \frac{\partial}{\partial p} \left( p^2 D_{pp} \frac{\partial F_o}{\partial p} \right) , \qquad (6.13)$$

where $D_{pp} = D_2 p^2 \exp(-x/L_2)$ in $x > 0$. For the case of monoenergetic ($p = p_1$) injection at the shock and with $\kappa_1$, $\kappa_2$ independent of $p$, the solution for large $p/p_1$ far downstream has the form:

$$F_o(x,p) \propto (p/p_1)^{-\mu} , \qquad (6.14)$$

with

$$\mu = \frac{1}{2} \left[ \{3-\eta(1+\eta)(R-1)/3\xi\}^2 + 4\eta(1+\eta)R/\xi \right]^{1/2}$$

$$+ \frac{1}{2} \{3-\eta(1+\eta)(R-1)/3\xi\} , \tag{6.15}$$

where $\eta = V_2 L_2/\kappa_2$ and $\xi = D_2 L_2^2/\kappa_2$. For small $\xi$ it can be shown that

$$\mu \to \frac{3R}{R-1} (1 - \frac{9}{2} \xi/\eta(1+\eta)(R-1)^2 + ...) , \tag{6.16}$$

and hence the spectrum is flatter at high energies than found in the case without second order Fermi acceleration ($\xi = 0$). In a more general situation where $\kappa_2$ and $D_{pp}$ depend on the charge/mass ratio of the particle species one would expect that the acceleration is selective.

## 7. Acceleration by Non-Planar Shocks

With one exception, namely the model of an interplanetary corotating interaction region given by Fisk and Lee (162), all attempts to deal with the acceleration of energetic particles by non-planar shocks have been restricted to cases of spherical symmetry. The essential difficulty is that the flow on either side of the shock is divergent (div $\underline{V} \neq 0$, in contrast to the one-dimensional case), so that energy changes occur due to adiabatic expansion and compression in addition to the acceleration associated with the shock. Furthermore, in a non-planar geometry, energetic particles may diffuse towards the shock from downstream and also escape upstream relatively easily, whereas in plane shocks convection usually dominates the behaviour at large distances, both upstream and downstream. It is not surprising then, that the energy/momentum spectra produced by non-planar shocks are not in general power laws and that the acceleration produced by the shocks should in many cases be partly undone by adiabatic expansion in the downstream region.

Approximate solutions to non-planar problems can be obtained if one is willing to assume that as far as the upstream medium is concerned, the shock can be treated as being locally plane (i.e. $\kappa_1/V_s \ll R_s$, the shock radius) and the downstream region has $\kappa_2 = 0$, so that convection and adiabatic energy changes dominate (163), or the downstream flow is incompressible so that only convection and diffusion occur (164,165) Several quite complicated problems, in some cases with time dependence, have been treated assuming that the energy/momentum spectrum preserves a particular power law form everywhere (57,58,3) but in view of the results obtained in section 5 this is hardly justifiable (it amounts to neglecting a particular solution). A partial correction of this deficiency can be attempted by combining two power laws above a certain energy and always ensuring that the spectrum is smooth (as shown in Figure 6b for example) (166); however, this procedure is not strictly correct and leads in general to non-conservation of particle numbers.

Two spherically symmetric problems have been solved exactly, namely the cases of a stellar wind with a stationary inwards-facing terminal shock surrounded by an extended region of essentially incompressible

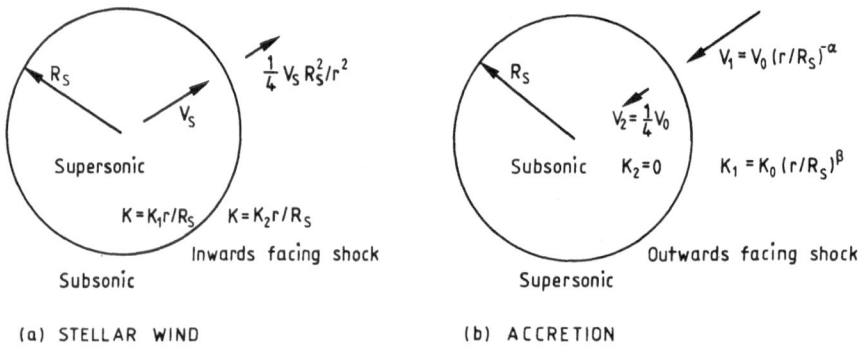

**Fig. 9.** Configurations assumed for solutions of problems corresponding to shock acceleration at (a) a stellar wind terminal shock (167-169), (b) an accretion shock (170).

flow extending to infinity (167-169) and a stellar accretion flow with a stationary outwards-facing shock, within which convection is dominant (170). The configurations are shown schematically in Figure 9. In both cases the energy/momentum spectrum has a given form and the radial current vanishes at $r = \infty$ and particles may also be injected at the shock ($r = R_s$).

The solution of the terminal shock problem for which $F_0(r,p) \rightarrow \delta(p-p_1)$ as $r \rightarrow \infty$ and $r^2 S = 0$ at $r = 0$ has the form

$$F_0(r,p) = g(r,p)\left[1-H(p-p_1)\right] + h(r)(p/p_1)^{-\mu}H(p-p_1) , \qquad (7.1)$$

where $g(r,p)$ and $h(r)$ are known functions and $\mu$ is a function of $V_1 R_s/\kappa_1$, $V_2 R_s/\kappa_2$ and $V_2/V_1$. The second term on the right of this expression is associated with shock acceleration, being a power law as in the plane case, but with a spectral index no longer dependent only on $V_2/V_1$. The first term on the right represents the effects of deceleration within the stellar wind region (i.e. modulation) which causes some particles to lose energy but in general also leads to a net acceleration (168). The cosmic ray energy flux $\int 4\pi r^2 STdT$ is positive and finite at infinity.

It has been proposed that stellar winds associated with O-stars could be an important source of galactic cosmic rays since (marginally) enough energy seems to be available (120,171,172). The mechanism has the advantage that it is stationary to a first approximation, so that there is no upper limit to the particle energies that can be achieved due to time dependent effects as described in section 6. However, there is a significant difficulty in that the magnetic field lines in the stellar wind region should be tightly wound Archimedian spirals at the position of the shock. Thus if $\Omega$ rad/sec is the rotational speed of the central O-star, the average angle between the field lines and the shock surface is $\phi \sim V_1/\Omega R_s$, which with $V_1 = 2000$ km/sec, $2\pi/\Omega = 10^6$ seconds and $R_s = 2$ parsecs, yields $\Omega R_s \sim 4 \times 10^{13}$cm/sec $\gg$ c; the shock is therefore almost everywhere perpendicular so that particles appearing at the shock, for example, cannot move upstream and be further accelerated. Furthermore, as a result of the tightly wound up magnetic field, the radial component of the diffusion coefficient in the subsonic region behind the shock is very small and hence the intensity of the galactic cosmic radiation must be substantially depressed, so that few particles can reach the shock from behind.

The accretion problem can be solved exactly provided once again the diffusion coefficient is chosen to be independent of energy/momentum. Taking $V_1 = V_0(r/R_s)^{-\alpha}$, $\kappa_1 = \kappa_0(r/R_s)^\beta$ and assuming injection at the shock such that $Q \propto \delta(r-R_s)\delta(p-p_1)$, solutions have been obtained for the case $\alpha + \beta = 1$ with asymptotic forms at high energies such that:

$$F_0(p,r) \sim A_1(r/R_s)^{-\eta\delta}(p/p_1)^{-\delta\lambda_1} , \quad \eta > \eta_c , \qquad (7.2)$$

$$F_0(p,r) \sim \{B_1-B_2 \log(r/R_s)\}(r/R_s)^{-m}\log(p/p_1)(p/p_1)^{-n}, \quad \eta<\eta_c, \quad (7.3)$$

where $\eta = V_0 R_s/\kappa_0$, $\eta_c = (1+\beta)(V_0+V_2)/(V_0-V_2)$, $\delta = 1-(1+\beta)V_2/(V_0-V_2)\eta$, $m = 1/2(1+\beta+\eta)$, $n = 3(1+\beta+\eta)^2/4\eta(1+\beta)$ and $A_1$, $B_1$, $B_2$ are constants. Although in general, the spectrum is essentially a power law, the spectral index is affected by the diffusion coefficient through $\eta$. For $\eta \to \infty$, the spectral index approaches that of the plane case ($\delta \to 1$) as expected, but for finite $\eta$ the spectrum is harder as a result of the additional acceleration associated with the convergence of the flow ahead of the shock. (In this case, the particle pressure becomes infinite at the shock, requiring a non-linear treatment.) For small $\eta$ the shock produces less efficient acceleration and the spectrum is very soft, such that $n \to 3(1+\beta)/4\eta$.

It has been possible to find exact analytic solutions for the accretion problem only for the case in which $\kappa_1$ is independent of momentum. However, an approximate asymptotic solution valid for large values of $p/p_1$ can be obtained for the case $\kappa_1 \propto (p/p_1)^\gamma$. The essential feature of this solution is that the spectrum becomes exponential:

$$F_0(p) \propto (p/p_1)^{-3/2} \exp\{-\frac{3}{2}\frac{(1+\eta)}{4\eta\gamma}(p/p_1)^\gamma\} \ . \qquad (7.4)$$

In all cases the accelerated particles are eventually convected through the shock since $r^2 S \to 0$ as $r \to \infty$; thus in contrast to the case of stellar winds, the configuration cannot provide a source of cosmic rays to the external world without some further modification, such as a "free escape" boundary somewhere outside the shock wave position.

It has been suggested (170,173) that accretion shocks around neutron stars could be an important source of galactic cosmic rays, at least for energies up to $10^6$GeV/nuc. Since many ($10^9$-$10^{10}$) neutron stars are likely to exist in the galaxy and a significant number of these ($10^4$-$10^5$) are in cool, dense clouds where the energy released by accretion can reach $\sim 4 \times 10^{36}$ergs/sec, it appears that enough power is available. It is also possible to generate the observed spectrum, although it is not obvious why any particular spectral index should be chosen. In view of the high background plasma densities involved it is also unclear whether or not the diffusion coefficient is sufficiently small that $4\kappa_0/V_0^2 \ll \tau_0$. In any case, an additional assumption is required to permit particles to leave the system, which must lead to a substantial reduction in efficiency as well as a change in the form of the spectrum. There is also of course the problem of angular momentum in the external medium which tends to further reduce the efficiency by reducing the solid angle in which an accretion flow can occur.

## 8. Acceleration of Cosmic Rays by Supernova Blast Waves

In this section we review our present understanding of cosmic ray acceleration by shock waves associated with expanding supernova remnants assuming that the cosmic rays play no part in the dynamics of the remnant. No exact formal solutions describing shock acceleration by blast waves exist since unsteady flow is involved and the problem is too difficult. One can obtain similarity solutions corresponding to the Sedov blast wave solutions (175) but this requires in general that $\kappa$ is a function of time, which may not be a very reasonable assumption. An approxi-

mate solution has been given by Krimsky et al. (163) however it is
rather oversimplified and the characteristic effects associated with
three-dimensional flows described in section 7 are absent; it is assumed
that $V_s R_s/\kappa_1 \gg 1$ so that the shock is effectively plane and $\kappa_2 = 0$ so
that the cosmic rays behind the shock simply expand adiabatically. Nevertheless, despite the oversimplification we can at least proceed in this
manner to consider how supernova remnants accelerate cosmic rays provided the limitations of the assumptions are recognized.

In order to obtain a feeling for the behaviour of supernova remnants in the HISM it is useful to consider the implications of the diagram given in Figure 10. Here we have assumed, following section 3, that
one supernova with average energy $10^{51}$ ergs occurs every 30 years in
a volume $V_g$ and that the HISM has density $n = 3 \times 10^{-3} \mathrm{cm}^{-3}$, temperature
$T_g = 7 \times 10^5 K$ and magnetic field strength $B = 1$ μGauss. We neglect the
effects associated with radiative cooling and the presence of dense
clouds (6,118) and assume simply that a supernova remnant can be represented by the Sedov solution with

$$R_s = 13(E_{51}/n_o)^{2/5} t_4^{2/5} \text{ parsecs} , \tag{8.1}$$

where $E_{51}$ is the supernova energy in units of $10^{51}$ ergs, $n_o$ the number
density of the HISM $(\mathrm{cm}^{-3})$ and $t_4$ the age of the remnant in units of $10^4$
years. Assuming that the supernovae are uniformly distributed, it is
possible to calculate the average number of shocks $(N_s(\geq V_s))$ with speed
greater than a given value passing any point per year and the fraction
$(F (\leq R_s))$ of the cosmic ray containment volume $V_g$ which is covered by
shocks with radii less than a given value:

$$\dot{N}_s(\geq V_s) = 4\pi R_s^3 \dot{N}_g/3V_g, \tag{8.2}$$

$$F(\leq R_s) = (4\pi\dot{N}_g/3V_g) \int_0^{R_s} \dot{R}_s^3 \, dt , \tag{8.3}$$

where $\dot{N}_g$ is the supernova rate in the galaxy. These quantities are shown
in Figure 10 with various abscissae determined from (8.1) and the plane
shock results (5.9) and (5.11). It is important to note that the Sedov
phase begins only after about $t_4 \sim 1$, since prior to this the mass
ejected by the supernova is not negligible compared with the mass of
the HISM overrun by the shock. The Sedov phase ends when the total enthalpy of the ambient HISM existing prior to the passage of the shock is
comparable to $E_{51}$ (i.e. at $R = R_c$) which is roughly the point at which
the Mach number of the shock $(M_1^2 = V_s^2/(c_g^2 + c_A^2))$ is unity. As shown in
Figure 10, this occurs at a radius $R_c \sim 180$ pc. Since the speed of the
shock cannot be less than the speed of sound, the Sedov solutions under-
estimate $N_s$ and $F$ when $R_s = V_s \rightarrow \sqrt{(c_g^2 + c_A^2)} \sim 185$ km/sec and at all
later phases. It is interesting to note, however, that with the condi-
tions assumed this occurs roughly when $F \sim 1$ so that the whole of $V_g$ is
in effect covered by supernova remnants of moderate strength as required to account for the existence of the HISM (6).

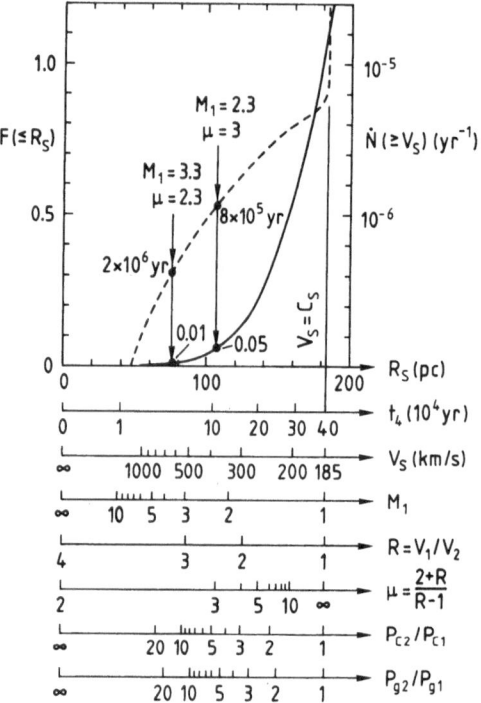

Fig. 10. The rate at which shock waves with speeds $\geq V_s$ pass a given point ($\dot{N}(\geq V_s)$) and the fraction of space in the galaxy covered by shock waves with radii $\leq R_s$ ($F(\leq R_s)$), as functions of shock radius ($R_s$), age ($t_4$) and shock speed ($V_s = \dot{R}_s$). We have assumed that the Sedov solution is valid and that the shock waves propagate into the HISM with $n_0 = 3 \times 10^{-3}$, $E_{51} = 1$, $T_g = 7 \times 10^5$ K, $B_g = 1$ μGauss and supernova rate $\dot{N}_g = 1/30$ per year. The corresponding values of the Mach number, the cosmic ray spectral index and the cosmic ray and plasma pressure jumps at the shock are also shown. The observed primary cosmic rays must be accelerated predominantly around the region where $\mu = 2.3$, which suggests that the important cosmic ray sources are supernova remnants with radii of the order of 70 parsecs, age $3 \times 10^4$ years and occupying about 1% of space. Secondary nuclei produced by interactions taking place throughout the galaxy have their spectral index changed only by shock waves corresponding to $\mu < 3$ and these are seen to occupy only 4% of space.

An immediate implication of Figure 10 is that the shock waves which are largely responsible for determining the cosmic ray source spectrum ($\mu \sim 2.3$) have $M_1 = 3.3$, $R_s \sim 75$ parsecs, $V_s \sim 650$ km/sec, $t \sim 3 \times 10^4$ years and they cover less than $\sim 1\%$ of space. The cosmic ray pressure enhancement behind such a shock is of the order of 14 (see Figure 5b); presumably, however, this is something of an overestimate since the plasma pressure enhancement is only of the same order ($p_2/p_1 \sim 5M_1^2/4$ if $M_1^2 \gg 1$ and $\gamma = 5/3$) and we have neglected the effects of cosmic ray pressure on the dynamics of the supernova remnant. Evidently, cosmic rays produced in earlier phases (when $\mu < 2.3$) cannot make a significant contribution to the ultimate cosmic ray population due to the overwhelming effects of adiabatic expansion, whereas cosmic rays produced in later phases suffer less from expansion losses but are not accelerated as efficiently. It is important also to note that shocks which alter the observed spectrum of secondaries ($\mu \leq 3$, $M_1 > 2.3$) occupy only $\sim 5\%$ of space so that post-acceleration by shocks affects the spectral form of no more than 5% of the secondaries. However, since essentially all of the HISM is covered by shocks (perhaps relatively weak), all secondaries undergo post-acceleration, which must therefore be taken into account in making inferences from their fluxes (as discussed in section 2). The argument (60) that the acceleration of galactic cosmic rays must occur in the HISM because only in this way a large fraction of space can be "processed" is misleading since it implicitly assumes that cosmic rays are accelerated from a seed population rather than emerging from the background plasma; indeed, the above discussion is essentially contradictory to such an approach.

According to Figure 10, the cosmic ray pressure enhancements to be expected are not insignificant in the sense that an enhancement by a factor of 2 or more corresponds to $F \sim 0.35$, $\dot{N}_s \sim 5 \times 10^{-6}$/yr. That is, for 35% of the time and once every $2 \times 10^5$ years, the solar system should be immersed in regions with noticeably enhanced cosmic ray intensity, which is probably not consistent with observations (112,113). This is an indication of the fact that a non-linear approach to supernova dynamics is necessary, leading to a reduction of the cosmic ray pressure enhancement at a given $R_s$ shown in Figure 10, to a reduction of the Mach number (since the effective speed of sound is 200-230 km/sec rather than 185 km/sec) and to corresponding changes in the volumes occupied by the most important cosmic ray producing shocks and the shocks which can affect the secondary spectrum.

Despite the above reservations it is important that the non-self-consistent analysis of cosmic ray acceleration be carried as far as possible to see what further implications it might have. This has been done Blandford and Ostriker (60,61) who made the following assumptions:

(a) the shocks can be considered plane and steady with the accelerated spectrum being determined as described in section 2;

(b) fresh particles are injected at the shock at a constant rate and with a monoenergetic spectrum ($p_1 \sim 100$ MeV/nucleon, say);

(c) the accelerated particles undergo adiabatic expansion by a factor F until the original plasma density or pressure is reached;

(d) specific shock propagation models are used (e.g. Sedov (175) or McKee-Ostriker (6));

(e) the subsequent propagation of cosmic rays takes place as in the simple leaky box model.

On this basis a conservation equation of the following form can be derived:

$$\frac{1}{\tau_s} \left[ \int \phi(x-x')Y(x')dx' - Y(x) \right] + Q_o \phi(x) = \text{Losses} + \frac{Y(x)}{\tau_c(x)} , \qquad (8.4)$$

where $Y \propto p^4 F_o(p)$, $x = \log(p/Z)$, $Q_o$ represents the injection rate, $\tau_s$ is a parameter determined by the supernova rate and $\tau_e(x)$ is the residence time of cosmic rays in the galaxy. $\phi(x)$ is the "redistribution function" given by

$$\phi(x) = \int_4^{q'} (q-3)D(q)F(q)^{q-3}\{\exp(4-q)x\}dq , \qquad (8.5)$$

where q is the spectral index characterizing the shock of a particular strength, $D(q)$ is obtained from the shock propagation model assumed and $q'$ is determined by the weakest shocks predicted in this model. For reasonable shock propagation models the form of $\phi(x)$ is such that around the injection energy the redistribution corresponds to diffusion in a manner similar to that described by Bykov and Toptygin (11,176,177) for multiple encounters with many weak shocks. In addition, $\phi(x)$ has a high energy tail corresponding to occasional encounters with strong shocks.

The essential result of the work of Blandford and Ostriker is that on the basis of the assumptions listed above, the spectral forms of primary and secondary cosmic rays can be adequately accounted for (see Figure 11). In particular, at high energies (> 10 GeV) the proton energy spectrum has the form $j(T) \propto T^{-(2+\delta+\mu)}$ where in the spectral index, $(2+\delta) \sim 2.2$ results from the initial acceleration by strong shocks and $\mu \sim 0.4$ from the energy dependence of the escape mechanism. The normalization of the spectra and the relative abundances of different primary species are not determined since the injection rate is arbitrary. However, the relative abundances and spectra of secondaries are determined once the $Q_o$ are fixed, just as in the simple leaky box models. The short-comings of this procedure are fairly obvious and are all connected with breakdowns of the validity of the five basic assumptions listed above. In particular, the underlying assumption that cosmic rays can be treated as non-interacting test particles having no effect on the dynamics of the background plasma is not sound as evident from Figure 10.

It is implicit in any model of this class that the energy of any given cosmic ray particle is not constant in time, which has important implications for the interpretation of observations of secondary particles in particular. Most fragmentation products, for example, should be produced in the ISM as a whole, since the product (volume x time x cosmic ray flux) is much greater than in moderately strong supernovae blast waves (R $\lesssim$ 100 pc) even taking into account the enhancement of cosmic ray flux in the latter. Nevertheless, the secondaries must be affected by the large number ($\sim 10^2$) of shock waves they encounter during the time they spend in the containment volume. Even if the forms of their spectra are unchanged, the secondaries must be accelerated and

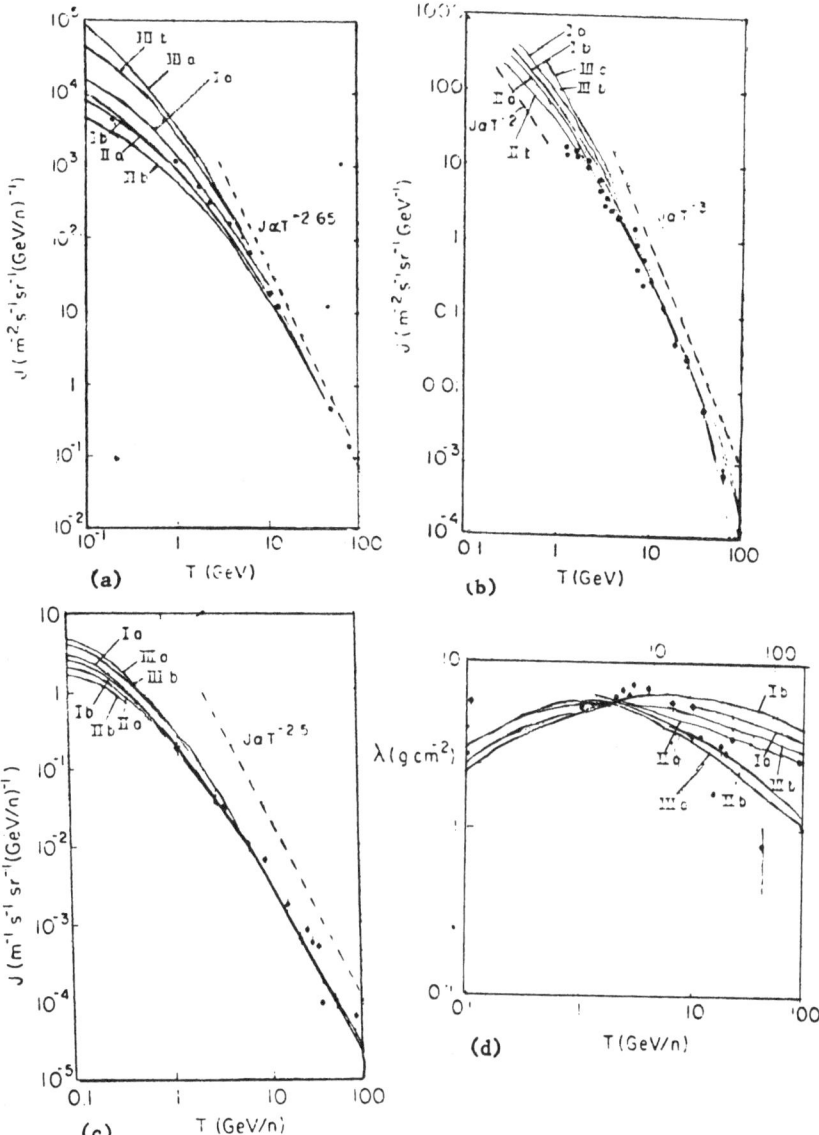

**Fig. 11.** Model galactic cosmic ray spectra calculated by Blandford and Ostriker (61), for (a) protons, (b) electrons, (c) heavy nuclei and (d) the apparent path lengths of secondaries. Model I corresponds to the supernova model of McKee and Ostriker (6), model II to the Sedov solution and model III is a solution with an assumed power law source and no redistribution. The escape laws used have a simple power law dependence on rigidity (a) and contain an extra rigidity independent term (b).

decelerated (i.e. the spectrum raised and lowered) as a result of such encounters following their formation; if the net energy changes significant, the path length must be correspondingly different from that deduced under conditions of propagation at constant energy. The Blandford-Ostriker model does not provide an especially good fit to the data concerning secondaries (178) but this does not mean that the model is wrong on this account; there are many improvements to be made and it should therefore be regarded as being a first step in the right direction.

Some secondaries, notably anti-protons, are particularly important in this regard, since their production cross sections are strongly energy-dependent (in contrast to fragmentation cross sections). Simple leaky box models with constant particle energy find great difficulty in explaining the presently available measurements of anti-protons, especially at low energies where the direct production rate is very low indeed (179-182). Although a detailed model is not yet available, it seems likely that the anti-protons can only be explained if their production occurs mainly in supernova remnants where the high energy (T $\gtrsim$ $10^2$ GeV) protons have intensities $\sim 10^4$ greater than normal (partly as a consequence of a flatter spectrum) and is followed by adiabatic expansion bringing significant numbers of anti-protons to energies of $\sim$ 100 MeV. Obviously such secondary particles can provide important information concerning the whole process of cosmic ray acceleration in supernova remnants, which is not obtainable from secondaries produced by fragmentation, for example.

## 9. The Cosmic Ray Diffusion Coefficient in the HISM

In order for galactic cosmic rays to be accelerated by supernova shocks up to any particular energy it is necessary that the acceleration time $\tau_a \sim 4\kappa_1/V_1^2$ be less than the time available for acceleration $(R_s/\dot{R}_s)$, otherwise the spectrum will exhibit a high energy cut-off. For Sedov blast waves, this condition can be written approximately as $4\kappa_1\dot{R}_s \lesssim R_s/\dot{R}_s$, hence

$$\kappa_1 \lesssim R_s\dot{R}_s/4 = 2\times10^{27}(E_{51}/n_o)^{2/5} t_4^{-1/5} \text{ cm}^2\text{sec}^{-1} . \qquad (9.1)$$

With $E_{51} = 1$, $n_0 = 3\times10^{-3}$, $t_4 \sim 1$, we require that $\kappa_1 \lesssim 5\times10^{27}$ cm$^2$/sec. Taking as an estimate $\kappa \sim cr_g/3(\delta B_w/B)^2$ with $r_g \sim 10^{-6}$ T(GeV)pc and $\delta B_w$ the amplitude of waves having wave length $O(r_g)$, we find that $\kappa_{||} < 1.5\times 10^{28}$ cm$^2$/sec corresponds to $\delta B_w/B \sim 10^{-3}$ (T GeV)$^{1/2}$ and hence $\delta B_w/B$ should range between $10^{-3}$ and 1 for 1-$10^6$ GeV protons.

The condition (9.1) is stronger than the requirement that the shock should appear approximately plane (namely $\kappa_1/V_1 < R_s$ or $\kappa_1 < R_s\dot{R}_s$ for isotropic diffusion). However, if the diffusion coefficient is highly anisotropic (as it probably is for T $\lesssim 10^6$ GeV/nuc), this requirement is always satisfied. Furthermore, it should be noted that $\kappa_1$ refers to the component of the diffusion coefficient normal to the shock and it could therefore be identified with $\kappa_{||}\cos^2\theta = \kappa_{||}/3$ if an average is taken over the (spherical) shock surface. Thus we require that $\kappa_{||} \lesssim 1.5\times10^{28}$ cm$^2$/sec.

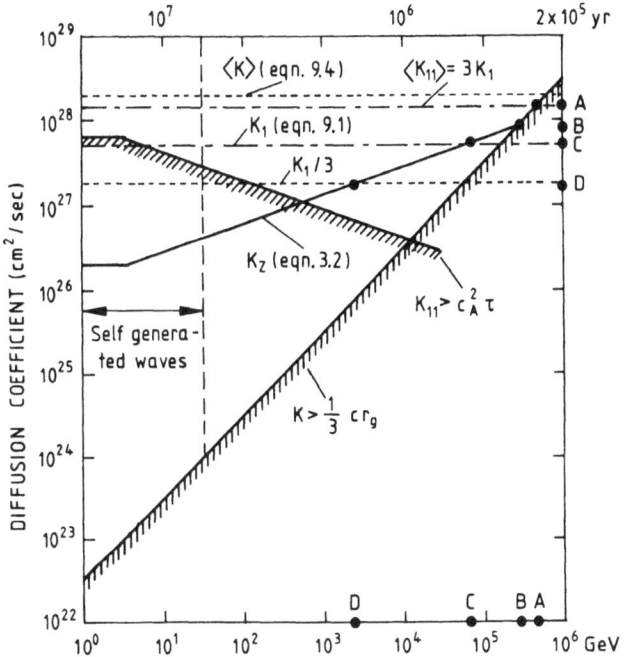

**Fig. 12.** Diffusion coefficients versus energy/nucleon and çonfinement time. The maximum diffusion coefficient which would allow the equilibrium spectrum to be achieved is $\kappa_1$ (equation (9.1)); this intersects the diffusion coefficient for escape from the galaxy at point C (T $\sim$ $6 \times 10^4$ GeV/nuc). If $\langle \kappa_{11} \rangle = 3\kappa_1$ is used, the intersection point yields an upper limit to acceleration at T $\sim 10^6$ GeV/nuc. Allowance for the energy dependence of the diffusion coefficient (equation (6.7)) yields an upper limit T $\sim 2 \times 10^3$ GeV/nuc (point D). The average diffusion coefficient $\kappa$, resulting from an equilibrium between shock amplification and non-linear damping (equation (9.4)), is slightly greater than $3\kappa_1$. The diffusion coefficient has a lower limit $\kappa = cr_g/3$ which yields an upper limit, using $\langle \kappa_{11} \rangle$, of T $\sim 5 \times 10^5$ GeV/nuc. Second order Fermi acceleration should not be important if the cosmic ray spectrum is to be determined by shock acceleration ($\kappa_{11} > c_A^2 \tau$); this could give rise to difficulties if the diffusion coefficient is as small as $\kappa_1/3$ or $\kappa_z$. Self-generated waves occurring near the shock can permit acceleration up to 10-50 GeV/nuc so that the above arguments should be ignored in this region of the diagram (185). It is evident that $\kappa_1$ or $\langle \kappa_{11} \rangle$, as shown, would be acceptable but $\kappa_z$ is too small in the range T < $10^3$ GeV/nuc. [Based on arguments presented by Ginzburg and Ptuskin (67), Cesarsky and Lagage (68) and Krimsky and Petukhov (183).]

These constraints on $\kappa_1$ and $\kappa_{||}$ are displayed in Figure 12 together with $\kappa_z(T)$, representing the coefficient for diffusion perpendicular to the plane of the galaxy (3.2). It would seem that there is no real difficulty in satisfying the requirements for shock acceleration by supernova shocks until energies of $\kappa$ order of $5\times10^5$–$10^6$ GeV/nuc are reached. However it should be noted that the form for $\tau_a$ used in obtaining (9.1) is rather crude and if we were to allow for the fact that the shocks are not infinitely strong (e.g. $V_1/V_2 \sim 3.3$) and the diffusion coefficient depends on energy (e.g. $\kappa_1 \propto T^{1/3}$ as in (3.2)) then a further factor $\sim 3$ is required (for acceleration from an injected population at very low energies) which corresponds to reducing the upper limit on the energy achievable by a factor $\sim 27$ to $2\times10^3$–$4\times10^4$ GeV/nuc. The essential problem here is that, apart from some conceptual difficulties, we do not know what the typical values of $\kappa_{||}$ might be in the HISM. The arguments given above concerning the anisotropy of the diffusion coefficient could equally well apply to $\kappa_z$, but on the other hand our estimate of the latter is entirely dependent on the most elementary leaky box model with $L \sim 200$ pc, $\tau_0 = 2\times10^7$ years and it is assumed that the diffusion coefficient so obtained is somehow representative of $\kappa_{||}$, which is of interest to us. In fact, the escape of cosmic rays from the galaxy may well be largely controlled by convective transport associated with the supersonic turbulence in the HISM (12,73) and by a galactic wind, so that the above difficulty, which has been emphasized by Ginzburg and Ptushkin (67), may not be real.

It has been pointed out that there is a further important constraint on the diffusion coefficient at high energies, namely the equivalent scattering mean free path should not be smaller than the particle gyro-radius so that $\kappa > cr_g/3$ (67,68,18). This constraint is also shown in Figure 12, where we see that it yields an upper limit on the energy of $\sim 3$–$5\times10^5$ GeV for protons and $\sim 1$–$3\times10^5$ GeV/nuc for other species if $B_g \sim 1$ μGauss. To reach these energies as a result of acceleration by a single shock, the magnetic field must be disordered on a length scale of 0.1–1 pc — which is perhaps not unreasonable. With the value of $B_g$ assumed, the upper limit quoted is not quite as severe as that obtained from considerations of $\kappa_z$, however if $B_g$ is as small as $10^{-7}$ Gauss, as argued by Hall (72,73), the upper limit to the energy achievable would again be reduced to $\sim 10^4$ GeV/nuc.

An additional constraint has been pointed out by Ginzburg and Ptuskin (67), namely that second order Fermi acceleration should not be important if the cosmic ray energy spectrum is to be determined by shock acceleration alone. For ultra-relativistic particles the time scale for second order Fermi acceleration by the Alfvén waves which cause pitch angle scattering is $\kappa_{||}/V_A^2$ (184) which is in general much greater than the shock acceleration time $4\kappa_1/V_s^2$ since $V_s \gg V_A$. However, second oder Fermi acceleration can operate over the whole period of containment in the galaxy, so the essential constraint is $\kappa_{||}/c_A^2 > \tau = \tau_0(T \text{ GeV/nuc})^{-1/3}$, again a lower limit on $\kappa_{||}$. This does not appear to be a serious constraint, however, since it approaches the upper limit required for shock acceleration only at energies less than $\sim 1$ GeV/nuc where there is no requirement that shock acceleration be dominant.

At low energies, the magnetic field irregularities required for scattering can be locally generated by cosmic ray streaming instabilities associated with the shock itself (4). According to Völk (185) how-

ever, if non-linear damping of the unstable waves is taken into account,
the upper limit on the energy to which particles can be accelerated re-
lying only on self-generated waves is only 10-50 GeV/nuc (see Figure 12).
Since linear effects strongly damp all magnetohydrodynamic waves other
than Alfvén waves propagating parallel to the magnetic field and the
latter are eventually damped by non-linear effects (186,187) it is
necessary to examine carefully the nature of the magnetic irregularities
required for particle scattering.

In view of the evidently violent nature of the HISM it seems difficult to imagine that the magnetic field is completely smooth on scales
of the order of 0.1-1 pc. Indeed we see from Figure 10 that weak shock
waves can be expected to pass any point in space with a frequency corresponding to such length scales. Weak shocks are not in themselves likely to be very effective in cosmic ray scattering since the large scale
magnetic field changes they produce are relatively small, however they
are capable of amplifying existing Alfvén (and other) waves (188) and
generating new waves as a result of the cosmic ray streaming they in-
duce (70,71). It is possible that the magnetic field in the HISM is more
or less random on length scales of the order of 1 pc, which would satis-
fy the requirement that $\kappa_{\shortparallel} \sim 1.5 \times 10^{28}$ cm$^2$/sec for T $\sim 10^6$ GeV. If this
turbulence could cascade to smaller wave lengths maintaining a Kolmogrov
spectrum then the diffusion coefficient would be easily small enough for
our purpose for energies less than $10^6$ GeV and it would have the energy
dependence required to account for the apparent escape lifetimes (189).
It is not clear however that the necessary cascading will take place at
a sufficient rate to overcome the effects of wave damping which become
progressively more important at small wave lengths (187,188).

For the case of non-linear Landau damping of Alfvén waves (186):

$$\frac{d}{dT}(\delta B_w^2) = -\delta B_w^4 / B^2 \tau_1 \;, \quad \tau_1 = \sqrt{(8/\pi)}/c_g \langle k \rangle \;, \tag{9.2}$$

where $\langle k \rangle$ is the wave number corresponding to the particle momentum of
interest. With values of the quantities involved which are appropriate
to the HISM we find that

$$\delta B_w^2 / \delta B_{wo}^2 = 1/\left[1 + (\delta B_{wo}^2 / B^2)(t/\tau_1)\right] \;, \tag{9.3}$$

where $\delta B_{wo}$ is the wave amplitude at t = 0. Consequently,

$$\langle \kappa \rangle = \frac{1}{3} c r_g \left\langle (B^2/\delta B_{wo}^2) + (t/\tau_1) \right\rangle \sim 2 \times 10^{28} \, t_4 \; \text{cm}^2/\text{sec} \;. \tag{9.4}$$

Thus if the waves are regenerated at intervals of the order of $10^4$ years,
an adequate diffusion coefficient can be maintained on the average. It
is important to note that in the case of non-linear damping, the temporal
behaviour is not an exponential decay, but inverse time; hence the decay
rate becomes progressively longer at later times and is not sensitive

to the initial value of $\delta B_w$, as long as $\delta B_{wo} \gg \delta B_w$. A more detailed
analysis of these points is at present being carried out (188).

Bell's suggestion that Alfvén wave generation occurs as a result of
instabilities induced by cosmic ray streaming ahead of shock fronts has
been criticised by Morrison et al. (189) on the grounds that, with the
effects of cosmic ray pressure included in the dispersion equation, the
phase speeds of the most unstable waves exceed the Alfvén speed by a
very large factor and may even be "supersonic". These authors quote an
approximate result for the phase speed of the most rapidly growing wave:

$$v_{ph} = c_A \left[ 1 + \frac{\pi}{5} \, (E_*/m) \, (kc/\Omega_*)^{1/2} (\epsilon \, \xi/c_A^2) \right]^{1/2} , \qquad (9.5)$$

where the cosmic rays are supposed to have a power law spectrum above a
cut-off energy $E_*$, $\Omega_*$ is the corresponding gyrofrequency, $\epsilon$ is the ratio
of the cosmic ray to plasma number density and $\xi$ the anisotropy. Taking
$E_* = 10^{-3}$ ergs, $\epsilon = 10^{-7}$, $kc/\Omega_* = 1$, $\xi \le 1$ and $c_A = 35$ km/sec, we obtain $v_{ph} = 2.4 \, c_A$. This is very much less than the minimum shock speed
in the HISM, namely $c_s \sim 200$ km/sec; hence the cosmic rays do not
stream supersonically and can therefore be accelerated as described in
section 5. If we take $c_A = 110$ km/sec corresponding to $B_g = 3$ μGauss,
we obtain $v_{ph} = 1.3 \, c_A$ which is again substantially less than the sound
speed in this case ($c_s = 230$ km/sec). If $B_g = 10^{-7}$ Gauss, then $v_{ph} \sim$
100 km/sec which is much greater than $c_A$, but smaller than the sound
speed $c_s = 200$ km/sec. [The formal analysis of Morrison et al. does not
differ significantly in substance from that of Zweibel (190) whose conclusions are essentially those presented here.] It should be noted that
the waves generated by streaming instabilities propagate in the direction of the magnetic field, which is not necessarily normal to the shock:
in this case the relative speed of the scattering medium is $V_1 =$
$V_s - v_{ph}\cos\theta$. It is certainly the case that such a reduction in relative
speed can reduce the effectiveness of shock acceleration (10) but it
does not appear to be sufficient to prevent it from occurring.

A different approach to the problem of generating the magnetic
field irregularities required to scatter cosmic rays has been taken by
Hall (72,73). He considers the effect of large scale supersonic turbulence in creating pressure anisotropies which cause firehose and mirror
instabilities which in turn produce small scale magnetic irregularities
capable of scattering cosmic rays. Using observations of electron density fluctuations in the galaxy obtained from pulsar scintillation data,
he estimates the plasma pressure anisotropies and the associated hydromagnetic wave field which might exist. The latter is then used to determine a cosmic ray diffusion coefficient which is in reasonable agreement with the shape of the cosmic ray path length distribution apart
from a scale factor which has not been adequately determined. In fact
this work is complementary to the other approaches to the production of
magnetic field irregularities we have outlined: it would be interesting
to have the effects of finite cosmic ray pressure and pressure anisotropies included in the analysis, together with an investigation of the
effects of linear and non-linear damping on the wave spectrum.

## 10. Non-linear Cosmic Ray Shocks

It has been noted at several points in this review that the linear or "test particle" approach which we have so far applied to the problem of cosmic ray interaction with shocks encounters certain difficulties. For example, it is evident from Figure 5 that if $\gamma_c = 4/3$ the cosmic ray pressure behind a strong shock may become very large if the compression ratio approaches 4; hence, given the fact that $p_{c1} \sim p_g$ in the HISM, the cosmic ray pressure produced by the shock may be the dominant pressure in the medium. This contradiction is also apparent in Figure 10, where the cosmic ray pressure in supernova remnants, which are of most interest for cosmic ray acceleration, is shown to be comparable with the gas pressure so that the cosmic rays must play a role in the dynamics of the expanding remnants. In connection with the problem of the diffusion coefficient discussed in section 9, it is evident that there is likely to be an intimate connection between the cosmic rays and the diffusion coefficient since cosmic ray anisotropies can produce the hydromagnetic waves which in turn scatter the particles and tend to quench the anisotropies which cause them.

These points and others which are perhaps more subtle, are all indications that the problem of cosmic ray acceleration by shocks is far more complicated than implied by the discussion given in the previous sections. That is, we cannot expect in general to be given the diffusion coefficient $\kappa$ and the velocity field in the medium $\underline{V}$ and be satisfied to calculate the behaviour of cosmic rays using linear transport equations as in sections 5 and 6, for example. In a realistic treatment it is necessary instead that we simultaneously determine the nature of the wave field (and hence the diffusion coefficient) and the reaction of the medium to cosmic ray and wave pressure gradients. Furthermore, we must consider the problem of cosmic ray injection, which opens a Pandora's box of questions concerning the state of the background plasma, the nature of collisionless shocks, pre-acceleration in plasma turbulence and so on (13,191-194). Although the literature concerned with the non-linear problem is not very extensive, the concepts and calculations involved are rather complicated and there is no space to give a detailed review here. Instead, we describe the general outline of the problem using Figure 13 as a basis, and provide a list of references.

In Figure 13, we show the system divided into three components, namely the plasma, the cosmic rays and the hydromagnetic wave field. The plasma may be described in terms of its distribution function, which is important in considering the problem of cosmic ray injection, and also in terms of hydrodynamic quantities such as pressure, density, velocity ($p_g$, $\rho$, $V$) and the mean magnetic field ($\underline{B}$). The cosmic rays can be described in terms of their distribution function (U, S) or in terms of the corresponding "hydrodynamic" quantities, namely the cosmic ray pressure and energy flux ($p_c$, $F_c$). The wave field can likewise be described in terms of its power spectrum as a function of wave number, or, where appropriate, the integral quantity $\delta B_w^2$, the mean square magnetic field fluctuation. The wave field determines the cosmic ray diffusion coefficient and, as a consequence, the nature of the coupling between the cosmic rays and the background medium.

The "linear" approach which we have reviewed in the previous sections, assumes that the quantities in the plasma and wave boxes of

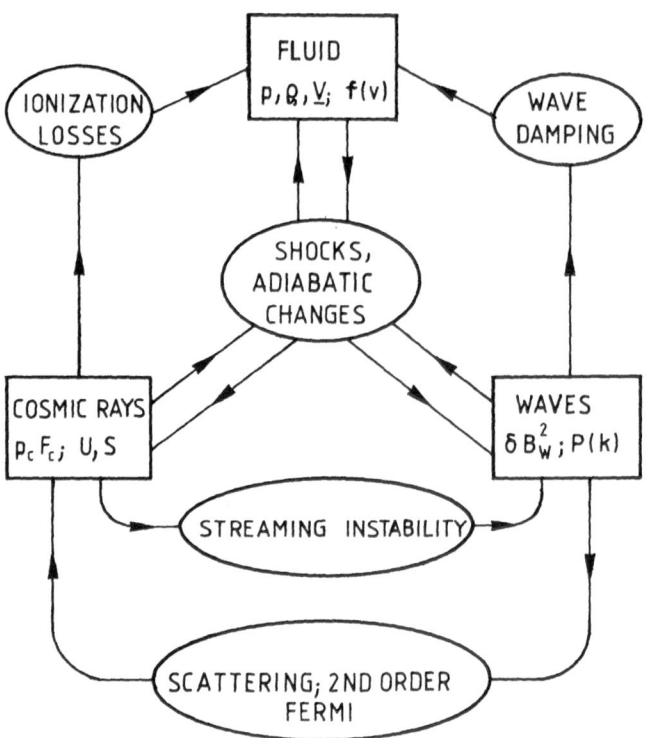

**Fig. 13.** A diagram showing schematically the non-linear interaction scheme described in the text. The arrows represent the direction of energy flow.

Figure 13 are determined independently and that the cosmic rays simply respond to the resulting diffusion coefficient and velocity field. We may arbitrarily assume a certain rate of injection of cosmic rays from the background plasma and also take into account cosmic ray energy change due to ionization losses, etc.

Ignoring the existence of cosmic rays to begin with, it is also possible to consider the interaction of waves with a given plasma velocity field, including shocks. It is found that the wave field must also gain and lose energy as a result of plasma compressions and rarefractions and, in particular, the waves can be strongly amplified by interacting with shocks (195,196). The effects of damping, which feed energy from the wave field back to the plasma may also be considered, as well as the direct effects of wave pressure and energy flux on the dynamics of the plasma as a whole (197-199).

A second non-linear combination can be considered in which the behaviour of the plasma as a whole is prescribed but the cosmic rays and wave field interact in such a manner that the waves which determine the diffusion coefficient are produced by cosmic ray streaming (4,69-71). The waves damp by giving energy to the cosmic rays, both directly and as a consequence of second order Fermi acceleration, and also by giving their energy up to the fluid as a result of various linear and non-linear plasma damping mechanisms.

The third non-linear combination of two of the components shown in Figure 13 comprises the cosmic rays and the plasma, with the wave field (and thus the diffusion coefficient) being given (3,12,13,191-194,200-205). The most straightforward procedure in this case is to treat the hydrodynamic aspects of the interaction first, describing the cosmic rays in terms of their pressure and energy flux and using a suitably defined average diffusion coefficient $(\overline{\kappa})$. Thus, for steady, one-dimensional flow:

$$\rho V = \text{constant} , \qquad (10.1)$$

$$p_g + p_c + \rho V^2 = \text{constant} , \qquad (10.2)$$

$$\rho V \left(\frac{1}{2} V^2 + \frac{\gamma}{\gamma-1} p_g/\rho\right) + F_c = \text{constant} , \qquad (10.3)$$

representing conservation of mass, momentum and energy respectively. The first moments of (5.1) and (5.2), integrated over T, yield:

$$\frac{dF_c}{dx} = V \frac{dp_c}{du} + \overline{Q}_1 , \qquad (10.4)$$

$$F_c = \frac{\gamma_c}{\gamma_c-1} p_c V - \frac{\overline{\kappa}}{\gamma_c-1} \frac{dp_c}{dx} , \qquad (10.5)$$

where $\overline{Q}_1 = \int_o^\infty TQdT$ can be neglected if cosmic ray injection occurs at very low energies. These equations can be integrated for given $\bar{\kappa}$ to yield shock-like transitions in which the total Mach number $M = V/c_s$ changes from supersonic to subsonic. The flow speed varies smoothly but an internal plasma shock occurs in some circumstances (3,201-203).

The properties of these transitions are such that for sufficiently large values of the upstream pressure ratio $(p_{c1}/p_1 \gtrsim 4.6$ for $\gamma_c = 4/3)$ the transitions are always smooth, with cosmic ray diffusion playing a role similar to heat conduction in an ordinary shock (12,13,200). For lower initial pressure ratios, an internal plasma shock must be introduced into the flow if the initial Mach number lies in a certain range but is otherwise also smooth (200-202). In the particular case where $p_g = 0$, smooth solutions are always the rule and the transition has the simple form:

$$V = \frac{1}{2}(V_1 + V_2) - \frac{1}{2}(V_1 - V_2) \text{ Tanh } x/L \quad , \tag{10.6a}$$

$$V_2/V_1 = (\gamma_c - 1)/(\gamma_c + 1) + 2/(\gamma_c + 1)M_1^2 \quad , \tag{10.6b}$$

where $L = 2\bar{\kappa}/V_1(1 - 1/M_1^2)$ and $M_1^2 = \rho_1 V_1^2/\gamma_c p_{c1}$. This solution (which is analogous to that obtained for the structure of a shock wave dominated by heat conduction rather than viscosity) shows the usual feature of being narrow if $M_1^2 \gg 1$ and broad when $M_1^2 \to 1$. The overall compression ratio $V_1/V_2 \to (\gamma_c + 1)/(\gamma_c - 1) \sim 7$ as $M_1^2 \to \infty$; in this case most of the kinetic energy of the fluid is converted to cosmic ray energy and the conversion efficiency approaches 98% as required to account for cosmic ray acceleration by shock acceleration in supernova remnants.

In order to complete the solution for this type of non-linear interaction it is necessary to determine the corresponding behaviour of the cosmic ray spectrum (i.e. U and S). That is, we must solve (5.1) and (5.2) with the flow speed determined from (10.1-10.5). This is in general difficult but a solution has been found in one case, where the flow speed is given by (10.6a) and the cosmic ray spectrum far upstream from the transition is a $\delta$-function (203,204). The downstream solution in this case is such that the peak of the spectrum shifts in a manner appropriate to adiabatic compression by a factor $V_1/V_2$ and a high energy tail develops such that $F_0(p) \propto (p/p_1)^{-\nu}$ where

$$\nu = \frac{3V_1}{V_1 - V_2}\left[1 + \frac{2}{\gamma_c + 1}\frac{V_2}{V_1 - V_2}\right] . \tag{10.7}$$

This index is larger than would be obtained for the same compression ratio in an unmodified shock (cf. 5.10), however this is not surprising since compression ratios larger than 4 are possible in this case and the cosmic ray pressure must remain finite (see Figure 5b). In fact, if $M_1^2 \gg 1$, $V_1/V_2 \to 7$ and $\nu \to 7/2(1 + (6/7)(1/6)) = 4$ which is to be compared with $\lambda_1 \to 4$ as $V_1/V_2 \to 4$; of course $p_{c2} \to \infty$ in either case.

On the basis of these results it appears that shocks associated

with supernova remnants can be very efficient in accelerating cosmic
rays. The transitions may be smooth in the very early and late phases,
but in general they contain an internal plasma shock which converts the
remaining flow kinetic energy largely into plasma internal energy. It is
not certain that smooth transitions really occur at early phases since
no dissipation of waves due to streaming instabilities has been included
and this could easily change the above result. The cosmic ray spectrum
at high energies should always be a power law but the index is not
necessarily that given by the linear theory, as suggested by the above
example. Finally, the cosmic rays may in some circumstances take up much
of the upstream kinetic energy available to the shock. This does not
mean that exceptional cosmic ray pressures are produced, rather the opposite, since the cosmic rays can at most absorb the energy available
whereas in the linear approach (Figure 5b) there is no limitation on the
energy given to cosmic rays. This is probably the resolution of the difficulty implied by Figure 10 which suggests that the long term variations of the cosmic ray intensity at the Sun should be much greater than
is observed (112,113,206).

In the above approach to the non-linear problem in which the diffusion coefficient $\underline{is}$ assumed known, it is necessary that a mean diffusion coefficient $\bar{\kappa}$ be given. Provided it is eventually possible to
solve for the cosmic ray spectrum, which determines the weighting in-
volved in $\bar{\kappa}$, this poses no real difficulty since $\bar{\kappa}$ can be merged with x
and only results in a distortion of the spatial form of the transition.
However, as far as the connection between cosmic rays and plasma is concerned (i.e. injection), the approach says nothing (200). An additional
theory must be constructed to determine how the cosmic ray and plasma
distribution functions merge, that is, to decide how much of the cosmic
ray pressure is to be provided by the acceleration of existing particles
or by the injection of fresh particles from the background. We are inclined to the view that this is perhaps the best approach since the
physics of the injection process may be very complicated, especially if
it involves the strong plasma turbulence associated with a collisionless
shock (12,13). On the other hand, it is interesting to consider a different approach where the details of the whole particle distribution
function are kept in view as far as possible and the problem not split
into two steps as described above. The difficulty with this procedure is
that the equation describing the distribution function is not easy to
handle and a set of integral-differential equations results if hydrodynamic equations are accepted wherever it is reasonable to do so.
Eichler (191) has obtained an approximate solution which represents some
progress in this direction but it will almost certainly be necessary to
fall back on a numerical treatment (193). An exact solution assuming
that the distribution function satisfies a Boltzman equation with a constant effective diffusion coefficient for all particles, has been obtained by Krimsky (194). This solution demonstrates that high energy
particles appear quite naturally out of the background plasma even when
the transition is completely smooth. A perturbation approach using the
linear or test particle solution as a basis has been attempted by Blandford (208) but the significance of the result is obscured by the fact
that the corresponding non-linear solution is not unique (200-202).

Although the results that have been obtained for the partially non-
linear configurations are interesting and instructive as far as the

central problem of cosmic ray acceleration is concerned, it is difficult
to avoid the conclusion that a completely non-linear treatment is necessary. That is, the entire complex represented in Figure 13 must be considered without making too drastic assumptions about any of its components. So far, it has been possible only to write down the relevant
set of equations (which is no easy matter in itself) and to obtain a
few rather specialised and simplified solutions (70,71,209). Very probably, it will eventually be necessary to resort to numerical methods.
However, we expect that the following features will emerge:
1) strong hydromagnetic turbulence will develop throughout the transition;
2) the shock will be smoothed out by cosmic ray pressure gradients;
3) much of the fluid kinetic energy will be transferred to cosmic rays;
4) a power law spectrum at high energies will result for plane, steady
transitions without losses;
5) damping of the hydromagnetic turbulence will lead to energy transfer
from the cosmic rays to the plasma so that the plasma may be efficiently
heated even in the absence of an internal collisionless shock.
The injection problem may eventually have to be treated separately but
it is to be expected that 1) a transition of finite width will tend to
favour acceleration of particles with large diffusion coefficients (191,
192) and 2) the turbulence generated in the transition, whatever form it
takes, will inevitably cause selective heating of the plasma so that
ions will be injected in a manner which depends on their current mass-
to-charge ratio (12,13). The latter may be the explanation for the observation that the relative elemental and isotopic source abundances of
galactic cosmic rays and of the energetic particles produced in large
solar flares are very similar (209), if both are the result of shock acceleration in media having "coronal" temperatures of the order of $10^6$ K.

## 11. Conclusions

Although the possibility of energetic particle acceleration in
shock waves has been investigated by numerous authors for more than
twenty years it is only comparatively recently that its more attractive
features, as far as the acceleration of cosmic rays is concerned, have
been generally realized. It must be accepted that the idea is not without its difficulties and perhaps unrecognized subtle features, but it is
surprising how much progress has been made. Whether or not all the theoretical and conceptual problems encountered so far can be resolved is
not clear but one can feel reasonably confident that the theory will not
be easy to reject.
    In our view, the immediate needs of the theory are: 1) to under-
stand the nature of hydromagnetic turbulence in the HISM as a whole;
2) to understand better the nature of fully non-linear cosmic ray shock
structures; 3) to investigate the expansion of supernova remnants without neglecting the cosmic rays as one of the most important components;
and 4) to understand better the nature of the galactic wind, its energy
sources and in particular the role played in its formation by cosmic
rays and extended supernova remnants. These are all very difficult problems and one should not expect rapid progress. In the meantime one is
free to speculate about the broader implications of shock acceleration
in the universe.

First we note that shock waves must be prevalent in the intergalactic medium, not only in obvious situations such as strong radio galaxies, but also in clusters of relatively normal galaxies. Shock waves originating in supernovae for example, can speed up and become strong when they encounter the dilute cluster medium and there is no reason to suppose therefore that there is not rather efficient acceleration of an extragalactic cosmic ray population in circumstances in which the losses due to universal expansion and other effects are minimized (12,13).

Secondly, shock acceleration must be considered as one of two distinct processes which give rise to solar energetic particles (the other process being magnetic field reconnection, especially at flare sites) (3,211,212). There is no lack of hydromagnetic turbulence in the solar corona and the shocks produced by flares can at times be extraordinarily strong (213). Except in the open magnetic field regions associated with coronal holes, solar flare produced shock waves must propagate for considerable distances through magnetically closed regions which could give rise to very efficient shock acceleration. These facts, together with the observation that the charge states of solar energetic particles often correspond to normal coronal and transition region temperatures (214) in contrast to very high temperatures observed in the vicinity of flares, is evidence in favour of shock acceleration. One would expect however that the gamma-ray emission associated with flares is the result of particles produced at the flare site, perhaps by magnetic reconnection, impinging on the lower atmosphere. It is interesting to note that, at least near the Sun, flare shock waves are likely to be highly non-linear in the sense discussed here.

## Acknowledgements

I wish to thank L.O'C. Drury, M.A. Forman, J.F. McKenzie, K. Richter, G.M. Webb, B. Wieser and H.J. Völk for their help in preparing this review.

## References

1) Lingenfelter, R.E.: 1979, 16th ICRC (Kyoto), 14, 135.
2) Krimsky, G.F.: 1977, Dok. Akad. Nauk., SSSR, 234, 1306.
3) Axford, W.I. et al.: 1976, EOS 57, 780; 15th ICRC (Plovdiv), 11, 132.
4) Bell, A.R.: 1978, MNRAS 182, 147 and 443.
5) Blandford, R.D. and J.P. Ostriker: 1978, Ap. J. 227, L49.
6) McKee, C.F. and J.P. Ostriker: 1977, Ap. J. 211, 148.
7) Johnson, H.E. and W.I. Axford: 1971, Ap. J. 165, 381.
8) Ipavich, F.M.: 1975, Ap. J. 196, 107.
9) Jokipii, J.R.: 1976, Ap. J. 208, 900.
10) Blandford, R.D.: 1979, AIP Conf. No. 56 (La Jolla), 333.
11) Toptygin, I.N.: 1980, Sp. Sci. Rev. 26, 157.
12) Axford, W.I.: 1980, IAU Symp. No. 94, 339.
13) Axford, W.I.: 1981, 10th Texas Symposium (in press).
14) Pesses, M.E. et al.: 1980, STIP Workshop (Smolenice); JHU/APL preprint 80-10.
15) "Upstream waves and particles": 1980, JGR 86, 4319.
16) Gold, T.: 1953, IAU Symp. No. 2, 103.
17) Blokh, Ya.L. et al.: 1959, 6th ICRC (Moscow), 4, 173.

18) Dorman, L.I.: 1963, Prog. El. Part. and CR Phys. 7, Section 4.7.
19) Dorman, L.I.: 1973, Comm. Astrophys. Sp. Phys. 5, 67.
20) Axford, W.I. and G.C. Reid: 1962, JGR 67, 1692.
21) Axford, W.I. and G.C. Reid: 1963, JGR 68, 1793.
22) Bryant, D.A. et al.: 1962, JGR 67, 4983.
23) Vernov, S.N. et al.: 1970, "Intercorrelated Satellite Observations
    Related to Solar Events", (Reidel), 53.
24) Richter, A.K. and E. Keppler: 1977, J. Geophys. 42, 645.
25) Hovestadt, D. et al.: 1981, 17th ICRC (Paris), 3, 451.
26) Barnes, G.W. and J.A. Simpson: 1976, Ap. J. 210, L91.
27) McDonald, F.B. et al.: 1976, Ap. J. 203, L149.
28) Hamilton, D.C. et al.: 1979, 16th ICRC (Kyoto), 5, 363.
29) Asbridge, J.R., S.J. Bame and I.B. Strong: 1968, JGR 73, 5777.
30) Lin, R.P., C.I. Meng and K.A. Anderson: 1974, JGR 79, 489.
31) Scholer, M. et al.: 1980, Geophys. Res. Lett. 7, 73.
32) Zwickl, R.D. et al.: 1980, Geophys. Res. Lett. 7, 453.
33) Zwickl, R.D. et al.: 1981, JGR 86,
34) Sarris, E.T. et al.: 1976, Geophys. Res. Lett. 2, 133.
35) Singer, S. and M.D. Montgomery: 1971, JGR 76, 6628.
36) Blokh, G.M. et al.: 1975, Cosmic Res. 13, 695.
37) Gosling, J.T. et al.: 1980, JGR 85, 744.
38) Anderson, K.A. et al.: 1972, NASA SP-315, 22-1.
39) Gosling, J.T. et al.: 1978, Geophys. Res. Lett. 5, 957.
40) Bonifazi, C., et al.: 1980, JGR 85, 3461.
41) Dorman, L.I. and G.I. Freidman: 1959, "Problems of MHD and Plasma
    Physics" (Riga), 77.
42) Shabansky, V.P.: 1961, JETP 41, 1107.
43) Parker, E,N.: 1963, "Interplanetary Dynamical Processes", (N.Y.,
    Interscience).
44) Wentzel, D.G.: 1962, Ap. J. 137, 135.
45) Wentzel, D.G.: 1964, Ap. J. 140, 1013.
46) Hudson, P.D.: 1965, MNRAS 131, 23.
47) Hoyle, F.: 1960, MNRAS 120, 338.
48) Schatzmann, E.: 1963, Ann. d'Astrophys. 137, 135.
49) Jokipii, J.R. and L. Davis, Jr.: 1964, Phys. Rev. Lett. 13, 739.
50) Jokipii, J.R.: 1966, Ap. J. 143, 961.
51) Van Allen, J.A. and N.F. Ness: 1967, JGR 72, 935.
52) Parker, E.N.: 1965, Plan. Sp. Sci. 13, 9.
53) Dolginov, A.Z. and I.N. Toptygin: 1966, JETP 51, 1771.
54) Dolginov, A.Z. and I.N. Toptygin: 1968, Icarus 8, 54.
55) Hasselmann, K. and G. Wibberenz: 1968, Z. Geophys. 34, 353.
56) Gleeson, L.J. and W.I. Axford: 1967, Ap. J. 149, L115.
57) Fisk, L.A.: 1969, Ph.D. Thesis, UCSD.
58) Fisk, L.A. and W.I. Axford: 1969, Trans. AGU 50, 307.
59) Morfill, G.E. and M. Scholer: 1977, Astrophys. Sp. Sci. 46, 73.
60) Ostriker, J.P.: 1979, 16th ICRC (Kyoto), 2, 124.
61) Blandford, R.D. and J.P. Ostriker: 1980, Ap. J. 237, 793.
62) Linsley, J.: 1980, Proc. IAU Symp. No. 94, 53.
63) Grigorov, N.L. et al.: 1971, 12th ICRC (Hobart), 5, 1746.
64) Yodh, G.B.: 1981, U. Maryland preprint.
65) Wolfendale, A.W.: 1977, 15th ICRC (Plovdiv), 10, 235.
66) Kiraly, P. et al.: 1974, Rev. del Nuovo Cimento 2, 1.
67) Ginzburg, V.L. and V.S. Ptuskin: 1981, 17th ICRC (Paris), 2, 336.

68) Cesarsky, C.J. and P.O. Lagage: 1981, 17th ICRC (Paris), 2, 335.
69) Lee, M.A.: 1981, 17th ICRC (Paris), in press.
70) McKenzie, J.F. and H.J. Völk: 1981, 17th ICRC (Paris), in press.
71) Völk, H.J. and J.F. McKenzie: 1981, 17th ICRC (Paris), in press.
72) Hall, A.N.: 1980, MNRAS 190, 335, 371, 385; 191, 739, 751.
73) Hall, A.N.: 1981, MNRAS, in press.
74) Urch, I.H. and L.J. Gleeson: 1972, Astrophys. Sp. Sci. 16, 55.
75) Ormes, J.F. and P.S. Freier: 1978, Ap. J. 222, 471.
76) Simon, M. et al.: 1980, Ap. J. 239, 712.
77) Cameron, A.G.W.: 1973, Sp. Sci. Rev. 15, 121.
78) Waddington, C.J.: 1977, 15th ICRC (Plovdiv), 10, 168.
79) Balasubrahmanian, V.K.: 1979, 16th ICRC (Kyoto), 14, 121.
80) Meyer, P.: 1980, IAU Symp. No. 94, 7.
81) Meyer, J.P., M. Cassé and H. Reeves: 1979, 16th ICRC (Kyoto) 12, 108.
82) Wiedenbeck, M.E. and D.E. Greiner: 1981, 17th ICRC (Paris) 2, 76.
83) Wiedenbeck, M.E. and D.E. Greiner: 1980, Ap. J. 239, L139.
84) Reeves, H.: 1980, IAU Symp. No. 94, 23.
85) Meyer, J.P.: 1981, 17th ICRC (Paris), 2, 265.
86) Ginzburg, V.L. and S.I. Syrovatskyi: 1964, "The Origin of Cosmic Rays", (Pergamon Press).
87) Cowsik, R. and M.A. Lee: 1977, Ap. J. 216, 635; 1979 ibid. 228, 297.
88) Giler, M. et al.: 1980, Astron. Astrophys. 84, 44.
89) Cowsik, R.: 1980, IAU Symp. No. 94, 93.
90) Taira, T. et al.: 1979, 16th ICRC (Kyoto), 1, 478.
91) Juliusson, E.: 1974, Ap. J. 191, 331.
92) Garcia-Munoz, M. et al.: 1979, 16th ICRC (Plovdiv), 1, 310.
93) Hagen, F.A., A.J. Fisher and J.F. Ormes: 1977, Ap. J. 212, 262.
94) Protheroe, R.J., J.F. Ormes and G.M. Comstock: 1981, Ap. J. 247, 362.
95) Ormes, J.F. and R.J. Protheroe: 1981, 17th ICRC (Paris), 2, 31.
96) Hayakawa, S.: 1969, "Cosmic Ray Physics", (Wiley, N.Y.), 551.
97) Sitte, K.: 1979, 16th ICRC (Kyoto), 2, 168.
98) Cowsik, R.: 1980, Ap. J. 241, 1195.
99) Fransen, C. and R.I. Epstein: 1980, Ap. J. 242, 411.
100) Raisbeck, G.M.: 1979, 16th ICRC (Kyoto), 14, 146.
101) Orth, C.D. et al.: 1978, Ap. J. 226, 1147.
102) Protheroe, R.J. and J.F. Ormes: 1980, IAU Symp. No. 94, 107.
103) Garcia-Munoz, M. et al.: 1977, Ap. J. 217, 859.
104) Webber, W.R. and J. Kish: 1979, 16th ICRC (Kyoto), 1, 389.
105) Garcia-Munoz, M. et al.: 1981, 17th ICRC (Paris), 2, 72.
106) Burton, W.B. and H.S. Liszt: 1980, IAU Symp. No. 94, 227.
107) Haslam, C.G.T. et al.: 1980, IAU Symp. No. 94, 217.
108) Scarsi, L. et al.: 1980, IAU Symp. No. 94, 279.
109) Wolfendale, A.W.: 1980, IAU Symp. No. 94, 309.
110) Kearsey, S. et al.: 1980, IAU Symp. No. 94, 223.
111) Issa, M.R. et al.: 1981, J. Phys. G. Nucl. Phys. 7, 973.
112) Forman, M.A. and O.A. Schaeffer: 1979, Rev. Geo. Sp. Phys. 17, 552.
113) Somayajulu, B.L.K.: 1977, Geochim. et Cosmochim. Acta 41, 909.
114) Eichler, D.: 1980, Ap. J. 237, 809.
115) Cox, D.P. and B.W. Smith: 1974, Ap. J. 189, L105.
116) McCray, L. and T.P. Snow: 1979, Ann Rev. Astron. Astrophys. 17, 213.
117) McKee, C.F.: 1981, 17th ICRC (Paris), in press.
118) Cowie, L.L. et al.: 1981, Ap. J. 247, 908.
119) Tamman, G.: 1974, "Supernovae and Supernova Remnants", (Reidel), 155.

120) Cassé, M. and J.A. Paul: 1980, Ap. J. 237, 236.
121) Biermann, L.: 1948, Z. Astrophysik 25, 161.
122) Parker, E.N.: 1958, Phys. Rev. 109, 1328.
123) Alexeyev, I.I. and A.P. Kropotkin: 1970, Geomag. Aeron. 10, 755.
124) Sarris, E.T. and J.A. Van Allen: 1974, JGR 79, 4157.
125) Pesses, M.E.: 1981, JGR 86, 150.
126) Chirkov, A.G. et al.: 1981, 17th ICRC (Paris), 3, 500.
127) Alexeyev, I. et al.: 1970, Is. Ak. Nauk. SSSR, Sov. Phys. 34, 2318.
128) Vasil'yev, V.N. et al.: 1978, Proc. 9th Leningrad Seminar, 159.
129) Chen, G. and T.P. Armstrong: 1975, 14th ICRC (Munich), 5, 1814.
130) Terasawa, T.: 1979, Planet. Sp. Sci. 27, 193 and 365.
131) Quenby, J.A. and S. Webb: 1973, 13th ICRC (Denver), 2, 1343.
132) Decker, R.B.: 1981, JGR 86, 4537.
133) Pesses, M.E.: 1979, 16th ICRC (Kyoto), 2, 18.
134) Sonnerup, B.U.O.: 1969, JGR 74, 1301.
135) Scholer, M. et al.: 1980, JGR 85, 4602.
136) Greenstadt, E.W. et al.: 1980, JGR 85, 2124.
137) Krimsky, G.F. et al.: 1979, 16th ICRC (Kyoto), 2, 39.
138) Yelshin, V.K. et al.: 1979, Geomag. Aeron. 19, 533.
139) Peacock, J.E.: 1981, MNRAS 196, 135.
140) Michel, C.F.: 1981, Ap. J. 247, 664.
141) Chandrasekhar, S.: 1943, Rev. Mod. Phys. 15, 1.
142) Jokipii, J.R.: 1971, Rev. Geophys. 9, 27.
143) Gleeson, L.J. and W.I. Axford: 1968, Astrophys. Sp. Sci. 2, 431.
144) Forman, M.A.: 1970, Planet. Sp. Sci. 18, 25.
145) Webb, G.M.: 1981, in preparation.
146) Forman, M.A.: 1981, 17th ICRC (Paris), 3, 467.
147) Fisk, L.A.: 1971, JGR 76, 1662.
148) Vasil'yev, V.N. et al.: 1980, Cosmic Res. 18, no. 3.
149) Forman, M.A. and G. Morfill: 1979, 16th ICRC (Kyoto), 5, 328.
150) Bulanov, S.V. and V.A. Dogiel: 1979, 16th ICRC (Kyoto), 2, 70.
151) Bulanov, S.V. and V.A. Dogiel: 1979, Sov. Ast. Lett. 5, 278.
152) Völk, H.J. et al.: 1979, 16th ICRC (Kyoto), 2, 38a.
153) Völk, H.J. et al.: 1980, IAU Symp. No. 94, 359.
154) Völk, H.J. et al.: 1981, Ap. J., in press.
155) Kaplan, S.A. and S.B. Pikelner: 1970, "The Interstellar Medium",
        (Harvard, U.P.), 244-247.
156) Forman, M.A.: 1981, Adv. Space Res. 1, 97.
157) Krimsky, G.F. et al.: 1979, 16th ICRC (Kyoto), 2, 75.
158) Tverskoy, B.A.: 1967, JETP 53, 1417.
159) Tverskoy, B.A.: 1969, Proc. 1st Leningrad Seminar, 159.
160) Tverskoy, B.A.: 1971, Proc. 2nd Leningrad Seminar, 7.
161) Webb, G.M.: 1981, in preparation.
162) Fisk, L.A. and M.A. Lee: 1980, Ap. J. 237, 620.
163) Krimsky, G.F. et al.: 1979, 16th ICRC (Kyoto), 2, 44.
164) Axford, W.I.: 1972, "Solar Wind", NASA SP-308, 609.
165) Völk, H. and M. Forman: 1981, Ap. J., in press; 17th ICRC (Paris).
166) Petukhov, S.I. et al.: 1981, 17th ICRC (Paris), 3, 460.
167) Webb, G.M. et al.: 1981, 17th ICRC (Paris), 2, 309.
168) Forman, M.A. et al.: 1981, 17th ICRC (Paris), 2, 313.
169) Webb, G.M. et al.: 1981, Astrophys. Sp. Sci., submitted.
170) Cowsik, R. and M.A. Lee: 1981, 17th ICRC (Paris), 2, 318.
171) Cassé, M. and J.A. Paul: 1979, 16th ICRC (Kyoto), 2, 103.

172) Montmerle, T.: 1979, Ap. J. 231, 95.
173) Katafos, M. et al.: 1981, 17th ICRC (Paris), 2, 348.
174) Lee, M.A. and L.A. Fisk: 1981, 17th ICRC (Paris), 3, 405.
175) Courant, R. and K.O. Friedrichs: 1948, "Supersonic Flow and Shock
     Waves", (Interscience).
176) Bykov, A.M. and I.N. Toptygin: 1979, 16th ICRC (Kyoto), 3, 66.
177) Bykov, A.M. and I.N. Toptygin: 1981, 17th ICRC (Paris), 2, 331.
178) Garcia-Munoz, M. et al.: 1981, 17th ICRC (Paris), in press.
179) Gaisser, T.K. et al.: 1980, IAU Symp. No. 94, 257.
180) Szabelski, J. et al.: 1980, IAU Symp. No. 94, 259; Nature 285, 386.
181) Tan, L.C. and L.K. Ng: 1981, 17th ICRC (Paris), 2, 210.
182) Buffington, A. and S.M. Schindler: 1981, Ap. J. 247, L105.
183) Krimsky, G.F. and S.I. Petukhov: 1981, 17th ICRC (Paris), 3, 503.
184) Jokipii, J.R.: 1977, 15th ICRC (Plovdiv), 1, 429.
185) Völk, H.J.: 1981, in preparation.
186) Lee, M.A. and H.J. Völk: 1973, Astrophys. Sp. Sci. 24, 31.
187) Cesarsky, C.J.: 1980, Ann. Rev. Astron. Astrophys. 18, 289.
188) McKenzie, J.F. et al.: 1981, in preparation.
189) Morrison, P.J. et al.: 1980, NASA-TM 82032.
190) Zweibel, E.: 1979, AIP Conf. No. 56 (La Jolla).
191) Eichler, D.: 1979, Ap. J. 229, 419.
192) Eichler, D. and K. Heinebach: 1981, preprint.
193) Ellison, D.C. et al.: 1981, J. Geophys., in press.
194) Krimsky, G.F.: 1980, Eur. CRC (Leningrad), in press.
195) McKenzie, J.F. and K.D. Westphal: 1968, Phys. Fluids 11, 2350.
196) McKenzie, J.F. and K.D. Westphal: 1969, Planet. Sp. Sci. 17, 1029.
197) McKenzie, J.F. and M. Bornatici: 1974, JGR 79, 4589.
198) Scholer, M. and J.W. Belcher: 1971, Solar Phys. 16, 472.
199) Chao, J.K. and B. Goldstein: 1972, JGR 75, 6394.
200) McKenzie, J.F. et al.: 1981, Astrophys. Sp. Sci., submitted.
201) Drury, L.O'C. and H.J. Völk: 1980, IAU Symp. No. 94, 363.
202) Drury, L.O'C. and H.J. Völk: 1981, Ap. J. 248, 344.
203) Drury, L.O'C. et al.: 1981, 17th ICRC (Paris), 2, 327.
204) Drury, L.O'C. et al.: 1981, MNRAS, in press.
205) Achterburg, A. and C.A. Norman: 1980, unpublished report.
206) Webb, G.M. and W.I. Axford: 1981, in preparation.
207) Blandford, R.: 1980, Ap. J. 238, 410.
208) Achterburg, A.: 1981, Astron. Astrophys. 98, 195.
209) Meyer, J.P.: 1981, 17th ICRC (Paris), 2, 265.
210) Axford, W.I.: 1977, "Study of Travelling Interplanetary Phenomena",
     (Reidel), 145.
211) Achterburg, A. and C.A. Norman: 1980, Astron. Astrophys. 89, 353.
212) Lee, M.A. and L.A. Fisk: 1981, 17th ICRC (Paris), 3, 405.
213) Woo, R. and J.W. Armstrong: 1981, Nature 292, 608.
214) Gloeckler, G. et al.: 1981, 17th ICRC (Paris), 3, 136.
215) Webber, W.R.: 1973, 13th ICRC (Denver), 5, 3499.
216) Grigorov, N.L. et al.: 1971, 12th ICRC (Hobart), 5, 1746.
217) Grigorov, N.L. et al.: 1971, 12th ICRC (Hobart), 5, 1760.

THE SOLAR MAXIMUM MISSION

G.M.Simnett
Department of Space Research
University of Birmingham
England.

ABSTRACT

During its period of full operation from February -
December, 1980 the Solar Maximum Mission spacecraft
has made significant inroads into many of the outstanding problems in solar physics. The array of six
major instruments, pointed by a 3-axis stabilized
platform to better than an arc-second, have studied in
a co-ordinated way solar flares, active regions,sunspots,
limb phenomena and coronal transients at wavelengths
ranging from the visible to the gamma ray region of the
electromagnetic spectrum. A seventh instrument has
monitored the solar constant. This review outlines the
main objectives of the mission and samples the interesting
data sets analysed so far.

## 1. Introduction

The Solar Maximum Mission (SMM) spacecraft represents the best co-
ordinated and most comprehensive set of instruments ever assembled with
the single purpose of studying the active sun. This NASA mission was
launched from Cape Canaveral, Florida, on February 14, 1980, and for the
rest of the year accumulated a wealth of data on active regions, flares,
sunspots and coronal transients complemented by the world-wide array of
ground based observatories under the auspices of the organizers of the
Solar Maximum Year. Analysis of the results is an exciting but time
consuming task and so far only a small fraction of the data has even
been studied and an even smaller fraction analysed exhaustively. In
this review we first outline the scientific aims of the mission; then
we describe briefly the instruments on the satellite, and the unique
way the mission was operated; finally we discuss some of the most
interesting results obtained so far.

## 2. The Scientific Goals

The sun, as our nearest star,offers us a laboratory in which we may
study atomic physics, plasma physics and magnetohydrodynamics under
conditions not even closely approached on earth. Solar flares represent
impulsive energy releases of $10^{29} - 10^{32}$ ergs, plasma temperatures
exceed $10^8$ K and charged particles are given energies only recently
exceeded by ingenious man-made accelerators. At such times the
elemental constituents of the solar atmosphere may be completely
stripped of their atomic electrons. The energy source is in some way

related to the sudden disruption of a large scale magnetic loop, such
as those frequently found in the vicinity of sunspots, where the typical
scale may be 5 x 10⁴ km and the field strength may be in the kilogauss
region.   What happens in the first instants of this cataclysmic
process is, like the Creation itself, still somewhat speculative, but in
the expansion that follows the formation of such high temperatures,
shocks are formed, energetic particles are accelerated and material is
ejected. Every conceivable type of electromagnetic radiation is emitted.
Each discipline has its own rather special interest in the physical
processes surrounding solar flares, but the common resolve is to understand exactly what is happening at the point of instability that triggers
a major flare.

Prior to SMM the most recent comprehensive attacks on problems in
solar physics were through the Skylab workshops, culminating in the
publication of "The Proceedings of the Second Skylab Workshop on Solar
"Flares", ed. P.A. Sturrock, 1980.   Skylab operated during the declining
part of the last solar cycle, and the experiments were particularly
suited to studying the later, slower, more thermal phases of flares.  In
contrast SMM was aimed at the peak of solar cycle 21 and several
experiments were tuned to the study of the impulsive phase of large
flares.  During operations in 1980 many medium to large flares were
studied  by the whole observatory, although the very large flares
remained elusive.

To put the SMM in perspective, it is useful to summarize what we
know already about solar flares, so that it will be apparent what new
information is necessary for advancement.  Large solar flares occur in
active regions containing sunspot groups, often groups with two dominent
spots of opposite magnetic polarity.   Sub-photospheric motions may
cause twisting of the magnetic fields, leading to regions with high
shear, such as those observed in the August 1972 flares (e.g.Zirin and
Tanaka, 1973).  The flare energy is supplied by the magnetic field, and
somehow magnetic field annihilation produces current  or current sheets
which both heat the ambient plasma-through Joule heating and accelerate
ions and electrons through mechanisms not totally understood to
energies leading to significant departures of their velocity distribution
from Maxwellian - hence the term non-thermal is used to describe them.

Large flares frequently appear in H$\alpha$ as two ribbons, on each side
of the magnetic neutral line threading the active region at photospheric
level.  These ribbons are likely to be the footpoints of arcades of
magnetic loops.   At photospheric level, the temperatures are such that
spectral lines are excited in the visible region of the spectrum, thus
making them suitable for study by ground based observatories.  As we
pass through the transition region, where the temperature rises from
4500 K - 10⁶ K, line emission in the UV to soft X-ray range is produced.
In the dynamic situation at the time of a flare, the normal local
temperature distribution is completely changed.  Hot plasma fills the
magnetic loops, such that they may suddenly be seen in emission in soft
X-rays, only becoming visible in H$\alpha$ as they cool.   Thus H$\alpha$ emission
while being easily studied from the ground, is really looking at the
final stages of the magnetohydrodynamic disturbance we call a flare
while the high temperature signatures are at wavelengths easily absorbed
by the earth's atmosphere.  It is for this reason that spacecraft
observations are so valuable.

Following the trigger, most of the flare energy is transferred
to non-thermal electrons, which interact with the ambient medium to
produce the hard X-ray burst and with the overlying magnetic fields to
produce the radio emission.   Theoretical studies have studied the
individual phenomena in great detail, but the lack of good temporal and
(3 dimensional) spatial information presents the theorist with far too
many free parameters to build a truly convincing theory of flares.   The
temperature, density, elemental composition and magnetic field are
additional parameters required to describe the plasma state fully. When
we recognize further that the solar flare is strictly a non-equilibrium
situation, involving mass motions, then the magnitude of the problem is
only too obvious.   The flare build up is not yet understood fully,
especially with regard to possible slow energy releases prior to the
flash phase of the flare; the acceleration of the protons is a problem;
and the role of the corona above the flare site is unclear.
      The prime objective of the SMM is to make simultaneous measurements
to define as many as possible of the above parameters with the hope
that by the end of the day solar flares will be less of an enigma than
they were.

3.   The SMM Payload

      SMM observes electromagnetic radiation in wavelengths ranging from
the $H_\alpha$ line at 6563Å to the gamma ray region of the spectrum, with
equivalent photon energies up to $\approx$ 10 MeV.   The regions in the visible
spectrum are, in fact, confined in the spacecraft observations to the
corona beyond1.25 $R_\odot$, but of course, in general excellent coverage of
the solar disc in this region is available from the world-wide network
of ground based observations.   Also, because of technical difficulties,
there was no coverage in the EUV, 1100 - 22.4Å.
      The instruments have been described in Solar Physics, 65, 1980, and
in references therein, but it is useful here to outline the principle
characteristics of each one, together with specific experiment goals.
The principle investigator(s) and participating institutions are given
after each title.

3a   The Gamma Ray Spectrometer (GRS) E.L. Chupp,   U.New Hampshire
      (Max Planck Institute, Garching; Naval Research Lab, Washington)

      This consists of 7 actively shielded high resolution NaI(Tl)
detectors operating between 0.3 and 9 MeV, with an energy resolution
better than 7% at 0.662 MeV and an effective area of 200 $cm^2$ at 0.3 MeV.
Time resolution may be as high as 2s. In addition there is a 7.6 cm thick
CsI (Na) crystal behind the spectrometer which responds to 10 - 100 MeV
photons (covering the $\pi^0$ decay) and in principle neutrons $\gtrsim$ 20 MeV.
The effective area of this system is 100 $cm^2$ for gamma rays and   7 $cm^2$
for neutrons.   Completing this instrument is an 8 $cm^2$ X-ray detector
covering the 10 - 140 keV range.
      One of the prime objectives is to establish the time sequence of
acceleration of energetic ( $>$ 30 MeV) protons relative to other flare
radiations.   By virtue of its improved sensitivity over previous
detectors it is able to identify better flares which accelerated high
energy nucleons or in which such nucleons were present.   It will measure

the intensity, energy and Doppler shift of narrow gamma ray lines, such as the prompt de-excitation lines of $^{12}C$ at 4.44 MeV, $^{16}O$ at 6.13 MeV and the neutron capture line at 2.223 MeV.

3b    The Hard X-ray Burst Spectrometer (HXRBS) K.J.Frost,
       Goddard Space Flight Center.

This is a well proven instrument, similar to that flown on OSO-5, consisting of an actively collimated CsI (Na) crystal, sensitive in the range 26 - 380 keV. The sensitive area is 71 $cm^2$ and one of the new features is the ability to accumulate data for short periods (few x 10s) with a time resolution of 1 ms.    In its standard operating mode 15 channel pulse height spectra are generated every 128 ms and the electronics are fast enough to accommodate counting rates in excess of $10^5$/s without spectral distortion.    Both this instrument and GRS have fields of view encompassing the whole sun and have no intrinsic spatial resolution.

One of the objectives of this experiment is to differentiate and identify the first and second stages of acceleration in flares.  In the first impulsive stage the millisecond time resolution enables a comparison to be made between the hard X-ray and other correlated emissions, such as microwave radiation, which originate at different electron densities, and therefore different heights.  The second stage acceleration produces an X-ray burst coincident with type II and IV radio bursts; the spectral, temporal and intensity characterisitics should be quite different to the impulsive burst.

3c    The Hard X-ray Imaging Spectrometer (HXIS) C de Jager,
       Space Research Lab, Utrecht (University of Birmingham)

This is a completely novel state-of-the-art instrument and is the first spaceborne detector capable of imaging X-rays in the 3.5 - 30 keV region.    It consists of a multiplate collimator divided in 576 sections, or subcollimators, each guiding X-rays to an individual point anode,Xe-$CO_2$ filled proportional counter.    Each subcollimator images a slightly different part of the total field of view. The full collimator array images a 6' 24" x 6' 24" area with 32" resolution, within which is an area of 2'40" x 2'40" imaged with 8" resolution. A schematic view of the imaging collimator and detector system is illustrated in Fig.1, which shows inset a cut-away section of the miniproportional counter array, a subcollimator pattern and the projection of the fields of view onto the sun.    X-ray images are accumulated simultaneously in each of six energy bands. The accumulation time for each set of images is variable between 1.5 and 7.5 s; this, and other complex instrument functions are under the control of a sophisticated microprocessor controlled data handling system.    One important function of HXIS is to provide the spacecraft with a flare alert which gives not only the intensity of the flare but the co-ordinate of the brightest point in soft X-rays (generally 3.5 - 5.5 keV) relative to the HXIS boresight.

The prime objective of HXIS is to try to resolve the controversy between thermal/non thermal origins of hard X-rays.    This is funda-mental to the estimate of the energy content of the electrons

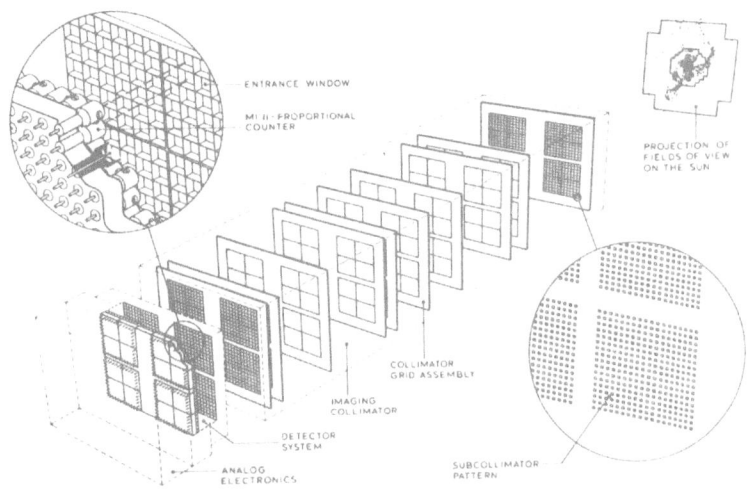

ENTRANCE WINDOW

MULTI-PROPORTIONAL COUNTER

PROJECTION OF FIELDS OF VIEW ON THE SUN

COLLIMATOR GRID ASSEMBLY

IMAGING COLLIMATOR

DETECTOR SYSTEM

ANALOG ELECTRONICS

SUBCOLLIMATOR PATTERN

<u>Fig.1</u> An open schematic view of the Hard X-ray Imaging Spectrometer(HXIS)

responsible for the hard X-rays; if they arise through heating of a
small volume of plasma to temperatures $\gg 10^7$ K, the energy content is
small compared to that required to produce the X-rays via a beam of
fast electrons.   The good spatial resolution will locate the hard X-ray
emission within the active region structure as determined by other
instruments; subsequent development will show how the "thermal" phase
of the flare relates to this.   HXIS can locate the source of the X-ray
bursts seen with better statistical accuracy by HXRBS.

3d    The X-ray Polychromator (XRP) L.W.Acton, J.L.Culhane, A.H.Gabriel;
      Lockheed;Mullard Space Science Lab. U. of London; Appleton Lab.

      The XRP consists of two complementary instruments, the Flat and Bent
Crystal Spectrometers (FCS and BCS) which investigate properties of
plasmas in the 1.5 - 50 million K range.   The principle of operation of
both the FCS and BCS is shown in Fig.2.   The FCS consists of seven
rotatable flat crystals which receive photons through a 14" x 14"
collimator which can be rastered over as much as a 7' x 7' area of the
sun.   Seven spectral lines may be analysed simultaneously, while by
rotating the crystals, the instrument becomes a scanning spectrometer.
covering the wavelength range 1.4 - 22.4Å.   The basic data gathering
interval is 0.256 s/pixel and the basic raster step size is 5".   Therefore the rate of accumulation of images is a function of the total
rastered area and the number of steps/pixel; a typical 3' x 3' raster
takes $\lesssim$ **90 s at 2 steps/pixel**.   Upon receipt of either a BCS or HXIS
flare alert, a raster of chosen size may be executed about the brightest
(or any other) point.
      The BCS is a dispersive Bragg spectrometer, reflecting different

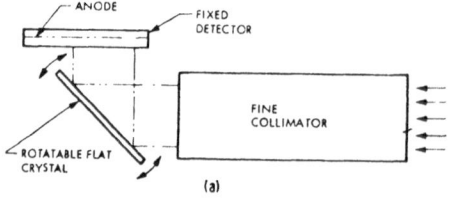

Fig.2 The operating principle of

(a) the Flat Crystal Spectrometer (FCS) and

(b) the Bent Crystal Spectrometer (BCS)

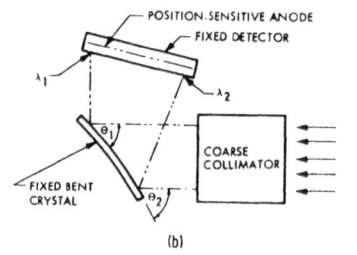

wavelengths into a position sensitive proportional counter. It has no imaging capability and accepts photons from a 6' x 6' field of view. There are eight individual crystals and counters, one of which studies primarily the Ca XIX line while the other seven concentrate on details of the iron line spectrum between 1.7 and 2.0Å. Spectral evolution during flares is so rapid that the important iron line region cannot be covered adequately by the scanning FCS. The XRP is under microprocessor control, which allows selection of wavelength and field of view for the FCS and permits an exchange of time resolution for spectral coverage for the BCS; the ultimate time resolution is 64 ms.

XRP is designed to study the physical conditions in the active region plasma both during a flare and in the pre-and post-flare stages. The spectral lines studied by the instrument provide a number of important temperature and density diagnostics which, when combined with the latest analytical techniques should determine the plasma state to a significantly better degree than previously. The broadness of the lines can establish the degree of turbulence, or the degree of thermal Doppler broadening depending on the cause. The iron lines can determine the degree of collisional excitation, and hence indicate the presence of non-thermal electrons. In addition to solar physics studies, XRP will contribute to spectroscopy and atomic structure physics.

3e  The Ultraviolet Spectrometer and Polarimeter (UVSP) E.A.Tandberg-Hanssen. Marshall Space Flight Center.(Goddard Space Flight Center)

This is a modification of the Colorado Univ. OSO-8 flight spare unit. The basic telescope is Gregorian with a rastered secondary mirror and an entrance aperture of 66.4 cm². Following the telescope is a 1 m Ebert spectrometer, covering 1150 - 1800Å in second order and 1750 - 3600Å in first order. The grating may be rotated for wavelength selection. Either of two rotatable waveplates may be inserted into the beam in the spectrometer for polarization studies, for which the grating serves as the analyser. There are 22 possible entrance/exit slit combinations feeding five separate detectors, which makes the UVSP extremely versatile

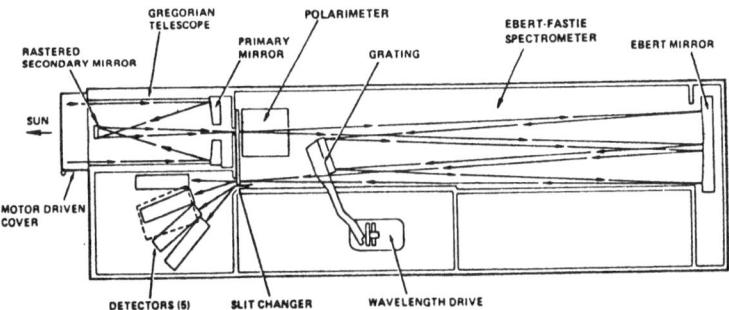

<u>Fig.3</u> Layout of SMM Ultraviolet Spectrometer and Polarimeter, showing the optical arrangement and major components.

but highly complex. The UVSP is illustrated schematically in Fig.3. Observations may be made with specific sets of 4 lines simultaneously or with both sides of 2 lines simultaneously for velocity and polarization measurements. The raster range may be any subset of a 256" x 256" field of view, which in turn may be offset from the spacecraft boresight if required. For co-ordinated studies UVSP will respond to the HXIS flare alert. For practical purposes the best resolution is 3". Time resolution is only dependent on the number of steps in the raster; an 8 x 8 raster takes 4.1s, resulting from a intrinsic time resolution of 64 ms. Instrument control is through a microprocessor.

The observational goal of the UVSP is to determine the physical properties of the solar flare plasma, namely temperature, density velocity and magnetic field as a function of space and time. The wavelength coverage is particularly suited to studies of the transition region and coronal lines, from atomic species only partially ionized. Use of the polarimeter permits determination of the 4 Stokes parameters- intensity, linear polarization (2) and circular polarization - from which the magnetic field vector can be deduced. The velocity measurements, from comparisons of the red and blue components of a spectral line split in the center, enable mass motions to be studied. Measurements in lines such as C IV formed at $\approx 10^5$ K, will study the flare build-up phase.

3f    The Coronagraph/Polarimeter (C/P). L.L. House,
      High Altitude Observatory.

The C/P represents a substantial improvement of the externally occulted Lyot coronagraph flown on Skylab. In addition to improved optics, it records coronal images on a vidicon detector rather than on photographic film. The image is transferred digitally onto a tape recorder, which has a capacity of $\approx 50$ complete high resolution pictures, and the recorder is read out to a ground station every few hours. The field of view is a quadrant of radial extent 1.6 - 6.0 $R_0$, or smaller under computer command, with a resolution of 10". Pointing is independent of the spacecraft, and the whole instrument is under the control of the SMM on-board computer. There is a range of spectral filters, including one to isolate the Fe XIV 5303Å coronal emission line;

polarization measurements on this are used to determine the direction
of the coronal magnetic field. The filters are also useful in
separating the K and F coronal constituents, which arise from scattering
of photospheric light off free electrons and heavy particles respectively.

The primary objective is to study the ejection of mass from the sun,
either after a flare or from an eruptive prominence. It was apparent
from prior observations that a coronal transient could contain kinetic
energy exceeding the net radiative output of a flare by over a factor of
2 (Hundhausen, 1972) and that the material was principally of coronal
origin (Gosling et al, 1975). Through careful monitoring of transients,
and correlation with other experiments, C/P hopes to gain further insights
into the physical processes initiating transients. The polarization
studies yield the best information on the magnitude of the coronal field
when combined with high (spatial) resolution metric radio observations.
This can determine if the ejection is driven primarily by mechanical
forces or magnetic forces. These studies will also provide a test of the
hypothesis that dense coronal structures outline the coronal field.

The Active Cavity Radiometer (ACRIM) R.C.Willson, Jet Propulsion Lab.

This instrument is designed to measure the total solar irradiance
or the solar constant with an accuracy better than 0.1%. It is part
of a long term goal to study solar irradiance over a 22 year magnetic
solar cycle. It consists of three independent, self calibrating
pyrheliometers, each of which views the sun through a $5^\circ$ full angle field
of view. The instrument is described in detail by Willson (1979).

## 4  Mission Operations

SMM was designed to make coordinated observations of solar
phenomena. HXIS, XRP and UVSP only view a small fraction of the sun, so
considerable care was taken to optimize the spacecraft pointing based on
the state of the sun and current scientific objectives. Either could
influence the other. Also, as these instruments were all highly
versatile their precise hour by hour configuration, via software loads
to the microprocessors, depended on the current scientific plans.

Early in the mission planning it was decided to conduct all
scientific operations and initial data evaluation through an Experiment
Operations Facility (EOF) at Goddard Space Flight Center. This contained
a forecasting office, operated by NOAA, using real time solar data from
both ground based observatories and a monitoring satellite, strong
resident scientific and technical teams from all instruments, a ground
based observatory co-ordinator and the facilities to generate command
loads via the Mission Control Center. The operational cycle was 24
hours, and a daily planning meeting in the EOF, attended by the
scientific staff, decided on the basis of the current state of the sun,
spacecraft and instrument status, requests from ground based observatories and guest investigators, and their own guidelines what the plan
should be for the next operational day. These plans included decisions
on the use of the HXIS flare flag. Following the planning meeting each
team would prepare a command load to configure its instrument to support
the following day's plan. Quick-look and real-time data came to the EOF
where it was evaluated by each scientific team using its own dedicated

computer. Limited capability to adjust instrument parameters existed in real time. There was also a capability (rarely used) to re-point the spacecraft within an operational day which could be exercised at the discretion of the chairman of the planning session. After the early part of the mission the spacecraft automatically tracked the target region to allow for solar rotation, rather than making pointing updates every 96 min.orbit.

SMM was operated in this way from Feb-Dec.1980, when a failure of the altitude control system removed the capability to point the spacecraft accurately at the sun.

## 5. Results

SMM has generated a wealth of data that will be analysed for many years and already the volume of exciting observations and conclusions exceeds the capacity of this review to accommodate. We outline some of the early results and the reader is urged to consult the references for details and discussion of the data.

Acceleration of high energy protons in flares is of intense interest to the Cosmic Ray community because of its relevance to the origin of high energy cosmic rays themselves. High energy protons signal their presence through gamma ray lines originating from the prompt decay of excited nuclei such as $^{12}C^*$ and $^{16}O^*$ which the protons produce through nuclear interactions. Other short lived radiative nuclei, such as $^{11}C$, $^{13}N$, $^{15}O$ produce prompt positrons, which, through annihilation with ambient electrons produce the 0.511 MeV line. Positrons from $\pi^+$ production are only significant if the accelerated protons exceed the production threshold of 293 MeV. Neutrons are also produced in these reactions, and as they thermalize in the denser regions of the solar atmosphere they are captured by hydrogen to form the $^1H (n, \gamma)^2H$ 2.223 MeV line. For very steep proton spectra, interactions with CNO are the dominant neutron source as the threshold is low (few MeV), whereas for flatter spectra, p $\alpha$ reactions, with a threshold of $\approx 26$ MeV, dominate. The ratios of the $\gamma$ ray line intensities and the time history of the delayed neutron capture line can be interpreted to give the time history of proton acceleration (e.g. Ramaty et al, 1975).

One such event was observed by the GRS and HXRBS on June 7, 1980 at 03.12 U.T. from AR 2495╱ at N17W70 (Chupp et al, 1981, Orwig et al, 1981, Kiplinger et al, 1981) and the X-ray intensity time history is illustrated in Fig.4. A series of multiply impulsive spikes, each with significant intensity fluctuations on a time scale of 100 ms, lasted about 70 s at the event onset. There are two further series of spikes, at around 03.14 U.T. and 03.16 U.T., with a 4th enhancement around 03.17 U.T.in the 26-52 keV channel. The width of the individual spikes is qualitatively similar in all sets.

Kiplinger et al (1981) have proposed that the pulse trains originate as repeated energy releases in a set of (7) adjacent loops. Note the modulation depth of typically a factor of 4 during the first series of spikes. Comparison of the 26-52 keV and 228-386 keV channels shows an underlying slowly varying component at the lower energies which is absent at the higher energies. This is presumably thermal emission

╱ Active regions are referred to by their NOAA classification

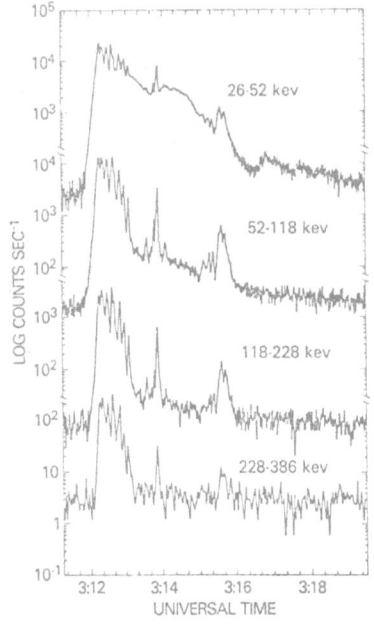

Fig.4 The multiply impulsive
hard X-ray burst,
June 7, 1980.

from hot flare plasma, superimposed on which are the impulsive, non-thermal bursts.

Interpretation of the GRS data required ions $\gtrsim$ 50 MeV to have been present at the impulsive phase. Fig.5 shows the intensity-time history of the 2.223 MeV line, which rises in coincidence with the hard X-rays, consistent with the production of fast neutrons between 03.12.14 U.T. and 03.12.54 U.T. This is compared in Fig.5 with a theoretical model (Wang and Ramaty, 1974; Wang 1975) in which neutrons are produced $3 \times 10^4$ km above the photosphere; the situation is simplified by having

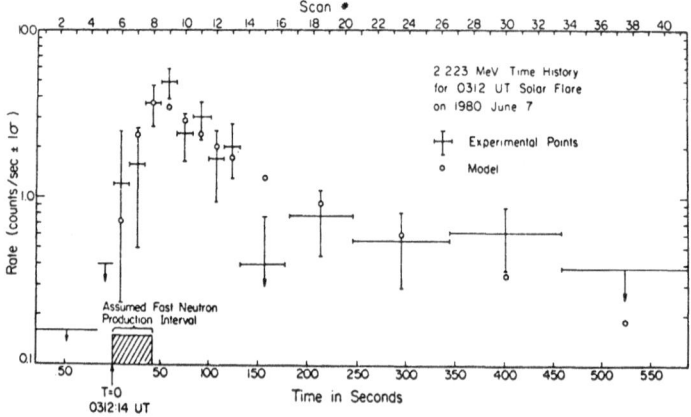

Fig.5 The time history of the 2.223 MeV line, with model predictions
June, 7, 1980.

two neutron populations at 0.5 MeV and 50 MeV. Forrest et al (1980) showed that the 4.1-6.5 MeV channel, covering the prompt de-excitation lines from $^{12}$C* (4.44 MeV) and $^{16}$* (6.13 MeV), also exhibited the deep modulation seen in the impulsive X-ray spikes, at least for the first 5 spikes. There was a slight tendancy for these MeV photons to be delayed ($\approx$ 1sec) from the X-ray burst.

The HXRBS data covering the peak of the most intense (4th) spike requires a temperature of $\approx 10^9$ K for the best thermal fit, while the best power law fit is $\alpha$ E $^{-2.7}$ (Kiplinger et al, 1981). There is a test for a thermal model of this high a temperature. If significant volumes of plasma are heated to these temperatures then thermonuclear reactions will certainly occur. From the X-ray data the emission measure, or the amount of material responsible for the emission, can easily be calculated. The thermal hypothesis may then be tested by looking for the resulting emissions based on thermonuclear theory.

This event requires, in the first seconds of the impulsive phase, the presence of electrons to $\approx$ 1 MeV and protons to at least 50 MeV. Thus the concept of 2nd stage acceleration for these populations seems to have been invalidated, or at least shifted to a much higher energy regime. One saviour of the two stage hypothesis would be to argue that the electrons and protons undeniably present were accelerated in a flare elsewhere on the sun, e.g. on the invisible part of the disc, and beamed in a modulated way onto the flare site, causing the impulsive emission and triggering the small, 1B "flare" at N17W70.

There was a source of such particles. An energetic particle event recorded on June 7, 1980 at ISEE-3 (von Rosenvinge, 1981) showed that the 18-70 MeV/n proton and $\alpha$ particle intensity starts rising around 00.00 U.T, some 3 hours before the gamma ray event to a peak around 06.00 U.T. At lower energies the event started even earlier, at $\approx$ 22.00 U.T., June 6, a signature inconsistent with prompt release from a flare favourably connected (magnetically) to earth. The rise and fall of the event at ISEE-3 was quite symmetrical, showing that the gamma ray event, despite its association with a flare at W70 longitude, did not inject particles into the interplanetary medium.

Of the gamma ray events detected in 1980, 4 had significant 2.223 MeV line emission, two of which also had 0.511 MeV emission. Of the latter, the stronger was from a limb flare (W91) on 21 June, when the intensity was comparable to that observed in the 4th Aug.1972 flare. By comparison, the 2.223 MeV line was about 6 times weaker; this may however, reflect the limb location as the prompt 0.511 MeV emission originates higher than the neutron capture line, which requires the neutrons to thermalize. Alternatively it could reflect a different initiating proton spectrum. Further analysis should resolve this.

One of the more intriguing events studied by the HXRBS was on March 29, 1980 at 09.17.10 U.T., which had a FWHM $\approx$ 10s (Dennis et al,1981) Using the FWHM as a criterion, it was the most intense short lived event > 50 keV ever reported; it did, however, possess substantial time structure. It was observed by the GRS out to $\approx$ 1MeV (Ryan et al,1981). Despite being qualitatively similar in almost every respect to one of the individual bursts from the June 7 event discussed above, there was no corresponding strong evidence for the presence of ions. The H $\alpha$ flare (Rust et al, 1981) was concentrated in a single bright loop. The time structure of the X-ray burst is interpreted by Dennis et al (1981)

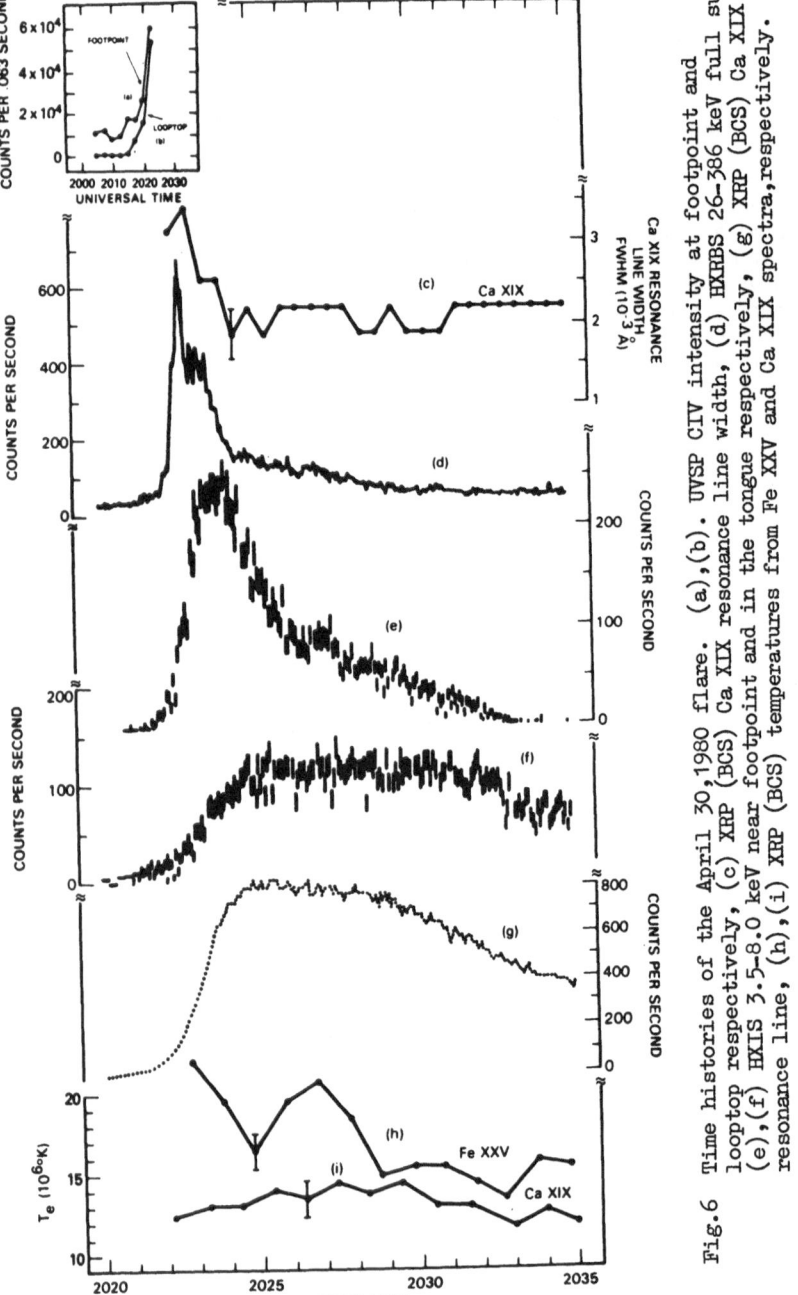

Fig.6  Time histories of the April 30,1980 flare. (a),(b). UVSP CIV intensity at footpoint and looptop respectively, (c) XRP (BCS) Ca XIX resonance line width, (d) HXRBS 26-386 keV full sun, (e),(f) HXIS 3.5-8.0 keV near footpoint and in the tongue respectively, (g) XRP (BCS) Ca XIX resonance line, (h),(i) XRP (BCS) temperatures from Fe XXV and Ca XIX spectra,respectively.

as representing the time profile of the electron acceleration, produced either by successive tearing mode instabilities or by triggering a closely packed arcade of loops (unresolved in H α),

Some of the more appealing data sets from SMM come from limb flares, where imaging instruments measure directly the altitude of the source. One such event was studied from AR2396, on the SW limb at S11W90, on April 30, 1980 starting at 20.17 U.T. about 4 minutes before the onset of the Hard X-ray burst (Fig.6d). A frequent characteristic of flares is that part of the region heats up prior to the hard X-ray burst, which is triggered when some critical parameter is exceeded. Some regions have substantial, long enduring pre-cursor activity before exploding, finally, in a major flare. Fig.6 (a-i) shows the time history of various para-meters as seen by HXRBS, HXIS, XRP(BCS) and UVSP (Ap.J.Lett,244 No.3, plate L3, 1981). UVSP made maps of the limb region in the 1548Å CIV line at 3" resolution during three consecutive orbits spanning the flare orbit (Woodgate et al, 1981). They concluded that prior to the impulsive phase, a small loop filled with matter moving at 20 kms$^{-1}$;then there was rapid brightening in CIV, Hα and hard X-rays followed by brightening in a higher set of loops, apparently where the small loop intersected a higher magnetic structure. This could be emerging flux causing energy release when it interacts with high, pre-existing flux of opposite polarity (e.g. Heyvaerts et al, 1977).

Not all observations support this completely.Fig.7 shows 3' x 3' rasters made by XRP(FCS) in OVIII,Ne IX and Mg XI resonance lines at two times just before the flare (Gabriel et al, 1981). The estimated limb position is shown, sun to the left. The maps clearly have enhanced emission along the limb, concentrated at the loop footpoint(s). The later Mg XI map has significant enhancement above the limb. Following a BCS flare alert, the FCS made small rasters over the hatched area in the second OVIII map. These data, and UVSP light curves (Fig.6a, b) from the footpoint and looptop respectively, show the footpoint bright early, and increase in brightness sooner that the looptop, where the model would predict field line merging. FCS rasters about the foopoint in the hot Fe XXV resonance line show the core gradually rise off the limb and head SW. Plasma temperature, from Fe XXV satellite j to resonance line ratio, is maximum coincident with the hard X-ray burst (Fig.6h), shown by HXIS (Van Beek et al, 1981) to be at the loop foot-point. Gabriel et al,(1981) calculated the cooling time to be ≈ 100 s but note the existence of the flare plasma for > 7 m, thus requiring continuous energy input.

HXIS imaged continuously the tongue that spread SW, and Fig.6e,f show the light curves of the footpoint and high altitude tongue respectively. The latter is ≳ 30000 km above the limb, and later in the event is the dominant X-ray source. Preliminary HXIS analysis gives the tongue a consistently higher temperature than the emission closer to the limb, even though it appears a few seconds after flare onset. There is a continuous energy input into the tongue which, from thermo-dynamics, cannot be via heat conduction. Analysis of this event is still in progress.

Line broadening due to turbulence was demonstrated for the first time by XRP. The Ca XIX spectra just prior to, and a few minutes after, the peak of the light curve (Fig.6g) are shown in Fig.8, A,B. The lines are considerably broadened during the impulsive phase, corresponding to

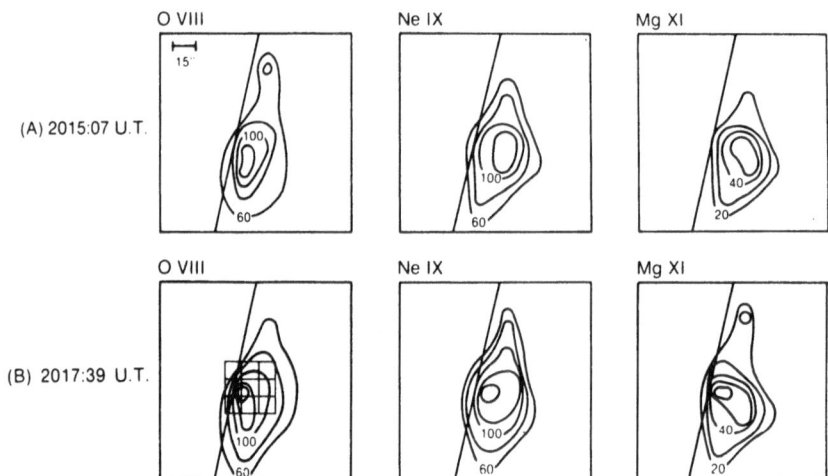

O VIII       Ne IX       Mg XI

(A) 2015:07 U.T.

O VIII       Ne IX       Mg XI

(B) 2017:39 U.T.

Fig.7  XRP (FCS) maps just before the onset of the April 30 flare.

velocities up to 180 km s$^{-1}$.  Fig 6c shows the time history of the line
width, which correlates almost exactly with the hard X-rays (Fig.6d).
Gabriel  et al (1981) believe the turbulence is either due to energy input
from the non thermal process causing the hard X-rays, or it is the source
of the fast electrons.

The 2B two ribbon flare on May 21, 1980, 20.50 U.T. from AR2456 at
S13W15 was observed by all SMM instruments.  It illustrates how close
collaboration between many teams has been co-ordinated to provide
comprehensive solar flare analysis, not only during the flare itself but
for many hours later.    This flare has been studied extensively by HXIS
and HXRBS (Hoyng et al, 1981).  They conclude, as in the April 30 flare,
that emerging flux played an important role in flare development; the
hard X-ray spikes coincided with intense Hα kernels and were probably
caused by electron beams; and the flare plasma needed a continuous energy
source to sustain it.

Accurate alignment of HXIS and Hα images was possible because UVSP
observed the flare in Mg II at 20.56 U.T.  Alignment between HXIS and UVSP
was known to within 8" and Mg II (formed at a comparable level to Hα) and
Hα images were sufficiently similar to allow exact overlay, Fig.9 shows a
time sequence of HXIS images at 3.5–8 keV and 16–30 keV corresponding to
the five times indicated on the HXRBS light curves, shown in Fig.10.
Images at 20.56.25 U.T. show, dotted and (●), the boundary and center of
the 8" x 8" f.o.v. respectively; the image at 20.55.55 U.T. shows the
position (dotted) of the active region filament.  At times 1 and 2, the
soft X-ray emission is predominantly along the filament, while the
16–30 keV X-rays are from spatially separated bright points, the most
northerly of which moves between times 1 and 2.    Emission from the pairs
of points is coincident within 3 s, a figure limited by counting statistics.
The weaker spikes 3,4 in  Fig.10 cannot be separated from the core of the
16–30 keV emission that rises in spatial coincidence with the 3.5–8 keV
emission.  At these energies at this phase in the flare, thermal emission
dominates. The weak feature, 5, cannot be separated.

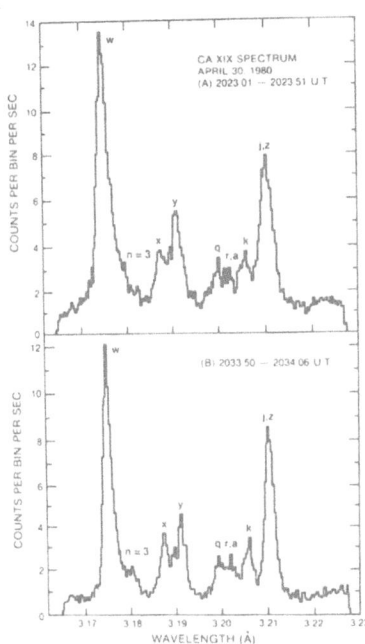

Fig.8  Ca XIX spectra,
illustrating line
broadening.

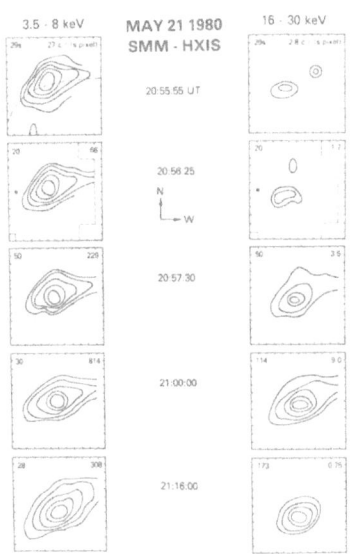

**Fig.9**  Time development of X-ray
images at 3.5-8.0 keV and
16-30 keV.

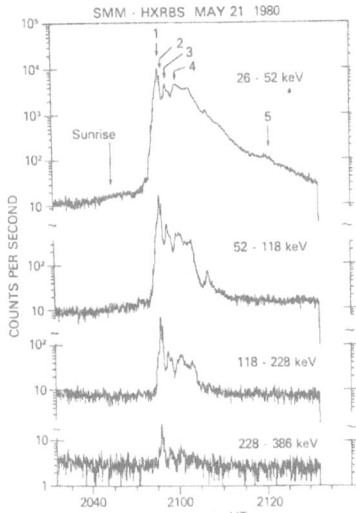

**Fig.10**  Hard X-ray intensity-time
history.

Fig.11a shows how the 16-30 keV X-ray points overlay the Hα photograph; it also identifies the region prior to the flare, showing the filament.    Point A is in the same place in Fig.11a, c, d. Comparison with a similar Hα picture (Hoyng et al, 1981) at 20.54.50U.T. shows the filament disruption and that bright Hα kernels appeared in coincidence with the hard X-ray bright points.   Bright post flare loops span the filament channel (Fig.11c).   Kitt Peak magnetograms (Fig.11b,d) taken at 16.26 U.T. and just before the flare at 20.20 U.T. show the emergence of a bipolar region, at A and B, Fig.11d.    An off-band Hα picture taken after the flare shows a new sunspot within 5" of A, hence the conclusion that emerging flux played a major role in the disruption of the filament and subsequent flare development. Analysis of the magnetic loop properties, together with the timing of the appearance of the hard X-ray footpoints led Hoyng et al (1981) to conclude that the X-rays must be produced at the footpoints by beams of electrons.

For more than 6h after the flare, HXIS imaged an extensive arch (Svestka et al, 1981).    Fig.12 shows a contour map of the 3.5-5.5 keV intensity integrated for 25 m about 03.30.45 U.T., May 22.    The dotted outline is the pre-flare emission contour at 20.52.54 U.T., May 21, while the black areas are sequential images showing the southward motion of the emission from the flare looptop between 20.57.30 U.T. and 23.05.57 U.T., an apparant projected speed of 1.5-2.0 km s$^{-1}$.    This speed is too low for the extensive arch to be a later manifestation of

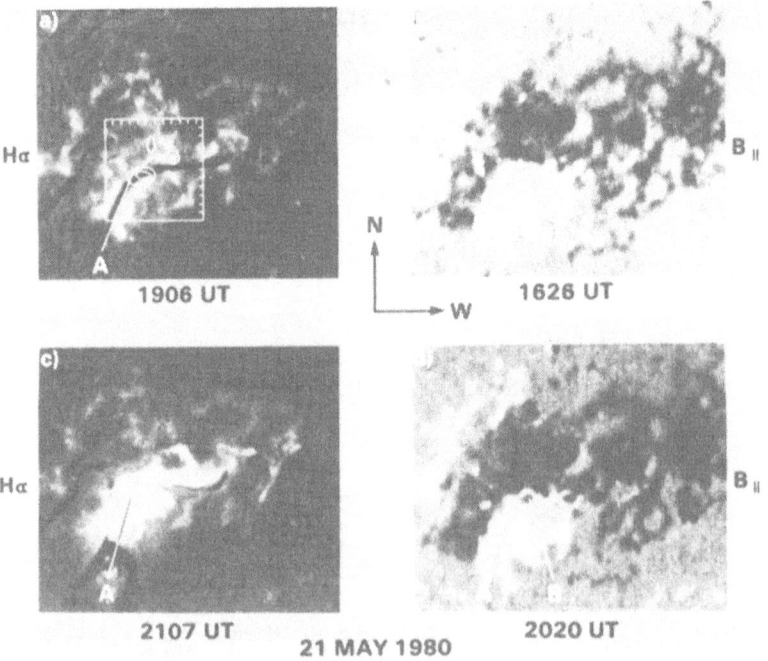

Fig.11  Hα photographs (a,c) and magnetograms (b,d) showing the location of the hard X-ray burst and the emerging flux AB (d)

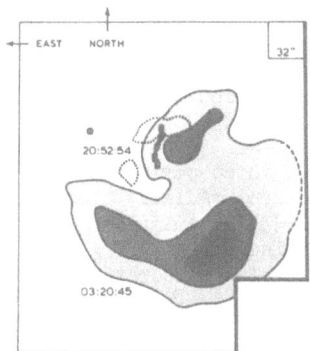

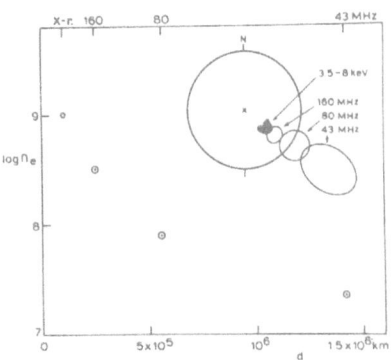

Fig.12  The immense X-ray
(3.5-5.5 keV) arch
early on May 22,1980.

Fig.13  Relative positions of the
X-ray arch and the Type 1
stationary noise storm. The
graph show electron density
$n_e$ derived from the emission
measure of the X-ray source
and from plasma frequencies
versus nominal height d.

of the flare looptops, and from careful analysis of the data, Svestka
et al,(1981) find evidence for the extensive arch, in the same position,
within 90 minutes of the flare.   The arch reaches to an altitude between
110,000 and 180,000 km and appears to have one footpoint near the flare
site and the other well to the SE.

During this time the Culgoora radio-heliograph observed a noise
storm at their three observing frequencies, 160,80 and 43 MHz.   Fig.13
shows the positions of the radio sources relative to the sun and the
X-ray arch, and also plots the electron density as a function of height
for the four components.  It is consistent with the data that the electrons
responsible for the radio noise storm and for the X-ray emission are all
part of the same population.  Their appearance after a major flare suggests
that they may represent particles accelerated in that flare and stored
in the corona, from which they gradually precipitate, producing the long
term emission.   Whether the structure deduced from Fig.13 is part of a
thin magnetic bottle, or whether it is itself one leg og a huge stationary
coronal arch, with the other leg, say behind the sun, cannot be resolved.

Moving now to the dynamic corona, one intriguing problem  is the
relationship between moving radio sources and coronal transients, and
their energy supply.   Is it a redistribution of flare energy, or is
energy removed from the coronal magnetic field?   The C/P observed a
large transient event on April 7, 1980 (Wagner et al, 1981) which
appeared to be initiated by a 1F flare at N32W67, AR2363 at 03.26 U.T.
Fig.14a shows the transient as a huge loop in the NW quadrant, extending
from near the equator to the pole, and its subsequent development. The
position of the initiating flare is shown (X), and the edge of the
coronagraph  occulting disc is at 1.5 $R_\odot$.   Following the transient,

coronal streamers traced the outer boundary of the former loop. Fig.14b shows the positions of the Culgoora radio sources. Circles indicate positions of stationary sources, which (at 43 MHz) were occasionally joined, early in the event, by an extended bright source. This pattern ceased once the sources at B started moving. The full line extending to B' marks the motion of the more southerly 80 MHz source while the dashed line extending to B" marks the motion of the 43 MHz source. Fig.15 is a radial altitude-time history of three parts of the event, namely the outermost loop, the loop enhancement radially above the flare including the moving 80 MHz source, and the 43 MHz emission. The duration of the stationary sources at B is indicated in Fig.15; heavier segments are times when they exceed a brightness temperature of $4 \times 10^8$ K and $6 \times 10^8$ K at 80 and 43 MHz respectively.

These observations show that the moving (Type IV) radio source was located very close to, if not on, the loop. The outermost part expands at $\approx 650 \pm 50$ kms$^{-1}$, while the motion of the brighter, western part is $600 \pm 30$ kms$^{-1}$, consistent with an origin in the flare at 03.26 U.T. Wagner et al (1981) interpret the fast appearance of 80 MHz emission at 1.7 $R_\odot$ as due to fast electrons generated in the impulsive phase of the flare; why the plasma emission is seen at only one of Culgoora's frequencies is unclear. One explanation is that a significant population of electrons are temporarily trapped in a pre-existing magnetic loop (c f May 21 above) extending to around the normal 80 MHz plasma frequency level. Fluctuations in the magnetic field could produce excursions of the trapped electron population out to the 43 MHz level. As the flare induced transient moves out, it takes this magnetic loop with it, thereby causing motion of the radio sources and cessation of the infilling at 43 MHz.

There was no Type II burst, normally indicative of shock formation when a disturbance moves faster than the Alfvèn speed. All transients

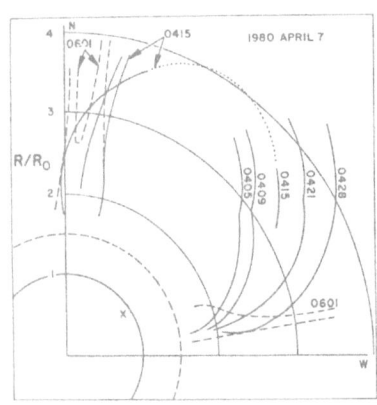

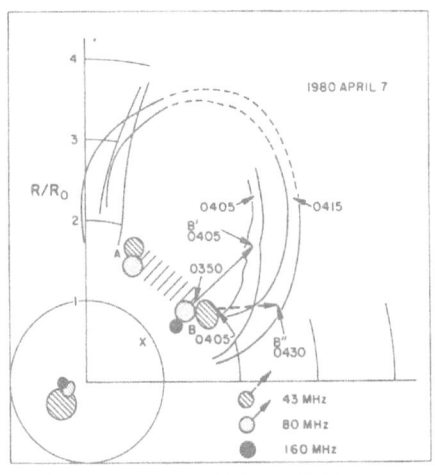

Fig.14 (a) Development of the transient in white light.

(b) relative position of the radio sources.

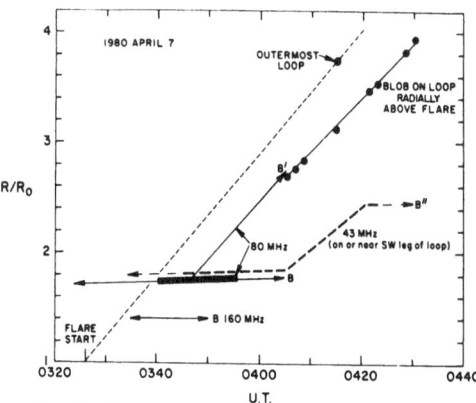

Fig.15  Altitude-time history
        of the loop and the
        radio sources

seen by Skylab were accompanied by Type II emission if the expansion
exceeded 500 kms$^{-1}$(Gosling et al, 1976).   If the Alfvén velocity were
not exceeded in this event, either the magnetic field is high, the
electron density low, or both.   From detailed analysis, Wagner et al
(1981) showed that the mechanical energy in the transient exceeded the
total radiated flare energy by over an order of magnitude.   Further
comparisons of the convected magnetic energy with the mechanical energy
show that the former dominates, confirming an earlier Skylab result.
Studies of transients thus indicate that a relatively minor disturbance
(the flare) can trigger an enormous transport of energy across the 5 $R_{\odot}$
level.   Our concept of flares themselves contains a similar element of
tremendous energy gain.   Whether transient formation contains similar
physics to the flare process, but with different field strengths,
densities, temperatures and timescales will need to await better understanding of both.   But like the flare, transient formation appears to
need the equivalent of emerging flux.

        Apart from collaborative results, SMM experiments have made
numerous individual contributions.   Gurman et al (1981) reported
significant transition region oscillations in small sunspots of periods
129 - 173 s, using observations of the line profile of the 1548.19Å
C IV resonance line.   The measurements were from a 3" x 3" pixel chosen
at the center of the sunspot umbra where the line intensity is minimum.
Fig.16a shows a time history (14.88s resolution) of intensity, $I_0$, the
shift of the line peak from the nominal line wavelength,$\lambda_0$,(corresponding
to velocity)and line width, d$\lambda$, from a sunspot in AR2522 on June 26,1980.
The data are corrected for spacecraft orbital motion and other low
frequency ( < 3mHz)effects.   The power  spectrum of these data is shown
in Fig.16b; a strong peak at 136 $\pm$ 10 s is present in all parameters,with
virtually no power elsewhere.   Maximum intensity occurs when the C IV
line is blueshifted, and the velocity lags the line width by $\approx$ 45°.
However, the phase correlations are not a frequent feature of the sunspot
studies, whereas the $\lambda_0$ velocity profile,i.e. the oscillations, are a
common feature.   They conclude they are witnessing upward propagating
acoustic waves in the umbral atmosphere.   The energy content is
$\lesssim$ 2 x 10$^3$ erg cm$^{-2}$s$^{-1}$, insignificant (by 7 orders of magnitude) when
compared with the "missing" radiative flux of sunspots.

        The XRP can study time evolution of X-ray spectra with exceptional
precision.   Fig.17 shows a subset of high resolution spectra(Culhane

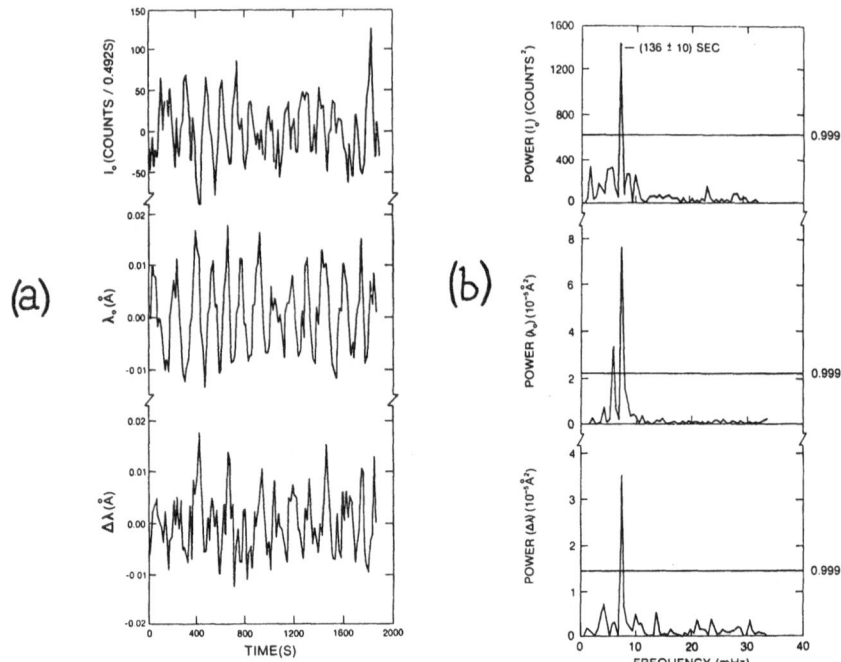

<u>Fig.16</u> (a) The time history of the intensity $I_0$, velocity ($\alpha \lambda_0$) and
line width, $\Delta \lambda$ over a sunspot. (b) Power spectra of the
time histories in (a).

et al, 1981) from a flare on April 7, at 05.38 U.T. from AR2372,together
with light curves of the Ca XIX and Fe XXV resonance lines and the Fe Kα
spectrum. Temperatures derived from these spectra show that the Fe XXV
spectrum originates at a significantly higher temperature (e.g. 22 versus
16 million K at 05.38. U.T.) than the Ca XIX spectrum; it also reaches
peak intensity earlier, and these results confirm the complex multithermal
nature of the flare plasma. One significant result to come from these
early spectral studies concerns the origin of the Fe Kα radiation. It has
long been suspected (Doschek et al, 1971) that fluorescence from the
intense X-ray continuum produced in a flare causes this radiation,
probably in cool photospheric Fe II. Detailed comparison by Culhane et
al (1981) with predictions of a fluorescence model (Bai,1979) for both this
event and the April 30 event (see above) confirm this conclusion. There
are other flares when, during the <u>impulsive</u> phase, the Fe Kα emission is
more consistent with inner shell collisional excitation, with the later,
decay phase, due to fluorescence.

The ACRIM, studying the constancy of the Solar Constant, has shown
that it is, in fact, rather variable. Willson et al,(1981) presented
the first 153 days data(Fig.18)corrected for orbital effects, temperature,
and normalized to 1 A.U. plus a small orbit factor. There is a large
drop, $\approx$ 0.2%, during early April and a smaller drop in late May. These
decreases are highly correlated with specific sunspot groups, that in
April due to AR2370 and 2372, and that in May due to AR2469 and 2470.

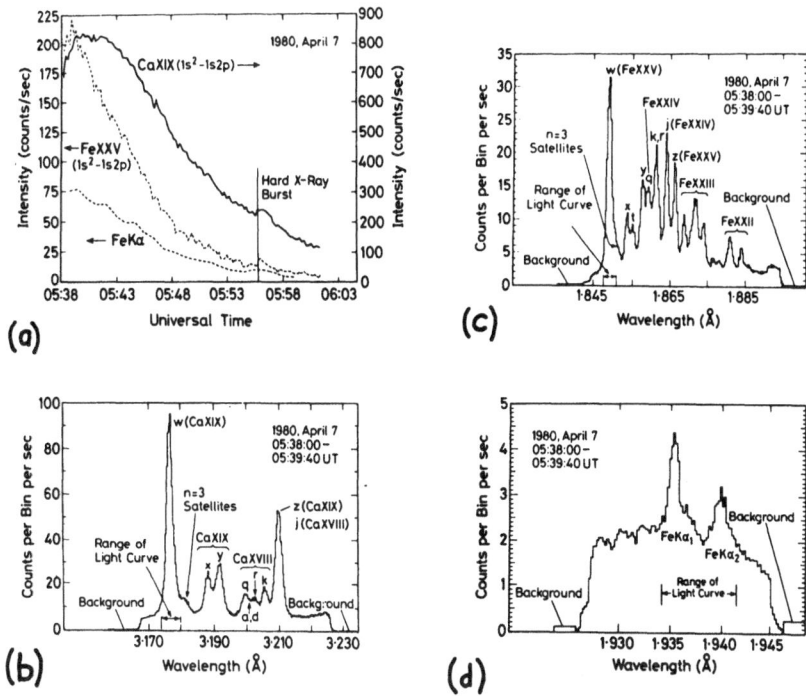

Fig.17 (a) Light curves of the Ca XIX, Fe XXV and Fe Kα lines for the April 7 flare. (b) (c) (d) high resolution spectra of Ca XIX Fe XXV and Fe Kα respectively.

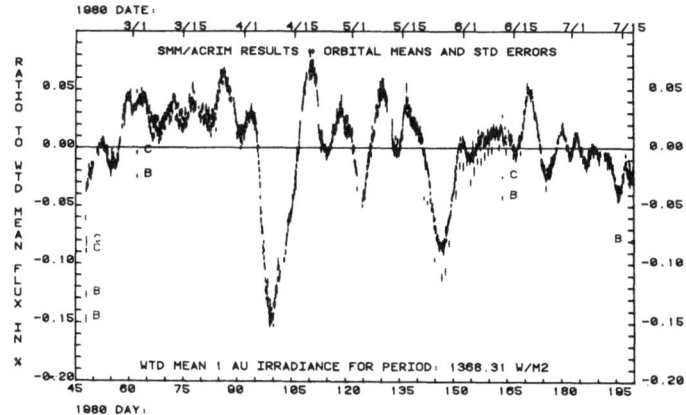

Fig. 18 The time history of the Solar Constant.

Note also the lack of significant high frequency fluctuations. Power spectrum analysis shows a high frequency cut-off $\approx$ 7 days,which supports the concept that development of sunspot groups and facular areas produces the variability.

While sunspots represent cool regions, white faculae are hot regions. Lack of perfect correlation of irradiance with sunspot number is believed to be caused by counteracting variations in faculae. The fact that the "missing" flux in sunspots does not always manifest itself through enhanced faculae emission shows that energy must be stored in the convection zone for $\approx$ 7 days or longer.

A major triumph of ACRIM is that the standard error of relative measurements is $\approx$ 1 part in $10^5$ for orbit averages. Thus it may be able to detect the 5 minute global solar oscillations, predicted by Isaak (1980) to be in the $10^{-5}$ region when considered as a fraction of total solar irradiance.

## 6.   Summary

The flare build-up and the impulsive phases have been well studied because the SMM chose a likely flare site,observed it continuously with all instruments and had a programmed, highly sophisticated response to a flare.   This contrasts strongly with the situation that existed previously, when observations were generally not co-ordinated, had poor time resolution and frequently (e.g. in the case of Skylab) started observing after flare onset.

This review has given a sample of the capability of the SMM and the interpretative skills of the scientists working with the data.   It has already made a big impact on solar physics and it is clear that over the next few years a substantial number of new developments will be forthcoming, carrying on where Skylab left off.

## Acknowledgments

The author expresses his deepest thanks to all the SMM experiment teams for supplying the material for this review.  The figures are reproduced from articles referenced in the text.  The ACRIM results were provided through the courtesy of the Jet Propulsion Laboratory,California Institute of Technology, Pasadena, CA.

## References

Bai, T, 1979 Solar Phys. 62, 113,
Chupp, E.T.  et al, 1981, Ap.J.Lett, 244, L171.
Culhane, J.L. et al, 1981, Ap.J.Lett, 244, L141.
Dennis, B.R.,et al, 1981, Ap.J.Lett, 244, L107.
Doschek, G.A.  et al,1971, Ap.J. 170, 573.
Forrest, D.J. et al, 1980, Bull AAS, 12, 890 (SPD Meeting, Taos, 1981)
Gabriel, A.H, et al, 1981, Ap.J. Lett, L147.
Gosling, J.T. et al, 1975, Sol.Phys. 40, 439.
Gosling, J.T. et al, 1976, Sol.Phys. 48, 389.
Gurman, J.B. et al, 1981, Ap.J. (in press).
Heyvaerts, J. et al, 1977, Ap.J.,216, 123.

Hoyng, P. et al, 1981, Ap.J. Lett. (in press).
Hundhausen,A.J., 1972, Coronal Expansion and Solar Wind,Springer-Verlag.NY.
Issak, G.R. 1980, Nature 283, 644.
Kiplinger, A. et al, 1981, Ap.J. (in press).
Orwig, L.E. et al, 1981, Ap.J.Lett. 244. L163.
Ramaty, R. et al, 1975, Sp.Sc.Rev., 18, 341.
Rust, D.M. et al, 1981, Ap.J.Lett, 244, L179.
Ryan, J.M. et al, 1981, Ap.J.Lett, 244, L175.
Svestka, Z. et al, 1981, Sol.Phys. (in press).
Van Beek, H.F. et al, 1981, Ap.J.Lett, 244, L157.
Von Rosenvinge, T.T. 1981, private communication.
Wagner, W.J. et al, 1981, Ap.J.Lett. 244, L123.
Wang, H.T. and Ramaty, R. 1974, Solar Phys. 36, 129.
Wang, H.T. 1975, Ph.D.thesis, University of Maryland.
Willson, R.C. et al, 1981, Science, 211, 700.
Woodgate, B.E. et al, 1981, Ap.J.Lett, 244, L133.
Zirin, H. and Tanaka, K, 1973, Solar Phys. 32, 173.

PLANETARY MAGNETOSPHERES:
THE IN SITU ASTROPHYSICAL LABORATORIES

S. M. Krimigis
Applied Physics Laboratory
The Johns Hopkins University
Laurel, Maryland 20707

## 1. INTRODUCTION

Investigation by spacecraft of several parts of the solar system has revealed a diverse variety of plasma processes, ranging from the relatively steady state conditions in the solar wind to the explosive release of large amounts of energy observed to occur in solar flares, and to the dynamic processes taking place within the magnetospheres of the Earth and planets. The magnetospheres represent the most accessible medium for in situ observations among all plasma environments within the solar system. The intrinsic value of studying planetary magnetospheres however is not their accessibility, but the fact that they represent the basic link between small-scale laboratory plasma physics and large-scale astrophysical plasmas; thus the study of magnetospheres is not only of fundamental importance in its own right, but offers the best hope as a testing ground for our models and theories which can then be applied to remote astrophysical objects, where our only source of information is the x-ray, radio, UV, $\gamma$-ray, and IR signatures.

A magnetosphere is the result of the interaction between a streaming, hot, magnetized, collisionless plasma and an intrinsically magnetized or electrically conducting body. The result of this interaction is a cavity, which is formed in the streaming collisionless plasma by the magnetic field or intrinsic conductivity of the central body, and is called "magnetosphere". The intrinsic or induced magnetic field plays a principal role in tying together the diverse phenomena observed to exist within the cavity, such as cold and/or hot plasmas, electric currents, plasma waves, and often trapped energetic particles. Magnetospheres are rarely at a steady state, in that there they respond to pressure variations in the streaming plasma outside the cavity, and often store energy which is then explosively released to produce a large variety of plasma phenomena.

So far magnetospheres of various kinds have been discovered in association with all six planets within 10 AU of the sun. These magnetospheres have formed in the streaming plasma (the solar wind), the characteristics of which are shown in Figure 1 (Russell, 1981) for the first 15 AU. The figure shows that the parameters of the solar wind change substantially between the orbits of Mercury and Saturn: the pressure by a factor of 100, the magnitude of the magnetic field by a factor of 1000 and the temperature of the plasma by a factor of $\sim$ 40. Yet despite these large changes in the streaming plasma, we will find that there exist remarkable similiarities which characterize the interaction between the solar wind and the magnetospheres of these planets.

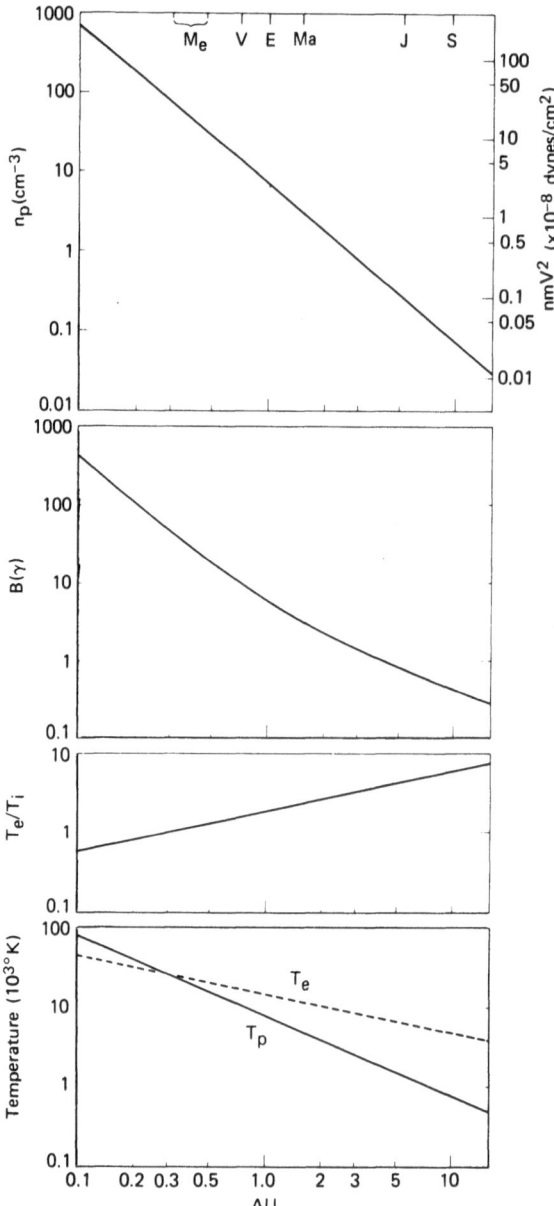

<u>FIG. 1</u> - Dependence of solar wind parameters on heliocentric distance, assuming observed dependences of density n, temperature T and field strength B (Russell, 1981).

In the following sections, the magnetosphere of Earth will be described first because it represents the one most extensively studied to date, and will serve as the baseline for characterizing the phenomena to be observed in other planetary magnetospheres. The magnetospheres of the inner planets (Mercury, Venus and Mars) will be examined next and then a description of the magnetospheres of the outer planets will be presented. The description of the magnetosphere of Jupiter will form the bulk of the paper, not only because of its enormous size and rich variety of physical phenomena within, but because it may represent the closest analog to the many astrophysical objects thought to possess magnetospheres. Several properties characteristic of magnetospheres in general will then be pointed out, and the significance of the findings will be discussed in the context of solar system and astrophysical plasmas.

## 2. MAGNETOSPHERE OF THE EARTH

Figure 2 gives a schematic view of the Earth's magnetosphere, and emphasizes aspects which are different than the typical diagram of the Van Allen belts. The figure shows those features of the plasma environment which are essential to both the steady state and dynamical aspects of the Earth's magnetosphere. The shaded region represents the magnetospheric bow shock, which is a characteristic feature of all magnetospheric interactions with the solar wind. Immediately inside the bow shock is depicted the magnetopause current, which in effect establishes the boundary between the solar wind plasma and the magnetosphere plasma. The solar wind plasma is shown to have direct access to the magnetosphere through the polar cusp and perhaps through the flanks, where it may enter directly into the plasma sheet. Closer in is shown the plasmapause and the plasmasphere where the electron density increases by a factor of $\sim 10^2$. The plasmasphere represents a relatively cold ionospheric plasma which is corotating with the Earth. The ring current circling the Earth is typically located between 3 and 6 $R_E$ and represents a large reservoir of stored energy in association with magnetic storms. Other current systems noted are the neutral sheet current and the tail current, which in conjunction with the intrinsic magnetic field of the Earth determine the overall magnetic structure of the magnetosphere.

Most notable of all, however, is the field-aligned current, which connects the ionosphere of the Earth to the distant magnetosphere. The Earth's field-aligned current system (e.g., Potemra, 1978), represents one of the most important discoveries of space plasma physics in that it demonstrated that currents could flow along magnetic field lines, even though such field lines are equipotentials. The discovery of field-aligned currents, whose existence was first suggested by Birkeland (1908), has had profound influence in the theory and understanding of magnetospheric processes and is likely to be of fundamental importance to the field of solar physics and astrophysics (e.g., Alfven, 1981).

The classical Van Allen belts, not shown here, represent a miniscule fraction of the energy density contained in the radiation belts. Most of the energy density resides in the range from $\sim$ 10 keV to $\sim$ 200

Earth's Magnetosphere

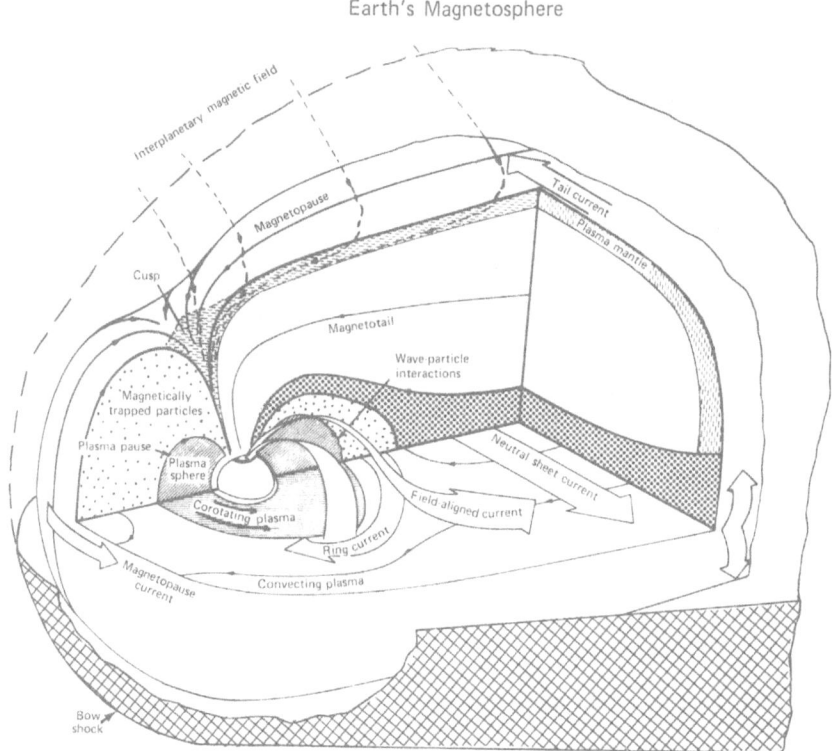

FIG. 2 - A three-dimentional view of the Earth's magnetosphere, adapted from Heikkila (1972) (see text for explanation).

keV. The presence of energetic trapped particles in the radiation belt, however, have raised the rather profound and universal question of the acceleration mechanism. At this point most magnetospheric physicists seem to agree that the primary acceleration mechanism operating in the magnetosphere at these energies is magnetic pumping, i.e., the conservation of the first adiabatic invariant $\mu = E_\perp/B$ (where $E_\perp$ represents the component of particle energy perpendicular to the magnetic field $B$) and the violation of the third invariant (for details see Alfven and Falthammar, 1963). Variations of magnetic pumping have been used to describe acceleration processes in several astrophysical problems (e.g., Kulsrud, 1979).

An example of energetic particle spectra in the Earth's plasma sheet is shown in Figure 3, and illustrates the wide energy range and the large intensity range of energetic ions within the Earth's magnetosphere. The points show a differential energy spectrum which extends smoothly from the highest energies (> 1 MeV) to the lowest energies (~ 2 keV). These spectra (Sarris et al., 1981), can often be represented by a hot, convected Maxwellian distribution (dashed line) joined smoothly to a power law at higher energies. The transition energy between the Maxwellian and the power law is given by E = (γ + 1) kT (Roelof et al., 1976), and occurs at an energy of ~ 100 keV in the present case, i.e. in excellent agreement with the observations. We shall come back to this type of spectral representation in the discussion of energetic ions in Jupiter's magnetosphere. Although the plasma shown in Figure 3 is relatively hot, the form of the spectrum seems to have this general shape even for cases when the plasma in the plasma sheet is somewhat colder (5-10 keV, Sarris et al., 1981).

Some of the most important advances in the last few years in understanding the magnetosphere have been made in the area of wave-plasma interactions. A summary view of a few plasma wave regions within the Earth's magnetosphere is given in Figure 4 (Colgate et al., National Academy of Sciences, 1978). These emissions extend from a few hertz to several hundred kilohertz and affect not only the dynamics of the trapped radiation belts, but also represent important dissipation of energy within the magnetosphere, such as the Auroral Kilometric Radiation (AKR), which radiates up to ~ $10^9$ watts from high latitude regions.

It should be noted that the Earth's magnetosphere is rarely in a quiet state. There are continuous auroral

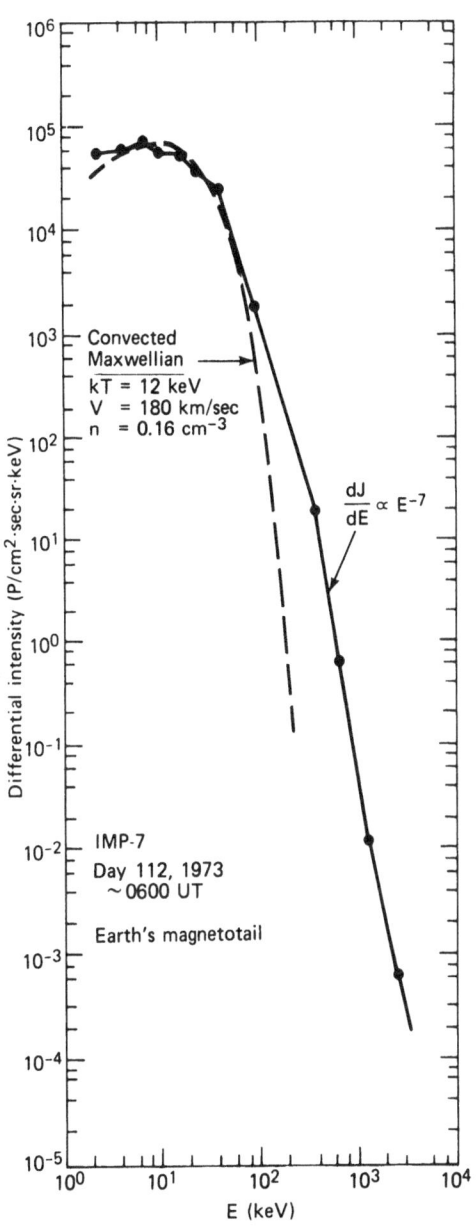

FIG. 3 - An observed energy spectrum in the Earth's plasma sheet (Sarris et al., 1981), and a Maxwellian fit at the low energy end with the indicated parameters.

displays on the Earth's polar caps which are the result of energetic particles precipitating from the plasma sheet; the total power involved in this process, including Joule heating of the ionosphere as $\sim 2 \times 10^{11}$ watts. There also exist magnetospheric substorms which release energy stored in the magnetotail every few hours at the rate of $\sim 10^{11}$ watts; some of this energy goes into the acceleration of energetic particles (Krimigis and Sarris, 1979). There is currently a lively debate within the magnetospheric community on the presence or absence of magnetic reconnection or merging (Figure 4) in the magnetotail, which is connected with the origin of both the mass plasma motions and the acceleration of energetic particles observed during substorms. It is important to point out that magnetic merging as a source of particle acceleration has been extensively discussed in connection with solar flares, and the magnetosphere offers us the best opportunity to study and understand this phenomenon in situ. Analogies between the magnetosphere substorm and solar flares have been ellaborated on in the literature (Obayashi, 1975; Akasofu, 1979).

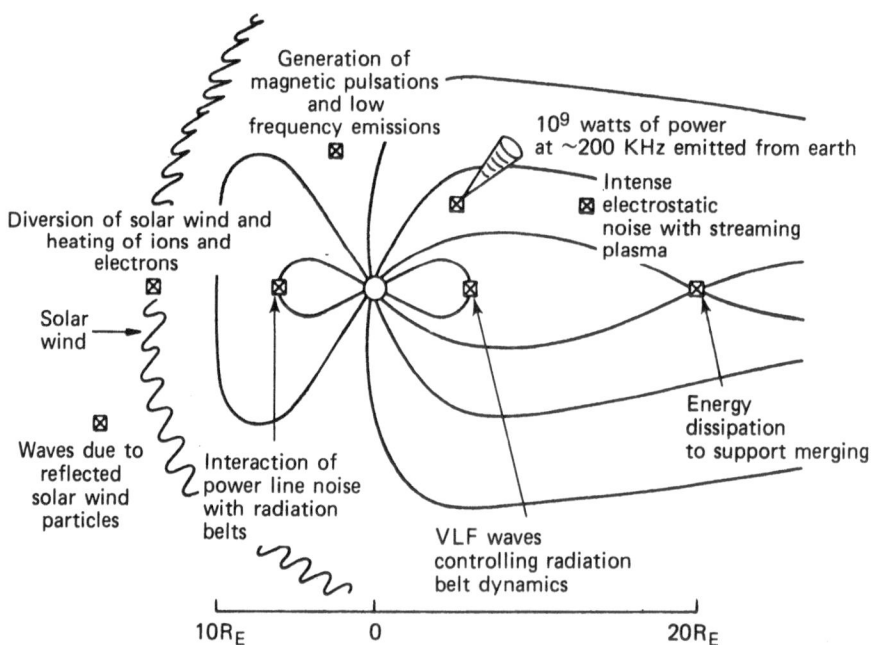

FIG. 4 - A schematic view of significant plasma wave regions within the Earth's magnetosphere (Colgate et al., 1978).

Finally, there exists a region upstream from the Earth's bow shock centered at $\sim 0900$ local time, where copious numbers of energetic ($\sim 2$ to 150 keV) ions are observed intermittently. There is substantial evidence that some acceleration may be taking place as well as leakage of ions trapped within the magnetosphere. Here, models of shock formation and plasma heating and acceleration could be tested for applications to other, non-accessible space plasma environments.

## 3. MAGNETOSPHERES OF MERCURY, VENUS AND MARS

During the past two decades, the plasma environments of the inner planets were investigated by spacecraft. Mercury was investigated in successive encounters by the Mariner-Venus-Mercury spacecraft and found to possess an intrinsic magnetic field and a well developed magneto-sphere. The modeled shape of the bow shock and magnetopause of Mercury is shown in Figure 5a (Ness, 1979), and displays a stand-off distance to the solar wind at $\sim 1.8$ $R_p$ (planetary radii). The magnetic moment of Mercury is $3 \times 10^{-4}$ that of Earth, as indicated on the figure. The mag-netosphere of the planet was found to be fairly well developed and con-tains both cool and hot electrons and probably large fluxes of more energetic electrons in the 70-300 keV range (for a review, see Ness, 1979). Because of the variability of these fluxes during the encount-er,it is believed that they represented transient events and did not correspond to spatial structures. The magnetosphere of Mercury is apparently too small to contain durably trapped energetic particle popu-lations.

The magnetosphere of Venus is shown in Figure 5b and has been investigated over the past two decades by both U.S. and Soviet space-craft. The most comprehensive view of the magnetosphere of Venus has come most recently from the Pioneer Venus data, the analysis of which shows the presence of a well defined bow shock at the standard distance of $\sim 1.3$ $R_p$ and an ionopause sitting just above the upper layers of the ionosphere of the planet (Slavin et al., 1980). Figure 5b shows calcu-lated bow shock and ionopause locations at Venus based on the Pioneer observations, given by Spreiter and Stahara (1980). The ionopause at Venus represents the first confirmed deflection of the solar wind by a planetary ionosphere and is well understood theoretically by gas dynamic calculations. An upper limit to the magnetic moment of Venus has also been deduced from the Pioneer measurements of $< 5 \times 10^{-5}$ that of Earth (Russell et al., 1980).

Although Mars has been investigated by a number of U.S. and Soviet spacecraft, the details of the interaction between the planet and the solar wind are rather unclear. A well-established bow shock has been observed, but the question of the presence of a possible magnetopause has not as yet been fully answered (for a review see Russell, 1979). The shape of the Martian bow shock is shown in Figure 5c and it is evident that it is farther away from the planet than the bow shock of Venus. Despite this fact, there is considerable evidence of ionospheric interaction with the solar wind (Vaisberg et al., 1976). The magneto-pause shown in Figure 5c has been scaled from the Earth's magnetopause on the basis of the bow shock-to-magnetopause ratio. As of this time, there has not been a spacecraft which came close enough to the planet to detect unambiguously an ionopause or a magnetopause.

The bow shocks of the terrestrial planets, scaled to a pressure of $3.5 \times 10^{-8}$ dynes/cm$^2$ at 1 AU is shown in Figure 6 (Slavin and Holzer,

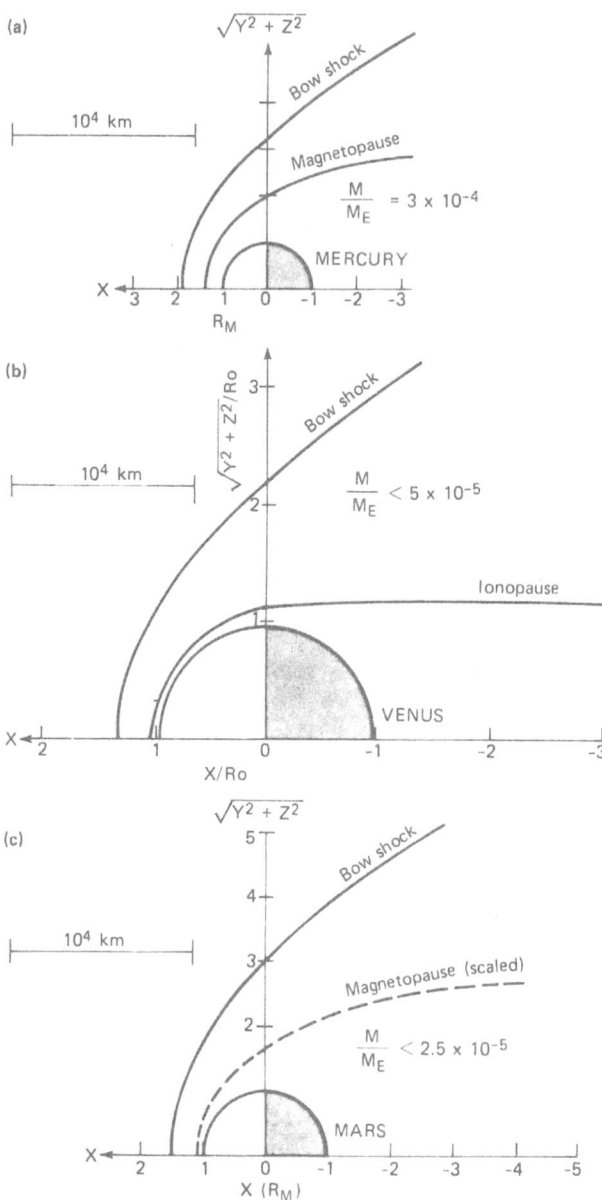

FIG. 5 – Average positions and shapes in terms of planetary radii of the bow shock, and magnetopause of (a) Mercury, (b) Venus, and (c) Mars; a scale of $10^4$ km is shown for absolute comparison.

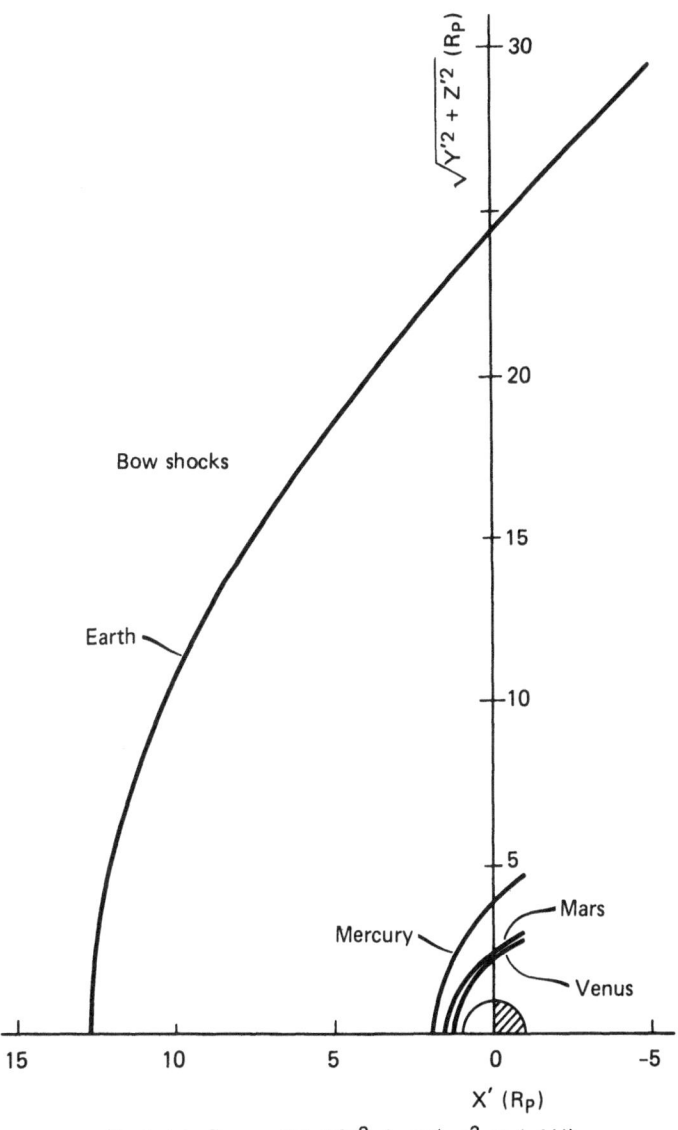

FIG. 6 — Comparative locations of the bow shocks of terrestrial planets normalized to the indicated solar wind pressure at 1 AU (Slavin and Holzer, 1981).

1981) in terms of planetary radii.  It is evident from the figure that the Earth's magnetosphere represents a very large obstacle to the solar wind compared to that of the other planets.  Mercury's bow shock has a similar shape to that of Earth, while the bow shock of Mars is further out than that of Venus, even though the atmosphere of Mars is much thinner.  It is possible that in the solar wind-atmosphere interaction the conductivity of the respective atmospheres could play a major role, or that Mars could well have a relatively small magnetic field.  This question remains to be settled from observations by future spacecraft.

To summarize, the magnetospheres of the terrestrial planets have provided us with two examples of solar wind-planet interaction: the case when the planet possesses an intrinsic magnetic field (Mercury, Earth), and the case where a field is probably absent but a conducting iono-sphere is present (Venus, Mars).  In both cases, well-developed bow shocks exist which can be modeled by gas dynamic calculations.  Heating of solar wind or ionospheric plasma is demonstrated in only the intrinsic magnetic field case, while scavenging of ionospheric ions, potentially important to atmospheric evolution, is taking place when no magnetic field is present.

## 4.  MAGNETOSPHERES OF THE OUTER PLANETS

Although the inner planets have helped elucidate a number of the basic problems associated with the planet-solar wind interations, it has been the giant outer planets which have presented us with the newest and most exciting physics.  The magnetospheres of both Jupiter and Saturn have been investigated by the Pioneer and Voyager spacecraft, while the probable magnetospheres of Uranus and Neptune will likely be investi-gated by the Voyager 2 spacecraft in the latter part of this decade.  Although Saturn is more spectacular than Jupiter in terms of physical beauty, Jupiter's magnetosphere is far more dynamic, complex, and rich in terms of the range of physical phenomena observed there.  A substan-tial portion of this review will be devoted to a description of the new aspects of magnetospheric physics revealed by the Voyager investigation of Jupiter's magnetosphere during 1979.

### Jupiter's Magnetosphere

Figure 7 presents a schematic summary of the main features of Jupiter's magnetosphere, as adapted from the paper by Scarf et al. (1981a).  First the size of the magnetosphere compared to the dimensions of the planet is impressive, extending to $\sim 100$ $R_J$ in the sunward direc-tion to perhaps as far as 4 astronomical units in the antisunward direc-tion (Krimigis et al., 1975; Scarf et al., 1981b).  The inner magneto-sphere is dominated by the Io plasma torus, centered at about the orbit of Io and containing hot and cold heavy ion plasmas with peak densities of about 3000 $cm^{-3}$ (Bridge et al., 1979a; Warwick et al., 1979).  Sulfur dioxide emitted by volcanoes and vents on Io becomes ionized and forms the heavy ion plasma torus.  There exist field aligned currents (just as in the case of Earth) which couple the Io torus to the ionosphere of Jupiter, and generate decametric and kilometric radiations which are shown to escape the magnetosphere.  Heavy ions from the Io torus move

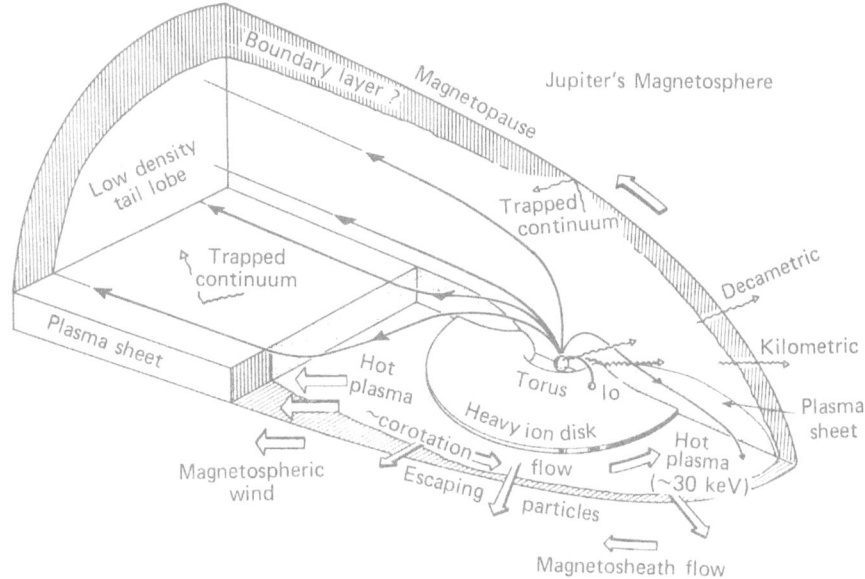

FIG. 7 - A schematic view of Jupiter's magnetosphere emphasizing the various plasma regions (adapted from Scarf et al., 1981a)

outward in a combination of diffusion and centifugal acceleration and populate the equatorial plasma sheet with a relatively cold plasma.

Further out ($\sim$ 35-40 $R_J$) the plasma becomes hot with characteristic temperatures of the order of 20-40 keV (Krimigis et al., 1979), and provides the primary pressure which both inflates the Jovian magnetosphere much beyond the distance which can be supported by the planetary magnetic field ($\sim$ 40 $R_J$), and pushes the planetary bow shock to distances as far as 110 $R_J$. The plasma more or less corotates all the way to the dayside planetary magnetopause (Krimigis et al., 1979; Carbary et al., 1981), although a marked departure from corotation has been noted in the middle ($10 \lesssim r \lesssim 30$ $R_J$) magnetosphere (McNutt et al., 1979). On the nightside, the plasma sheet is generally thinner ($\sim$ 5 $R_J$, Behannon et al., 1981), and the plasma flow changes from the corotation to the anti-sunward/anti-Jupiter direction at a distance of $\sim$ 130-150 $R_J$, (Krimigis et al., 1979; Carbary et al., 1981). The escaping plasma on the nightside forms a "magnetospheric wind" consisting primarily of heavier ions (sulfur and oxygen), (Krimigis et al., 1979) which begins inside the magnetopause and extends much beyond the magnetosheath of the planet. The estimated ion loss rate through the wind is $\sim 2 \times 10^{27}$ sec$^{-1}$, corresponding to an energy loss rate of $\sim 2 \times 10^{13}$ watts (Krimigis et al., 1981a). The hot magnetospheric plasma ions undergo charge exchange within the magnetosphere and appear as energetic neutrals in the inter-

planetary medium (Kirsch et al., 1981a). Finally, a well-formed
magnetotail exists (Ness et al., 1979a), whose lobes are populated by
trapped continuum radiation (Gurnett and Scarf, 1981), and whose length
extends to perhaps 6200 $R_J$ downstream from Jupiter (Scarf et al.,
1981b).

The presence of Jupiter's magnetosphere was detected in the interplanetary medium long before the spacecraft approached Jupiter. Figure
8 shows the trajectories of the Voyager 1 and Voyager 2 spacecraft
within ~ 1000 $R_J$ (0.45 AU) from Jupiter. Each black dot denotes the
observation of field-aligned ion beams first observed ~ 800 $R_J$ upstream
and at least as far as ~ 1200 $R_J$ downstream from Jupiter (Zwickl et al.,
1980). Figure 8 also shows two mass histograms of ions in the range 0.6

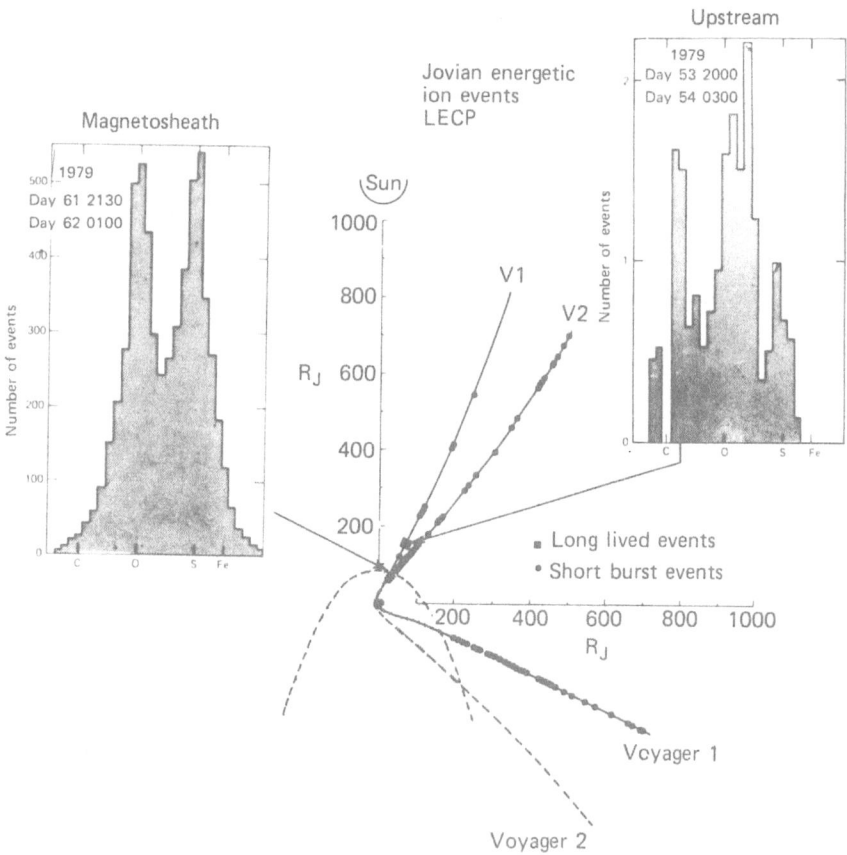

FIG. 8 - Location of upstream Jovian ion increases for Voyager-1 and
2; mass histograms (on the right) suggest that the origin of these ions
was the magnetosphere (left) (adapted from Zwickl et al., 1981).

to 1 MeV/nucleon; the one on the left was obtained in the magnetosheath while the one on the right was measured during one of the events upstream from the planet. It is evident from the histogram on the right that the upstream events contain substantial enhancements at oxygen and sulfur and that the composition is similar to the that observed in the magnetosheath as well as in the magnetosphere of the planet (Zwickl et al., 1981; Hamilton et al., 1981). These observations suggest that the upstream ions are definitely of Jovian origin and are most likely leaking out of the magnetosphere rather than accelerated at the planetary bow shock and foreshock regions. This observation is significant in the current discussion of ions present at the Earth's foreshock (e.g., Scholer et al., 1979), which are thought to be accelerated at the Earth's bow shock. The Jovian observations would suggest that leaking of already accelerated magnetospheric ions may be a more likely possibility.

The enormity of Jupiter's magnetosphere is illustrated by the data presented in Figure 9 which covers a period of $\sim$ 40 days beginning at distances $\sim$ 140 $R_J$ upstream and extending to > 350 $R_J$ downstream from the planet. The three curves in the top panel represent (from top to bottom) low energy ions ($Z \geq 1$), protons and medium ($Z \geq 6$) nuclei. It is evident, as mentioned in connection with the previous figure, that low energy ions are present in the interplanetary medium long before the first encounter of the planetary magnetopause. Following the last magnetopause crossing, the intensities stay relatively constant until the spacecraft comes into the inner part of the magnetosphere ($\sim$ 20 $R_J$), at which point the intensities increase rapidly until closest approach at 10.1 $R_J$. (The two vertical lines denoting the scale change indicate an intentional decrease in the instrument field-of-view by factors of 30 to 100, so as to keep the instrument operating in a linear range through closest approach). Following closest approach, the basic asymmetry of the magnetosphere between the day and night begins to manifest itself, with the first encounter of the nightside magnetodisc occurring at $\sim$ 22 $R_J$; these encounters continue in a periodic fashion with a 10-hour period to $\sim$ 130 $R_J$. Although there is gross similarity among the three particle species presented in the figure, there exist differences in the detailed profiles which are discussed elsewhere (Krimigis et al., 1981a). The 10-hour periodicity is generally disrupted at $\sim$ 150 $R_J$ with a large heavy ion spike (to be discussed in more detail in connection with Figure 16), and large increases in the intensities of heavy ions at both higher and lower energies. This change in the periodicity represents the point where the plasma flow changes from corotation to the tailward direction, and has been interpreted as the onset of the magnetospheric wind (Krimigis et al., 1979).

The second panel of Figure 9 shows the intensity profiles of low energy electrons which are generally similar to those of energetic ions presented above. The third panel shows the proton/alpha particle ratio in the indicated energy interval; the ratio displays large values during some of the discrete upstream increases and continues to increase closer to the planet, with maximum values of $\sim$ 300. Generally the ratios within the magnetosphere are substantially larger than what is typically observed in the solar wind or solar energetic particle events ($\sim$ 10-

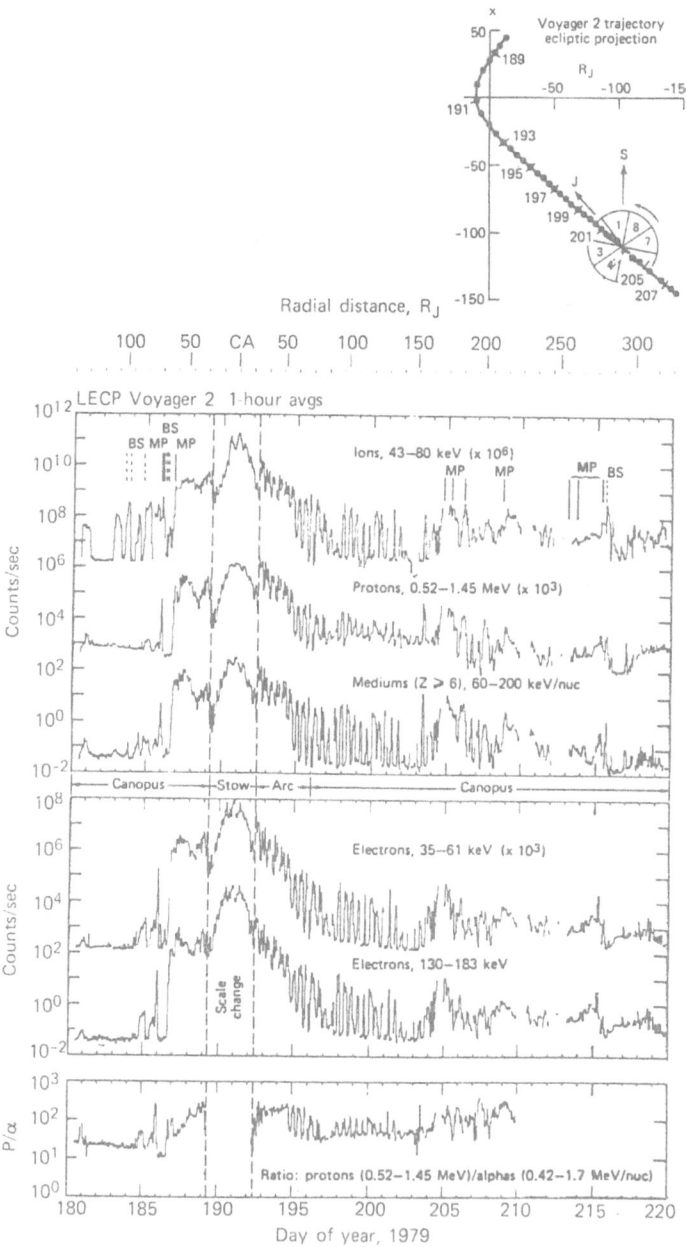

**FIG. 9** - An overview of particle intensity profiles during the Voyager-2 encounter of Jupiter as measured by the Low Energy Charged Particle (LECP) experiment (adapted from Krimigis et al., 1981a).

30). Note that the ratio in the magnetospheric wind region (i.e., after day 203) closely resembles that of the inner part of the magnetosphere and is at least an order of magnitude higher than values expected in the solar wind.

A most revealing view of the structure of the magnetosphere is provided by the first order ion anisotropies, which can be taken as indicative of convective plasma flow direction (Krimigis et al., 1979; Carbary et al., 1981). A summary of the overall anisotropy profile at low energies for both Voyagers 1 and 2 is presented in Figure 10. Here the projection of the first order anisotropy vector onto the ecliptic plane are drawn on the spacecraft trajectories from ~ 150 $R_J$ upstream to ~ 325 $R_J$ downstream of the planet (Krimigis et al., 1981a). The model magnetopauses from both encounters (Ness et al., 1979b) are shown for reference. We note that there exist two general flow directions in the Jovian magnetosphere, one which points in the direction of planetary corotation and obtains throughout the dayside and on the nightside to distances of ~ 130-160 $R_J$; beyond that distance and before crossing the magnetopause (identified by the plasma instrument of Bridge et al., 1979b), the flow direction changes to that of an anti-sunward/anti-Jupiter direction and continues to large distances away from the planet. These results are the basis of the schematic representation of corotation and anti-sunward flow presented in Figure 7.

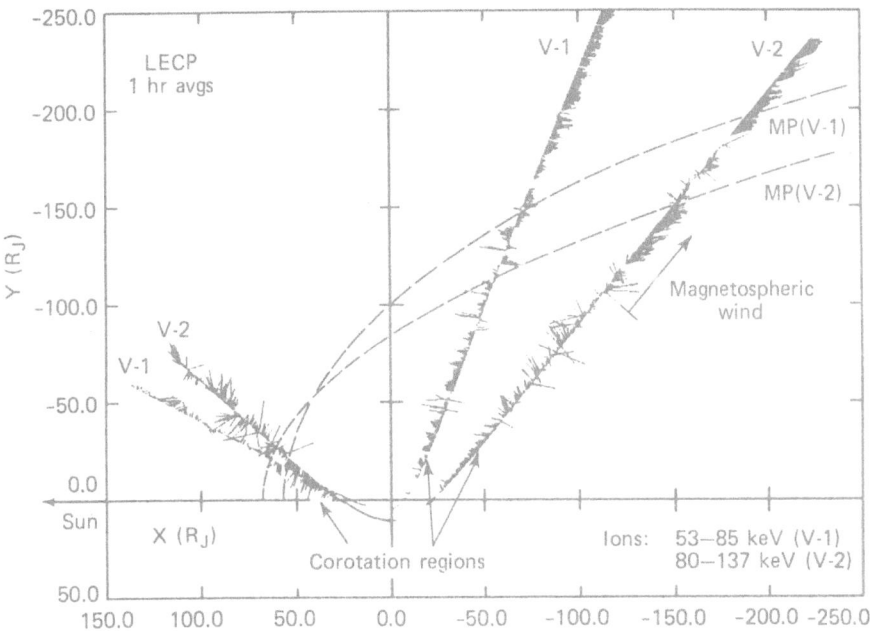

FIG. 10 - A projection of the low energy ion first order anisotropy to the ecliptic plane at Jupiter for the Voyager-1 and 2 encounters; the first-order anisotropies are indicative of underlying plasma flow pattern in the magnetosphere (adapted from Krimigis et al., 1981a).

In addition to the overall intensities and anisotropies, the form of the energy spectra and the ion composition represent essential inputs in the determination of density and pressure profiles, and in discussions of particle acceleration mechanisms. To illustrate the ion ($Z \geq$ 1) spectral form at low energies, we show in Figure 11 a typical spectrum measured in the outer magnetosphere by Voyager 2. It is evident that the shape of the spectra does not conform with a power law ($dj/dE \propto E^{Y}$) over the entire energy range, but rather exhibits significant flattening at the low energy range. The spectrum has been measured in a direction ~ 90° to the anisotropy vector, so as to be characteristic of the spectrum in the plasma rest frame (Ipavich, 1974). Above 200 keV, the shape of the spectrum seems to be consistent with a power law with the indicated exponent. Assuming that at 90° the detector is responding to protons (Krimigis et al., 1981a), we have obtained a fit to the three lowest energy points by assuming a convected Maxwellian distribution (Krimigis et al., 1979). The parameters providing the best fit are a convective velocity of ~ 700 km/sec, a temperature of 29 keV and a density of $8 \times 10^{-4}$ cm$^{-3}$. Similar temperatures have been obtained for most of the magnetosphere for both encounters and represent the basis for labelling the plasma in the outer magnetosphere in the schematic Figure 7 with a temperature of 20-40 keV (Krimigis et al., 1981a).

The composition characteristics of most of the Jovian magnetosphere at higher energies is presented in Figure 12. Figure 12a shows a histogram of light elements which reveals the presence of molecular hydrogen ($H_2$, $H_3$) in addition to the expected peaks of hydrogen and helium (Hamilton et al., 1980). The presence of $H_3$ which is expected to be a consituent of the Jovian upper ionosphere is taken as evidence that the ionosphere is a significant plasma source, at least for the outer part of the Jovian magnetosphere. The ionospheric source of hydrogen may well be the cause of the large values of the p/$\alpha$ ratios shown in Figure 9. Figure 12b shows a histogram of heavier ions (Krimigis et al., 1979; Hamilton et al., 1981) which reveals the presence of peaks at the location of carbon, oxygen, sodium and sulfur, and is characteristic of the composition throughout the Jovian magnetosphere. As pointed out previously O, S, and Na originate at Io. Figure 12c shows the dependence of the elemental abundance ratios (normalized to He) on radial distance from the planet. It is evident that there exist substantial changes in the relative abundances especially for the oxygen and sulfur ratios as the spacecraft moved to closest approach. These changes, however, are found to be related to the spectral changes in the energy/nucleon representation (Hamilton et al., 1981).

To investigate the spectral dependence noted above, a set of spectra for the various species in the outer magnetosphere is presented in Figure 13a. As is evident from the figure, all species have different slopes so that the relative abundances depend on the energy at which the ratios are evaluated. Hamilton et al. (1981) have shown that the most appropriate representation of the spectra in Jupiter's magnetosphere is on the basis of energy/charge. Their result is shown in Figure 13b which uses the identical data as those presented in Figure 13a, and also extends the energy range to lower energies where the

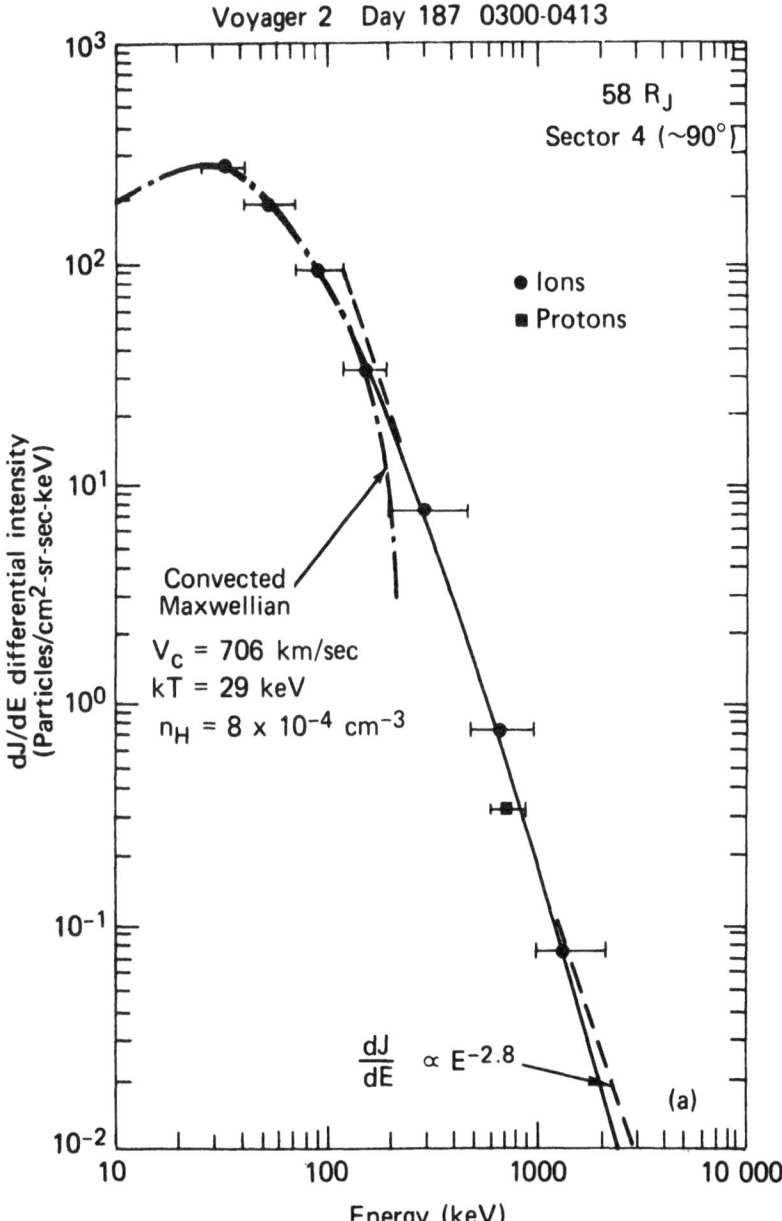

FIG. 11 — An example of the energetic ion spectra observed within Jupiter's magnetosphere; a convected Maxwellian with the indicated parameters fits well at low energies, and a power law obtains at higher energies (adapted from Krimigis et al., 1981a).

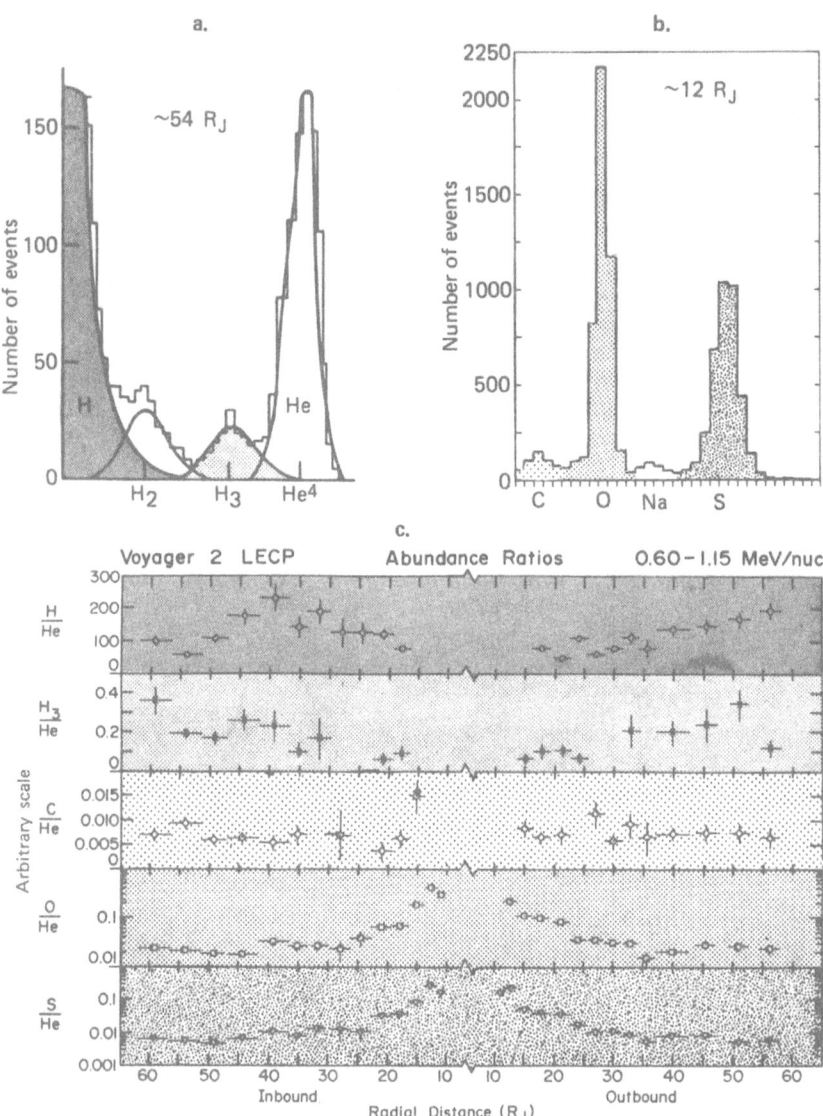

FIG. 12 - Elemental composition measurements by the LECP instrument within the Jovian magnetosphere. (a) mass histogram showing the presence of molecular hydrogen, $H_2$ and $H_3$. (b) mass histogram representative of the composition throughout the magnetosphere. (c) variation with distance to the planet of the abundances of the elements indicated relative to He (adapted from Krimigis et al., 1979 and Hamilton et al., 1981).

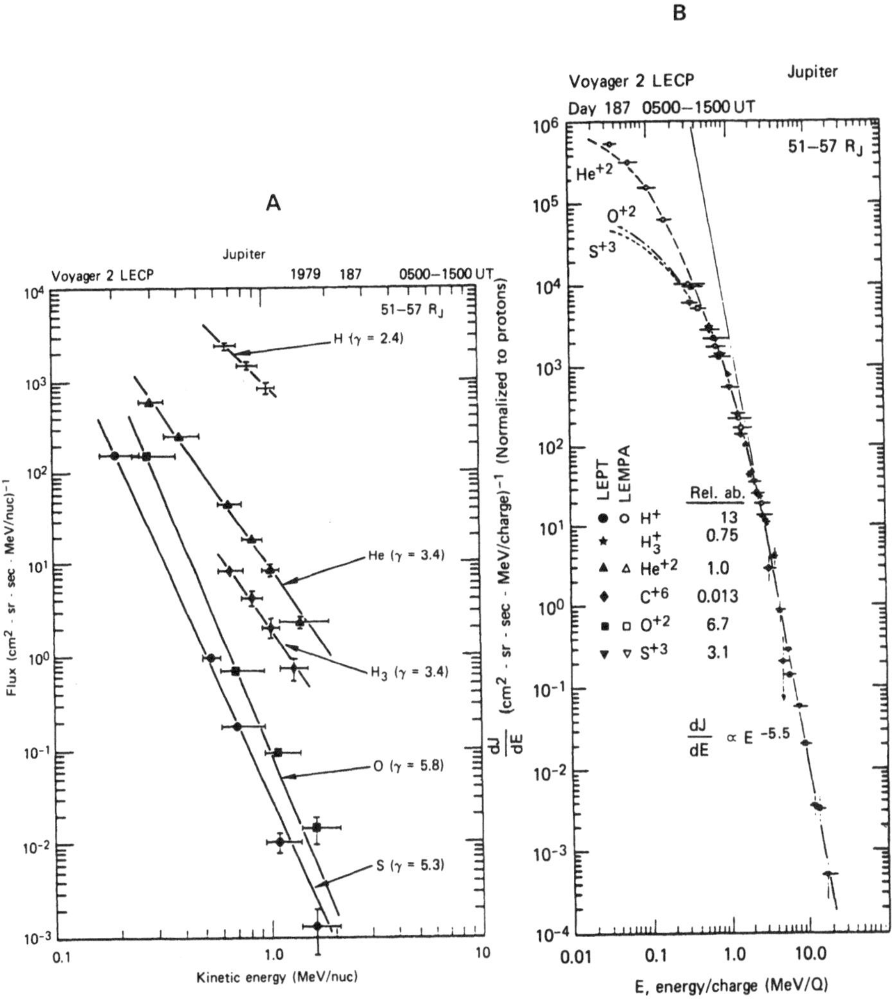

B

Voyager 2 LECP

Day 187 0500-1500 UT

Jupiter

51-57 R_J

$He^{+2}$

$O^{+2}$

$S^{+3}$

LEPT LEMPA  Rel. ab.
● ○ $H^+$   13
★   $H_3^+$  0.75
▲ △ $He^{+2}$  1.0
◆   $C^{+6}$  0.013
■ □ $O^{+2}$  6.7
▼ ▽ $S^{+3}$  3.1

$\dfrac{dJ}{dE} \propto E^{-5.5}$

E, energy/charge (MeV/Q)

$\dfrac{dJ}{dE}$ (cm² · sr · sec · MeV/charge)⁻¹ (Normalized to protons)

A

Jupiter

Voyager 2 LECP   1979  187   0500-1500 UT

51-57 R_J

H (γ = 2.4)

He (γ = 3.4)

$H_3$ (γ = 3.4)

O (γ = 5.8)

S (γ = 5.3)

Flux (cm² · sr · sec · MeV/nuc)⁻¹

Kinetic energy (MeV/nuc)

FIG. 13 - (a) Energy spectrum of the indicated species plotted versus energy per nucleon (velocity). (b) Differential intensities of the same species but plotted as a function of energy per charge (from Hamilton et al, 1981).

species are not identified uniquely.  Here the intensities have been evaluated at a given energy/charge and the curves representing each element were displaced vertically by a factor indicated in the figure as "relative abundances".  A charge state of +2 was assumed for oxygen and +3 for sulfur, primarily based on the charge states inferred from the UV experiment data (Broadfoot et al., 1979).  It is evident that in this representation, the energy spectra of all species fit on a common curve, and the relative displacements form the basis for estimating the relative number densities for all species.  We note from the relative abundances that the number of $Z \geq 2$ ions is comparable to the number of protons.  Thus, by inference from the energetic ions, the Jovian plasma is unique among magnetospheric plasmas in that it is dominated by ions heavier than hydrogen.

Coming to the inner magnetosphere, the basic aspects of the radiation belts of Jupiter were established by the Pioneer measurements and have been reviewed extensively in the literature (e. g., Thompsen, 1979), so that they will not be discussed here in any detail.  The new aspects are associated with the discovery of the plasma torus, the million ampere field-aligned current at Io (Ness et al. 1979c) and the inclusion of plasma wave and radio wave detectors on the Voyager experiment payload.  Figure 14 (Scarf et al., 1981a), gives a schematic of the plasma torus indicating the connection to the observed UV aurora region (emission of $\sim 10^{14}$ watts), and the expectation that the aurora is due to trapped particles which are precipitated because of their interaction with the plasma torus.  On top of the figure is shown an inset from a frequency-time diagram which shows the presence of hiss and chorus emissions, similar to those observed in the Earth's aurora region (Gurnett and Scarf, 1981).  The right hand vertical axis is labeled with the electron parallel energy which would be in cyclotron resonance with parallel propagating whistler waves.  The energy deposition from precipitating electrons of $E > 100$ keV is estimated to be $\sim 3$ ergs/cm$^2$/sec over the latitude range 65°-70°, with a total power dissipation of $\sim 10^{13}$ watts (Thorne, 1981).  This, however, is far less than the observed auroral UV emission, and several authors suggest that precipitating ions are the predominant energy source for auroral emissions (Goertz, 1980; Scarf et al., 1981a; Thorne, 1981).

The general association of the energetic ion profile with plasma waves is illustrated in Figure 15, which consists of the ion phase space density observed by Armstrong et al., 1981 (bottom panel), and the plasma wave measurements of Gurnett and Scarf (1981).  It is evident from the figure that the ion phase space density begins to decrease at the edge of the Io torus and drops by well over 3 orders of magnitude inside the orbit of Io, in close association with the presence of plasma wave activity, especially in the range from 1 to 10 kHz.  Although the waves specifically responsible for the loss of the ions have not been identified, there exists broadband impulsive emission with little frequency dispersion in the range of 2-12 kHz as indicated in the inset at the top of Figure 15.  Thorne (1981), concludes that precipitation of these ions in the few hundred keV range can account for $\sim 10^{14}$ watts, which is comparable to the auroral energy requirements.

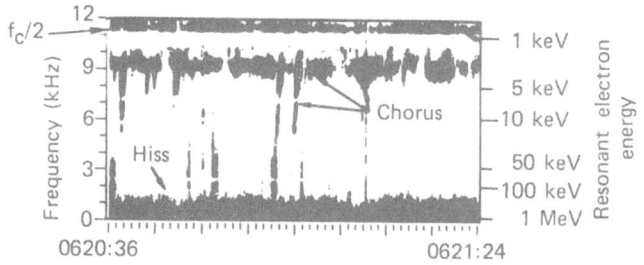

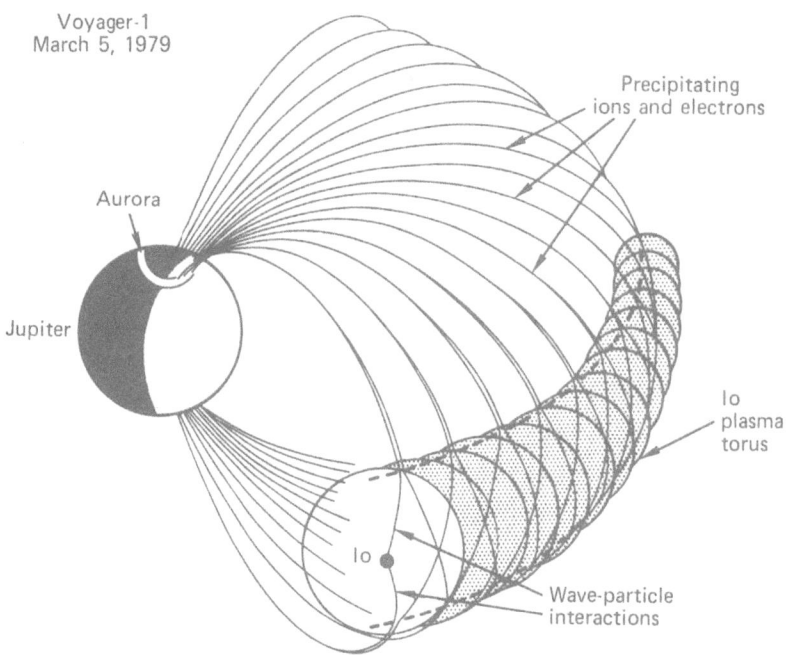

FIG. 14 - A schematic representation of the Io torus indicating the presence of Alfven waves and particle precipitation. The plasma wave frequency-time spectrogram on top indicates the presence of waves which may be interacting with the particles to cause precipitation (adapted from Scarf et al., 1981a).

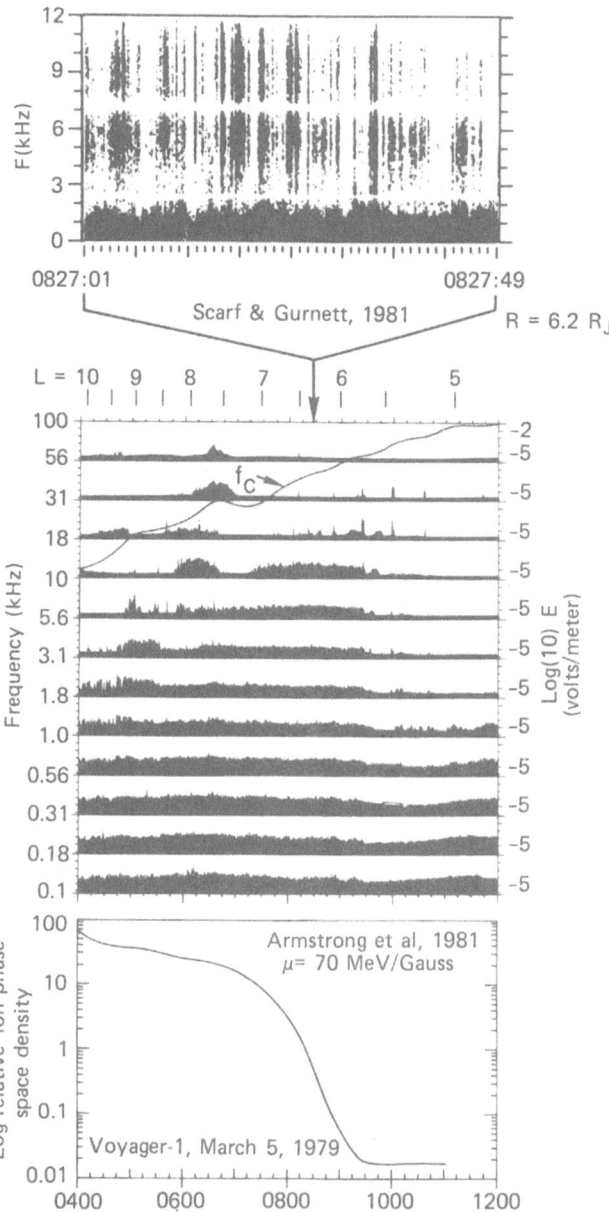

FIG. 15 - Phase space density of energetic ions across the Io plasma torus shown at bottom (Armstrong et al., 1981), on the same scale as plasma wave activity in the indicated frequency intervals. Note the the absence of activity above 1 kHz after the decrease in phase space density (adapted from Gurnett and Scarf, 1981).

At higher frequencies ($\geq$ 7 MHz) Jupiter has been known as a prodigious producer of radio waves. The instrumentation on Voyager has extended these measurements down to a few kHz and obtained polarization and total power profiles. The total radiated power over the range $\sim$ 30 kHz to $\sim$ 40 MHz is $\sim$ 4 x $10^{11}$ watts (Alexander et al., 1981), while at kilometric wavelengths it is $\sim$ 3 x $10^{8}$ watts, i.e., comparable to that at Earth.

Foremost among the many remarkable aspects of Jupiter's magneto-sphere is the hot, high beta (thermal energy density exceeding magnetic energy density) plasma in its outer parts, and the large number of energetic ions and electrons. Several acceleration mechanisms have been proposed including magnetic pumping (Nishida, 1976; Goertz, 1978), and charge exchange of ions in the plasma torus with consequent escape and acceleration in the outer magnetosphere (Cheng, 1980). Both mechanisms are probably operating within the magnetosphere, but others may also be important. A characteristic example of an acceleration process with a strange signature has already been reported in the literature (Krimigis et al., 1980); a beam of energetic ions with peak energies well over 100 keV were observed in the nightside magnetosphere at the point where plasma flow changes from corotation to the tailward direction.

The energy spectrum from this particular event is shown in Figure 16, based on an inference from observations at higher energies that the detector was primarily responding to sulfur ions (Krimigis et al., 1980). The data, plotted in the energy/nucleon representation, show a spectrum peaked at an energy of $\sim$ 8 keV/nuc, representing a total ion energy of $\sim$ 265 keV. This spectrum can be fit extremely well with a convected Maxwellian moving in the direction away from Jupiter at a speed of 1195 km/sec with a kT of $\sim$ 5 keV. Note that the thermal spread ($\sim$ 0.16 keV/nuc) is about as good as what one gets out of some labora-tory accelerators. This spectrum persisted for several · hours and, suggests that the entire plasma distribution must have passed through a potential drop of $\sim$ 130 kilovolts, if it is assumed that the sulfur ions are doubly ionized. The spectrum also includes a non-thermal tail at higher energies described reasonably well by a power law, a feature not uncharacteristic of spectra in both Jupiter's and the Earth's magneto-spheres. Alternatively, the non-thermal tail could be due to hotter, lighter ions such as O, He, and H. It is evident that such spectra will continue to challenge our current understanding of acceleration mechan-isms, not only in Jupiter's magnetosphere but in space plasmas in general.

Finally, Figure 17 shows a rather well known plot from Pioneer 10 (Pyle and Simpson, 1977) to remind us all of the fact that Jupiter is the source of what we used to call "cosmic ray" electrons. This work, coupled with the upstream and downstream ions observed hundreds of Jovian radii away from Jupiter, as discussed previously, should cause us to re-examine the hypothesis that Jupiter could perhaps be a significant source of low energy cosmic rays in the solar system (Krimigis et al., 1975). These ions (and neutrals, see Kirsch et al., 1981a) could be subsequently accelerated to higher energies via interactions with the solar wind (Fisk, 1976). To the best of my knowledge, a detailed

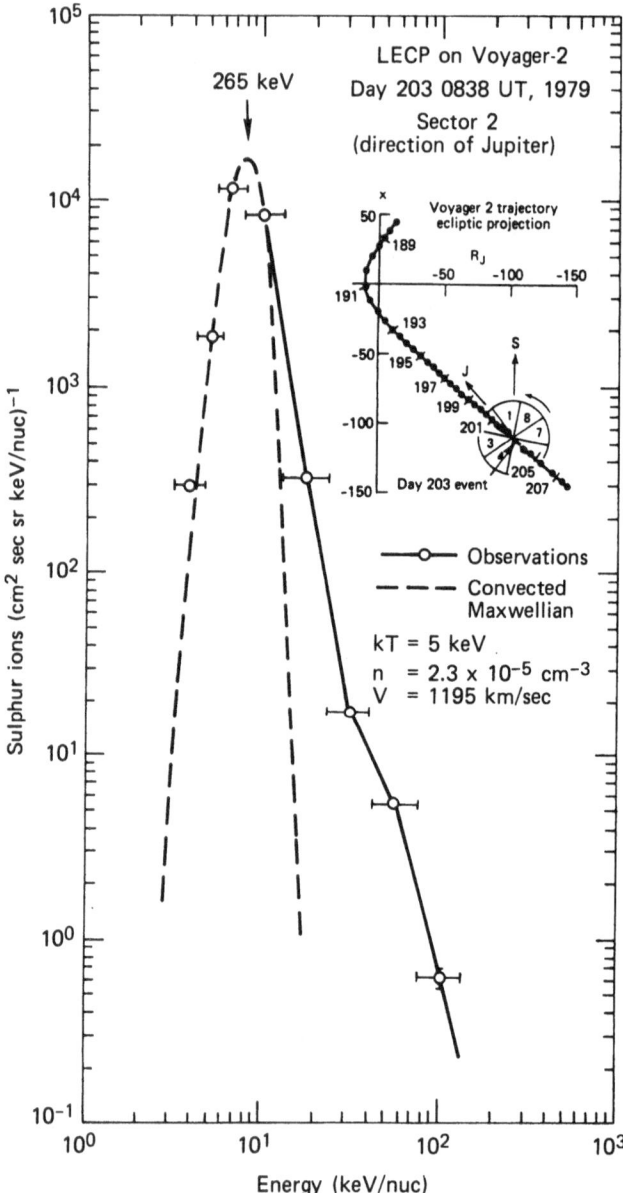

FIG. 16 — Differential energy spectrum of ions at the Jovian corotation boundary, deduced to be sulfur (Krimigis et al., 1980); the spectrum is represented well by a convected Maxwellian with the indicated parameters.

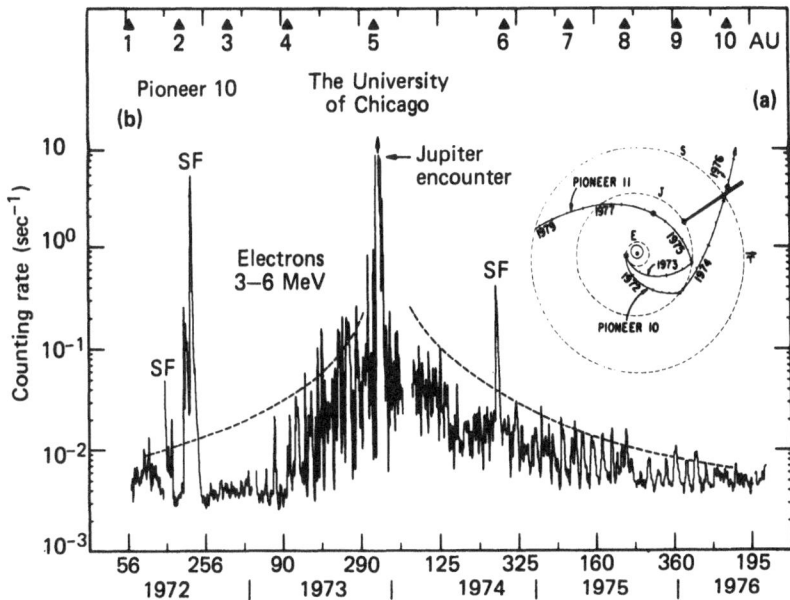

FIG. 17 — A vivid demonstration of the strength of Jupiter as a source
of high energy electrons (Pyle & Simpson, 1977).

examination of Jupiter as possible source of these ions has not been
carried out as yet.

The Magnetosphere of Saturn

     Saturn's magnetosphere has been investigated by two spacecraft so
far, Pioneer 10 in September of 1979 and Voyager-1 in November of 1980;
Voyager 2 will encounter the planet in late August of 1981. A substantial amount of knowledge about Saturn's magnetosphere has been collected
from the two encounters, some of which is summarized in Figure 18, based
primarily on the measurements of low energy ions and electrons ($>$ 30
keV) (Krimigis et al., 1981b). The magnetosphere is substantially
smaller than that of Jupiter, even though Saturn's radius is only marginally smaller (60,000 km compared to 71,400 for Jupiter) and the rotation period of both planets is similar. One of the notable aspects of
energetic particle observations in Saturn's magnetosphere is due to the
fact that Titan is sometimes inside (as was the case during the Voyager
1 encounter) and sometimes outside the magnetosphere (as was the case
during the Pioneer 11 encounter). The region around Titan's orbit
contains relatively soft protons (Krimigis et al., 1981b) which are
generally convected close to the corotation velocity of the planet
(Bridge et al., 1981; Maclennan et al., 1981). A most surprising aspect
of the pitch angle distribution of electrons and ions in the magneto-

Saturn's Magnetosphere

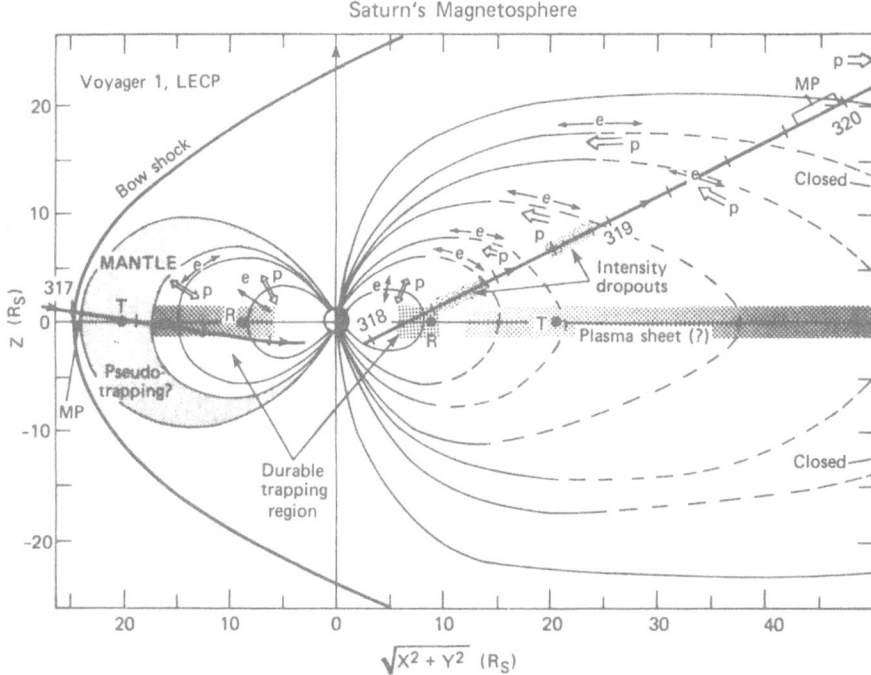

FIG. 18 - A schematic view of Saturn's magnetosphere, emphasizing the pitch angle distributions of ions (double arrows), and electrons (single arrows). The trajectory of the spacecraft is indicated by the thick heavy line (Krimigis et al., 1981b).

sphere of Saturn is that, although both electrons and protons have pancake (i.e., perpendicular to $\vec{B}$) pitch angle distributions inside the orbit of Rhea (indicated by arrows on the field lines), the electron distribution outside this orbit is more or less field-aligned both on the dayside and nightside magnetosphere. Further, the pitch angle distributions seem to be strong functions of energy (for details see November 1980 issue of Journal of Geophysical Research). Another unexpected feature of Saturn's magnetosphere is the presence of energetic particles on the nightside tail lobes, suggesting that field lines in the high latitude magnetotail are closed. This is in sharp contrast to the lobes of both Earth and Jupiter, where field lines are generally open and devoid of energetic particles. The plasma sheet at Saturn extends inside the orbit of Tethys, but is cut-off at ~ 4 $R_S$ (Frank et al., 1980). On the nightside, the plasma sheet extends to at least outside the orbit of Dione and perhaps farther, although as seen from Figure 18, the Voyager 1 trajectory was not optimal for determining the nightside plasma sheet (Bridge et al., 1981).

Just as in the case of Jupiter, a torus has been discovered to exist within the magnetosphere of Saturn and it extends from outside the orbit of Titan to about the orbit of Rhea as seen in Figure 19. The density of neutral hydrogen in the torus is thought to be uniform at $\sim 10$ cm$^{-3}$ (Broadfoot et al., 1981); the hydrogen source is thought to be the upper atmosphere of Titan through the interaction of hot electrons from the magnetosphere. Energetic particles trapped within the magnetosphere also interact with the neutral hydrogen via charge exchange and escape the magnetosphere (Kirsch et al., 1981b).

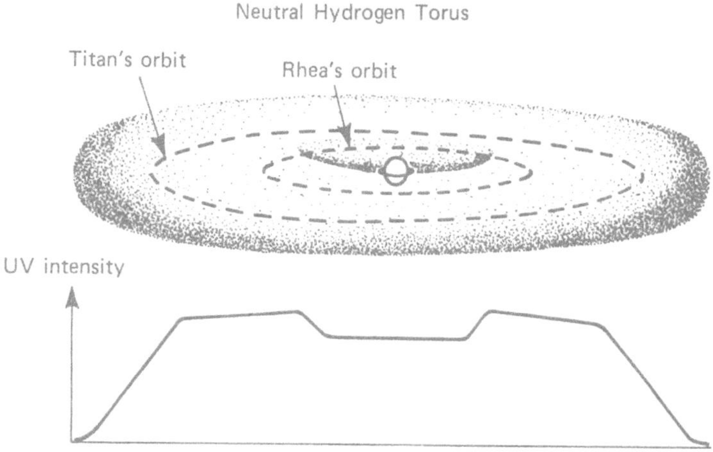

FIG. 19 - A schematic representation of the Titan torus (top), and the intensity profile of the UV signal observed by the Voyager-1 instrument (Broadfoot et al., 1981).

The satellites of Saturn drastically affect the trapped radiation within Saturn's magnetosphere as discovered by the Pioneer investigators (Van Allen et al., 1980; Simpson et al., 1980; McDonald et al., 1980). Figure 20 (Krimigis et al., 1981b), shows the phase space density of energetic ions as a function of radial distance for both inbound and outbound trajectories of Voyager-1. As noted previously, in this type of plot the phase space density should be constant as a function of radial distance, if diffusing particles from the outer part of the magnetosphere do not suffer any loses. This is clearly not the case at Saturn, where losses begin to appear on the inbound trajectory somewhere inside 12 $R_S$ and significant changes in the profile occur at the orbit of Rhea, Dione and Tethys. The overall decrease in phase space density from 15 $R_S$ to $\sim 4$ $R_S$ is some $\sim 10^5$. There is some speculation that the losses inside the orbit of Dione are due to the presence of tenuous ring-type material orbiting the planet. Inside 4 $R_S$ there are very few energetic ions, but there do exist substantial fluxes of high energy ($> 80$ MeV) protons close to the planet which are thought to be due to neutron albedo interactions of cosmic rays with the planetary atmosphere (Fillius and McIlwain, 1980).

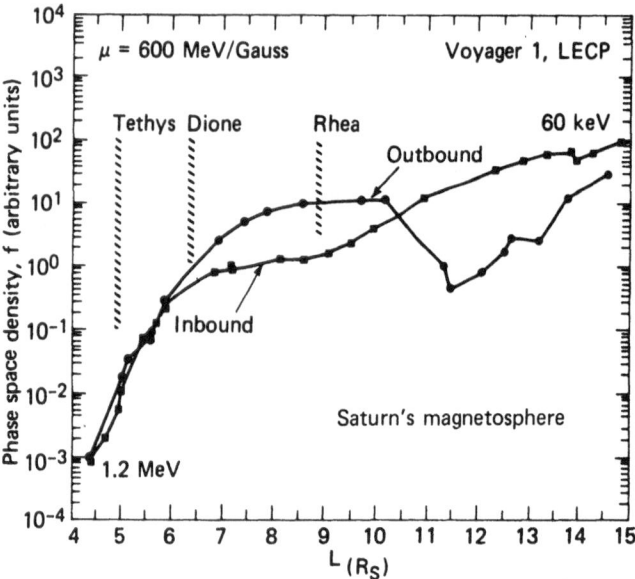

<u>FIG. 20</u> - The phase space density of energetic protons in the magneto-
sphere of Saturn (Krimigis et al., 1981b).

The composition of energetic particles in Saturn's magnetosphere is
not nearly as diverse as that of Jupiter's. Figure 21 shows a pulse
height matrix obtained on the inbound leg of the Voyager-1 encounter in
the indicated energy range and is compared to a similar pulse height
matrix from the Voyager-Jupiter encounter. There is a very pronounced
peak at the location of molecular hydrogen ($H_2$) and the location of $He^4$,
in contrast to the case at Jupiter however, no $H_3$ is present. We note
that although $H_3$ is expected to be an ionospheric constituent at Jupiter
(Atreya et al., 1974), no $H_3$ is expected to be present at the ionosphere
of Saturn (Waite et al., 1979). In fact the proton-to-helium ratios in
ᴸne magnetosphere of Saturn are as large of 5,000, i.e., larger than any
value observed in the magnetospheres of the other planets. This has
been interpreted as an indication that the primary source of plasma is
internal to Saturn (rather than the solar wind) and is likely to be the
upper atmosphere of Titan (Krimigis et al., 1981b).

A most important aspect of the magnetosphere of Saturn is the emission of kilometric radiation, a characteristic shared by both Jupiter
and the Earth. Figure 22 shows the frequency-longitude diagram of kilometric radiation (Kaiser et al., 1981) indicating that the radiation is
primarily concentrated at a particular local time relative to the Sun-
Saturn line. The power levels of Saturn Kilometric Radiation (SKR)
range from about $10^8$ at 50% occurrence probability to about $10^{10}$ watts

at the 1% level of occurrence probability (Kaiser et al., 1981). A more surprising aspect of SKR, first noted by Gurnett et al. (1981), was the modulation by the orbital phase angle of Dione. Figure 23 shows the intensity of SKR at 56.2 kHz before, during and after the Voyager encounter in November 1980. It is evident that the intensity is modulated with a period of ~ 3 days, which compares favorably with the orbital period of Dione at 2.74 days. This modulation has since been confirmed at higher frequencies as well by Desch and Kaiser (1981) and is reminiscent of the control of kilometric and decametric radiation exerted by Io in the Jovian magnetosphere. This could well mean that Dione is a plasma source in Saturn's magnetosphere, as is suggested by the Pioneer 11 plasma observations (Frank et al., 1980), and is consistent with the Voyager plasma observations (Bridge et al., 1981).

## Magnetospheres of Uranus and Neptune

The magnetospheres of Uranus and Neptune have not as yet been investigated by spacecraft so very little is known about their properties. Voyager-2 is scheduled to encounter the magnetosphere of Uranus on January 24, 1986 and, if successful, continue on to encounter Neptune on August 24, 1989. Thus, by the end of this decade all magnetospheres of the solar system with the exception of the planet Pluto will have been investigated by spacecraft.

The magnetosphere of Uranus is a particularly interesting case because its rotation axis is inclined 8° to the ecliptic plane, so that twice in its orbit of 40 years about the sun the pole of the planet is pointing to within 8° of

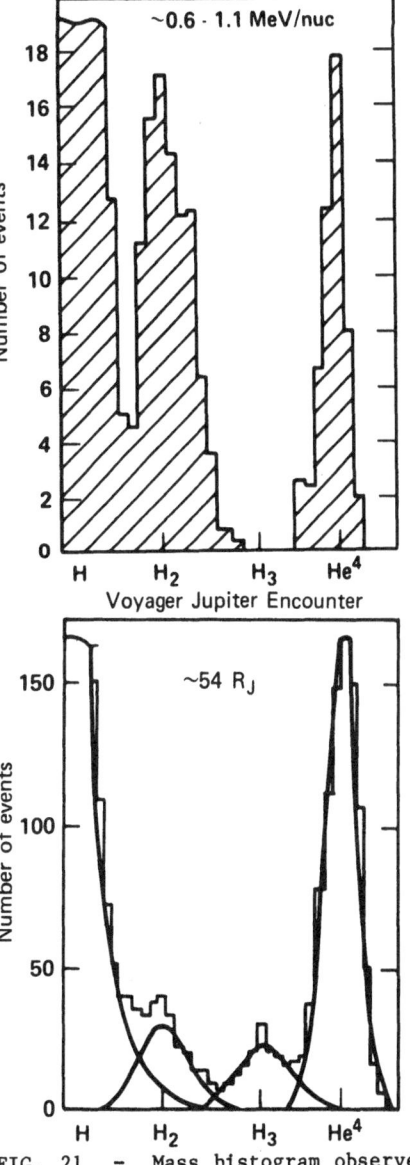

FIG. 21 - Mass histogram observed at Saturn by Voyager-1 (Krimigis et al., 1981b), compared to a similar histogram obtained in the magnetosphere of Jupiter.

Occurrence rate of SKR

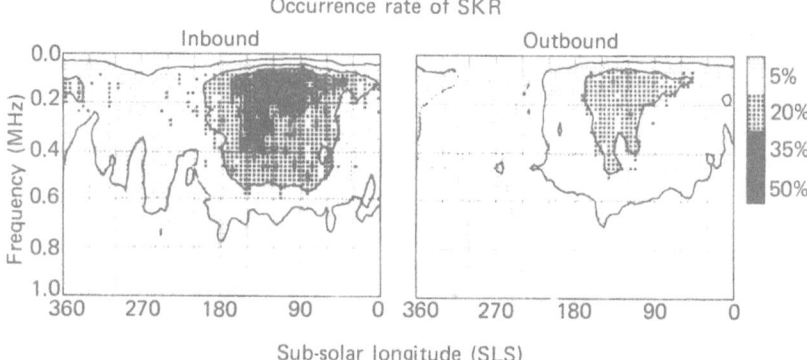

**FIG. 22** - The occurrence rate of Saturn's kilometric radiation above a level of $10^{-20}$ watt/cm$^2$ kHz is shown as a function of subsolar longitude and observing frequency of the Voyager-1 inbound and outbound observing positions (Kaiser et al., 1981).

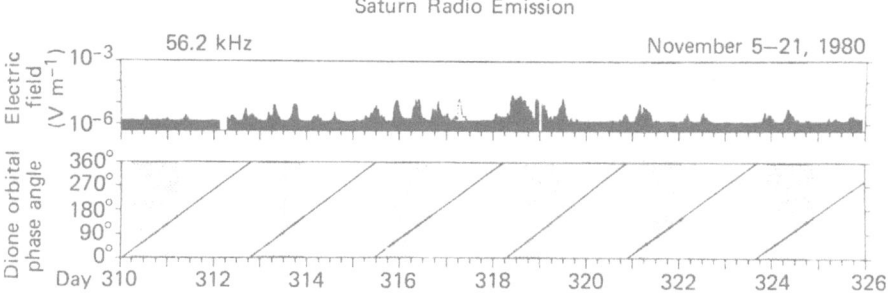

**FIG. 23** - A plot showing the long-term quasi periodic modulation of the Saturn radio burst intensities; the orbital position of Dione is superposed on the plot, and the association is rather obvious (Gurnett et al., 1981).

the solar direction. Although not much is known about the presence of a magnetic field at Uranus it is likely that it possesses one, even though its radius is a little less than half of that of Saturn and its rotation period is ~ 15.5 hours. A possible configuration of this "pole-on" magnetosphere is shown in Figure 24 (Siscoe, 1975). Investigation of this magnetosphere will provide a rather severe test to our understanding of magnetospheric structure and dynamics.

Neptune is similar in size to Uranus and has a period of 15.8 hours. We now know from the Pioneer 10 observations that the solar wind extends at least to 25 AU and is likely to exist much past the orbit of

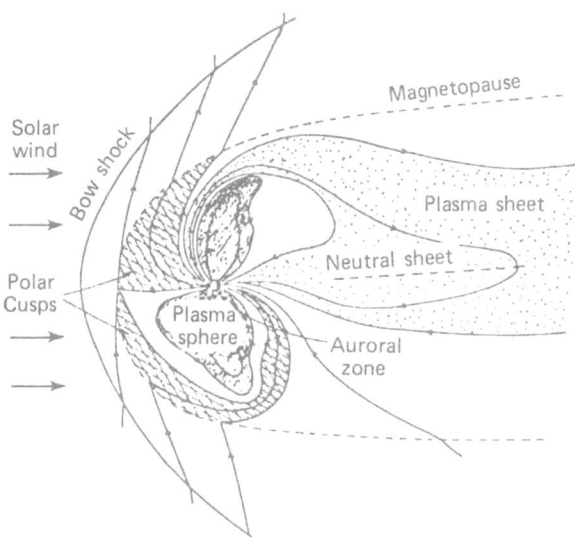

FIG. 24 - A possible configuration of a Uranian magnetosphere when the planetary axis is pointing in the general direction of the Sun (Siscoe, 1975). This is the expected orientation during the Voyager-2 encounter in January 1986.

Neptune, at a distance of 30 AU. It is likely that Neptune also possesses a magnetosphere, but it is rather pointless to speculate at this point on its characteristics.

## 5. GENERAL CHARACTERISTICS OF MAGNETOSPHERES

Despite the large diversity of interactions between the solar wind and planetary magnetospheres, there do emerge a number of characteristics which are shared by most magnetospheres, including plasma sources, some commonality in energy and plasma wave spectra, and radio waves.

### Plasma Sources

Figure 25 gives a schematic view of the plasma sources in the magnetospheres of Earth and Jupiter. In the Earth's magnetosphere (upper panel), both the solar wind and the ionosphere are important sources. Both plasmas are heated to several tens of keV and populate various regions of the magnetosphere. In the case of Jupiter (lower panel) the plasma sources are the solar wind (presence of $C^{+6}$), the ionosphere ($H_3$), and the Galileo satellites, especially the volcanoes and surface of Io (S, O, Na). Here the plasmas are again heated to several tens of keV and are present throughout the magnetosphere. Saturn's magnetosphere also has internal plasma sources, namely the upper atmosphere of

**Earth**

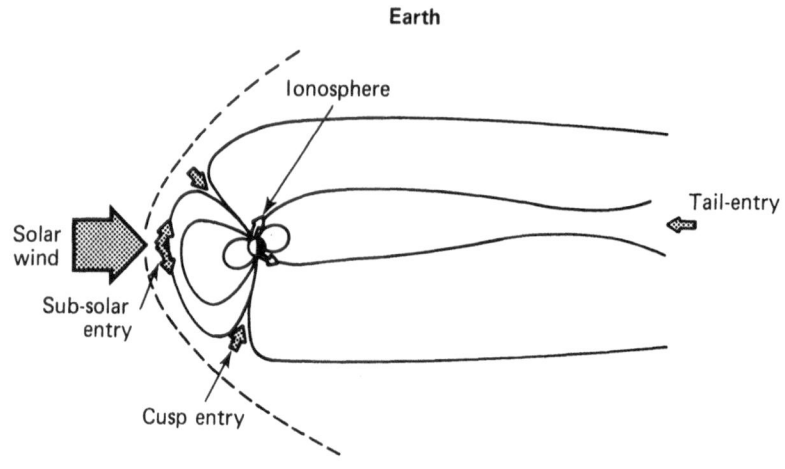

**Jupiter**

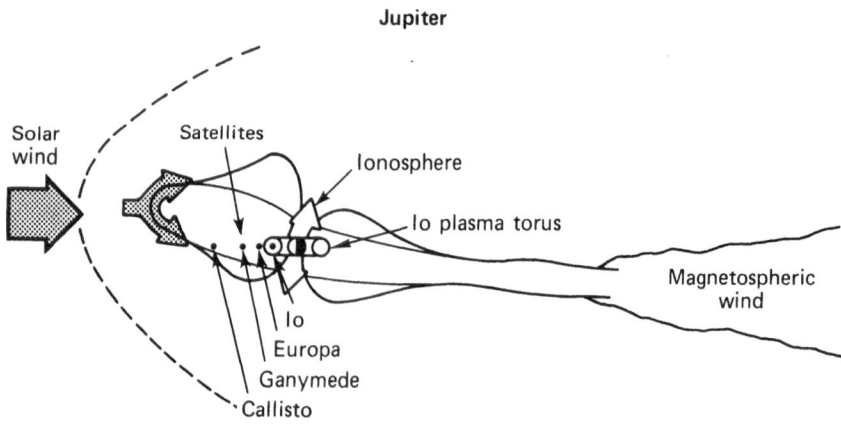

FIG. 25 — A comparison of plasma sources in the magnetospheres of Earth and Jupiter. In the case of Earth (top) both the solar wind and the ionosphere are important plasma sources, while in the case of Jupiter (bottom) the planetary satellites are apparently the most important plasma source.

Titan and possibly the ionosphere of the planet itself; no firm evidence of a solar wind tracer ion in Saturn's magnetosphere is yet available.

In the case of Mercury's magnetosphere, it is likely that the plasma is heated solar wind plasma since the planet possesses no measurable atmosphere. In Venus' magnetosphere, there have been observations of cold plasma which is mostly of ionospheric origin. Similar statements can be made about the inferred magnetosphere of Mars.

In summary, plasma sources of planetary magnetospheres are the solar wind, the ionospheres of the planets, and planetary satellites. Planetary satellite sources are only important in the outer planets, since they reside within the magnetospheric envelops. Ionospheric sources are dominant in those cases where an intrinsic magnetic field is likely to be absent (Venus, Mars), while the solar wind source is apparently dominant at Mercury and possibly Earth.

## Spectra

As indicated in the discussion of the magnetospheres of Earth (Figure 3) and Jupiter (Figure 11), energetic ion spectra tend to fall off as a power law in energy at relatively high energies and flatten or peak as a Maxwellian distribution at lower energies. The spectra at Saturn

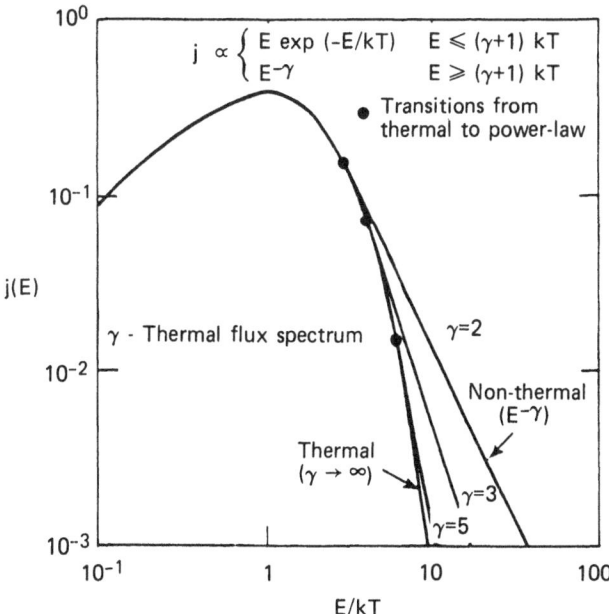

FIG. 26 - A "γ-thermal "distribution, with the energy normalized to kT (Krimigis and Roelof, 1981); such distributions are found to fit the data well at Jupiter (Krimigis et al., 1981a).

are similar in shape (Krimigis et al., 1981b), although a full analysis has not been completed. This spectral form can be generalized and expressed as γ-thermal distribution as shown in Figure 26 (Krimigis and Roelof, 1981). The spectrum expressed in the rest frame of the plasma can be described by

$$j(E) \propto \begin{cases} E \exp(-E/kT) & E \leq (\gamma + 1) \, kT \\ E^{-\gamma} & E \geq (\gamma + 1) \, kT \end{cases} \tag{1}$$

and is continuous across the transition point at $E = (\gamma + 1) \, kT$, from the Maxwellian to the power law. Although there exist other distributions, which have similar shapes (e.g., the $\kappa$-function, Olbert, 1968) the γ-thermal distribution is computationally quite easy to work with.

## Radio Emissions and Plasma Waves

As noted previously, kilometric radiation is present in three of the planets which possess significant magnetic moments namely the Earth, Jupiter and Saturn. Although the mechanism for kilometric radiation is presently not known, it is likely that the physical origin for this radio emission is common in all three cases.

In addition, plasma wave spectra obtained at the bow shocks of several planets show similar shapes. An example of these is shown in Figure 27 which presents spectra from the bow shocks of Venus, the Earth, Jupiter and Saturn (Scarf et al., 1981c). Here the electron plasma frequency ($f_{pe}$), the ion plasma frequency ($f_{pi}$), the electron cyclotron frequency ($f_{ce}$) and the Buneman mode characteristic frequency are marked for each spectrum. The spectral shape at Venus is not very well determined because the instrument only has four channels. Atlhough

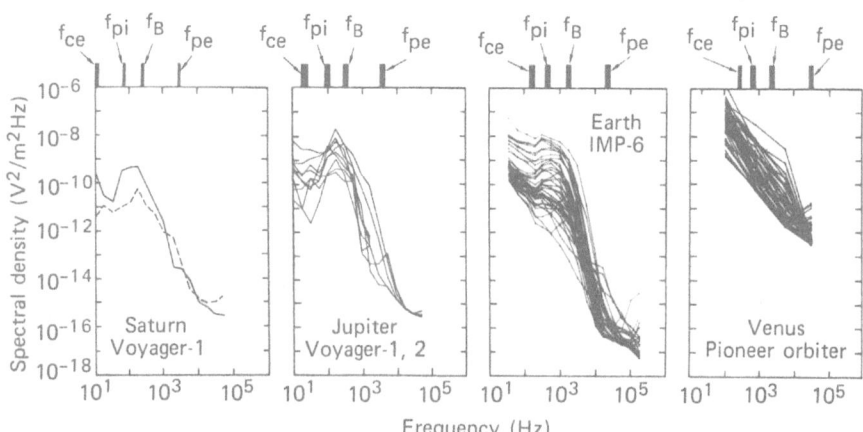

FIG. 27 - The plasma wave spectra at Saturn, Jupiter, Earth and Venus, adapted from Scarf et al., (1981c). The symbols denoting the frequencies at the top of the figure are explained in the text.

there are significant differences among the spectra, there are important similarities such as the presence of peaks close to $f_{pi}$ and $f_B$ at Earth, Jupiter and Saturn. This fact suggests that the basic processes at all bow shocks, over a range of some 10 AU, are very similar.

## Sizes and Scaling

The radii of the planets and the dimensions of their respective magnetospheres are shown in Figure 28 (Siscoe, 1979). Here the magnetospheres of Uranus and Neptune have been inferred by use of scaling rela-

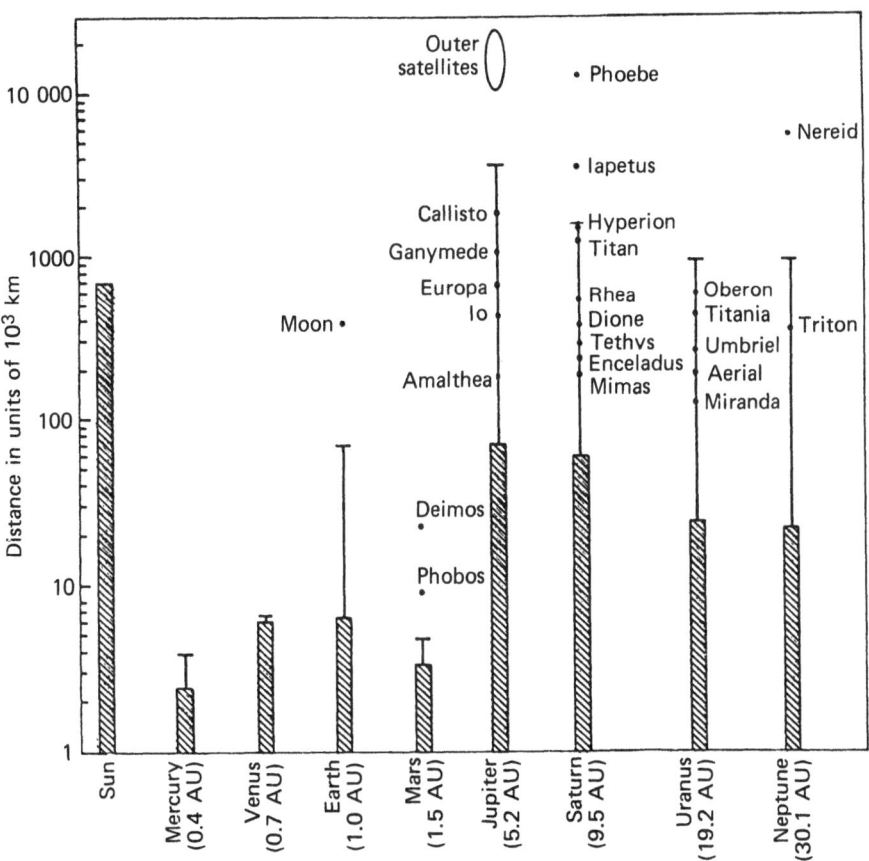

FIG. 28 — Dimensions of the radii of the sun and the planets (crosshatched), the radii of the orbits of their satellites (dots), and the characteristic dimensions of their magnetospheres (bars). The dimensions of the magnetospheres of Uranus and Neptune are estimated (adapted from Siscoe, 1979).

tions. From the obvious grouping of the inner and outer planets, we note that the magnetospheres of the outer planets are typically an order of magnitude larger than the planets themselves while the magnetospheres of the inner planets with the exception of Earth are generally quite small. The most obvious difference, however, is seen in the fact that the satellites of the inner planets lie outside the planetary magnetospheres while most of the satellites in the outer planets lie within the magnetosphere and as we know from the case of Jupiter and Saturn, interact strongly with the magnetosphere. More important, the magnetospheres of the outer planets are rotationally dominated and in that sense, very much different from the earth. Fairly sophisticated scaling laws attempting to deduce the size of magnetospheres in relation to other planetary and solar wind parameters have met with mixed success. For a full review of this work see Siscoe (1979).

An early attempt at predicting planetary magnetic moments using the so-called magnetic Bode's law (e.g., Kennel, 1973) has not been very successful in fitting the observations. Briefly, the magnetic Bode's law is a plot of the log of the magnetic moment of the planet against the log of its angular momentum. On this plot, the points are supposed to lie on a straight line. A recent re-evaluation of this relationship has been put together by Russell et al. (1980) and is shown in Figure 29. It is evident from the figure that the straight line drawn to connect the well known points of Earth and Jupiter has not been very successful in predicting the magnetic moments of Saturn, Venus or Mars. It has been suggested that the small magnetic moment of Mars is due to its relatively small size, while the magnetic moment of Venus, whose physical size is similar to Earth, is due to its slow rotation period. It is probably safe to conclude that there is no currently accepted theory of planetary dynamos which can account for the observations.

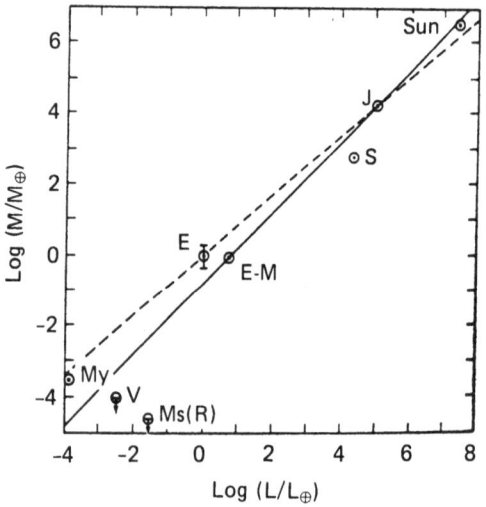

FIG. 29 - Magnetic moment of each planet normalized to the terrestrial moment (8 x $10^{25}$ Gauss $cm^3$), is plotted against angular momentum normalized to the terrestrial angular momentum. My, Mercury; V, Venus; Ms(R), Mars (Russell, 1979); E, Earth; E-M, Earth-Moon system; J, Jupiter; and S, Saturn. (adapted from Russell et al., 1980). This is known as Bode's law of planetary magnetism.

## 6. IMPLICATIONS FOR ASTROPHYSICS AND THE HELIOSPHERE

Ever since the discovery of the Earth's magnetosphere, astrophysicists have speculated that such magnetospheres may be present around many astrophysical objects, ranging from pulsars to entire galaxies. Figure 30, taken from a recent NASA brochure, compares the magnetospheres of the solar system with those inferred from various astrophysical objects. The figure suggests that the magnetosphere of a pulsar is similar in size to that of Earth, although its magnetic field is expected to be a factor of $10^{12}$ larger. The magnetosphere of Mercury, by comparison, is quite small compared to either of the two. Similarly, the magnetosphere of Earth is rather small on the scale of Saturn's, while Jupiter's magnetosphere is extremely large even compared to the dimensions of the sun. Even the magnetosphere of Jupiter, however, is totally insignificant when compared to the inferred magnetosphere of radio galaxy NCC1265 (Strom et al., 1975). Many of the astrophysical magnetospheric models have incorporated such standard concepts as field aligned currents, plasma sheets, tilted dipoles, etc. and the resulting physics has been both interesting and elegant. (For some of these concepts and the associated acceleration mechanisms, see IP Conference Proceedings No. 56, Particle Acceleration Mechanism in Astrophysics, AIP, New York, 1979).

I would like to call to your attention, however, a quotation from the summary of the aforementioned conference, by Gene Parker: "....all of us who are interested in the physics of the distant active objects must be interested in magnetospheric physics, in interplanetary physics and in solar physics while we continue to develop general ideas on the distant objects. The physics of the local activity may seem discouraging because of its complexity, but an extensive knowledge of it is the only firm basis for pursuing the physics of particle acceleration, and the constant humiliation which it provides helps to maintain our scientific equilibrium in judging what we have accomplished in the distant objects" (pages 431-432).

Since this is a cosmic ray conference, we could perhaps speculate on what the magnetospheres can teach us about cosmic ray spectra. We have seen that a "$\gamma$- thermal" distribution (Equation (1)) is often a good fit for magnetospheric particle distributions over a large energy range. Figure 31 shows an attempt to fit the cosmic ray spectrum with this spectral form. One would have to admit that the fit is rather good, and reproduces a number of the spectral features which we have debated in these conferences over the past 15 years. The diurnal anisotropy (the basis of the Compton-Getting) factor is an inherent part of the convected Maxwellian, and is of the same (small) magnitude as we have always calculated. Perhaps nature accelerates particles in such a way that, after a few time scales, the end result may be a $\gamma$-thermal distribution. Of course, if the shape of the cosmic ray spectrum is $\gamma$-thermal, then we may not find a sharp upturn at low energies in the nearby interstellar medium, which would play havoc with our ideas on modulation.

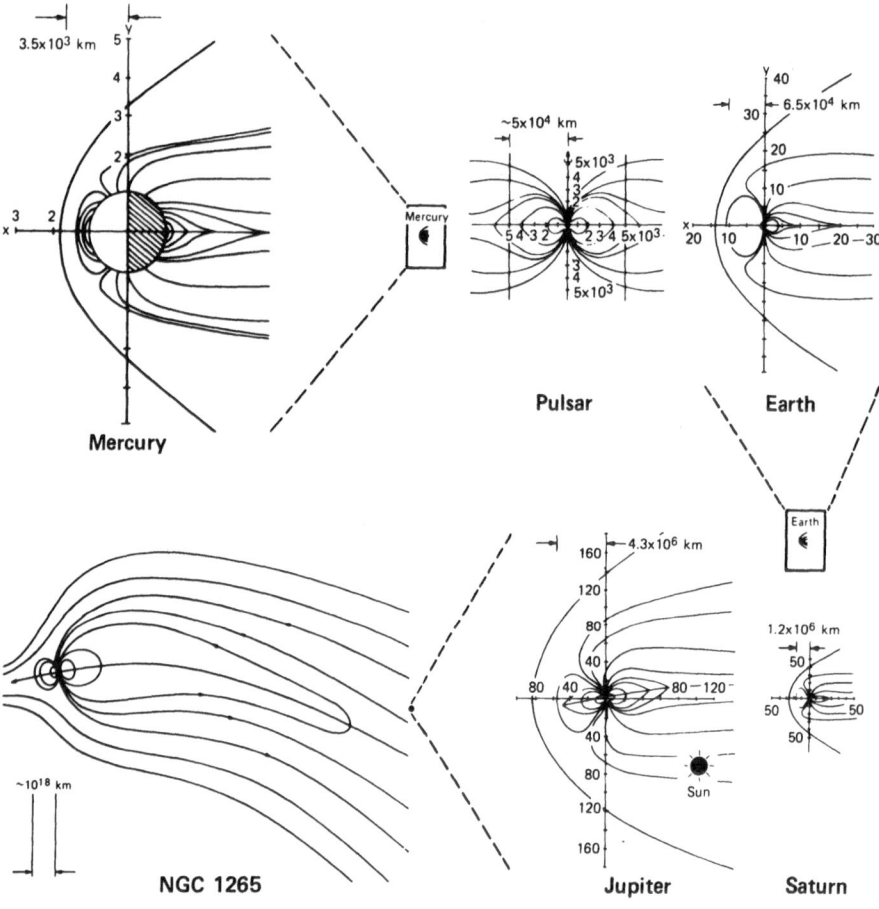

FIG. 30 — Relative sizes of the magnetospheres of some planets, and those of inferred astrophysical magnetospheres (courtesy of J. K. Alexander, NASA/GSFC).

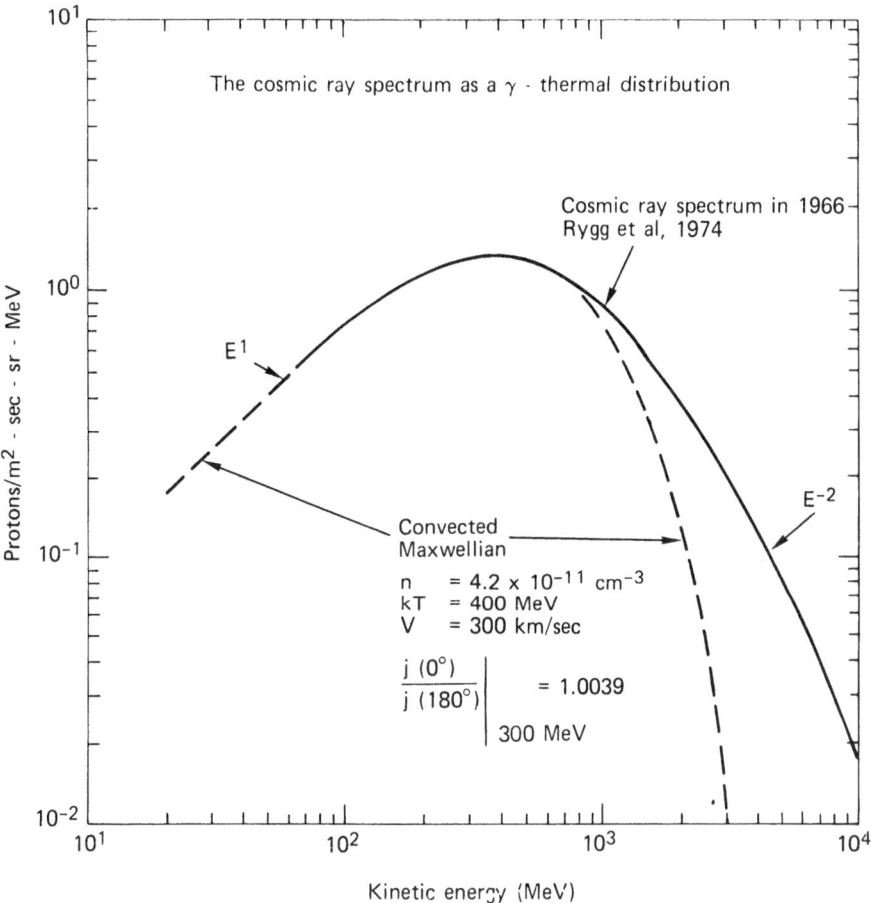

The cosmic ray spectrum as a γ - thermal distribution

Cosmic ray spectrum in 1966
Rygg et al, 1974

$E^1$

$E^{-2}$

Convected
Maxwellian

n   = 4.2 x 10⁻¹¹ cm⁻³
kT  = 400 MeV
V   = 300 km/sec

$$\left.\frac{j\ (0°)}{j\ (180°)}\right|_{300\ MeV} = 1.0039$$

Kinetic energy (MeV)

Protons/m² - sec - sr - MeV

FIG. 31  -  A "γ-thermal" distribution fit to the cosmic ray spectrum
with the indicated parameters.

    This thought brings up the point of what magnetospheres might tell
us about the shape of the heliosphere. After the discovery of the so-
called interplanetary sector pattern in the interplanetary magnetic
field (IMF), it was suggested by Alfven (1969) that the Earth found
itself above or below a massive solar plasma sheet similar to that of
the Earth's magnetosphere. The early of work of Rosenberg and Coleman
(1969) indicated some persistent IMF patterns in favor of this
hypothesis. It was not, however, until 1978 (Smith et al., 1978) when
Pioneer 11 moved above the ecliptic plane at large distances from the
sun that this hypothesis was placed on a firm observational footing. It
is now generally accepted that the sun possesses a general dipolar
magnetic field as indicated in Figure 32 (Hundhausen, 1977). Figure 32a

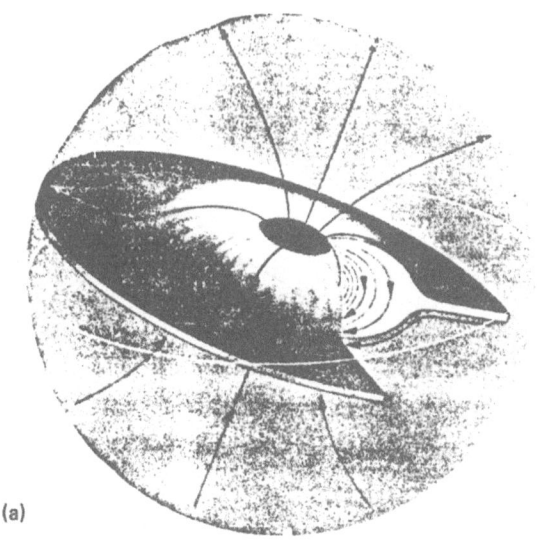

(a)

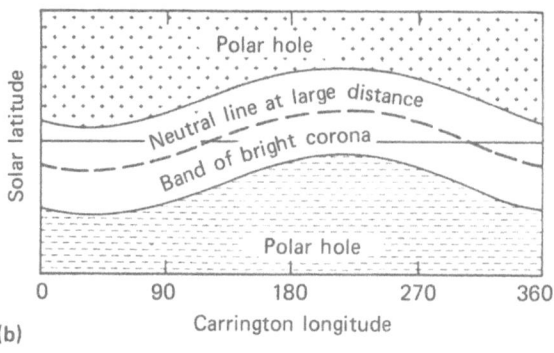

(b)

FIG. 32 - (a) An artist's drawing of the three-dimensional magnetic and density structure of a hypothetical solar dipole model. (b) The form of the coronal brightness map that would be observed if the corona had the three-dimensional structure illustrated in (a) (adapted from Hundhausen, 1977).

gives an artist's drawing of a three-dimensional magnetic density structure close to the sun implied by the dipole model of Pneuman and Kopp (1971). Figure 32b gives a general form of the brightness map that would be observed if the corona had the simple three-dimensional density structure of 32a. The tilted dipole would lead to a sinusoidal band of bright corona shown in the figure with a neutral line at large heliocentric distances rising above the bright structure. A two-sector, two-stream interplanetary structure should arise from the intersection of this structure with the ecliptic plane.

It does not take a large step to go on from this concept and suggest that the entire heliosphere can be viewed in the context of a magnetosphere. A speculative attempt at sketching the features of a heliomagnetosphere is given in Figure 33. There are a number of parallels between the magnetospheres that we have studied and some of the features which have already been established or inferred in the case of the sun. For example, plasma is present, perhaps with the highest density in the equatorial plane, and it is exceedingly dynamic. A dominant polarity of the interplanetary field is strongly supported by the Pioneer 11 excursion off the ecliptic plane. The existence of solar "polar caps" as regions of "open" field lines and low density plasma has been inferred from coronal holes. Efficient acceleration of ions from the background solar particles in association with interplanetary disturbances is observed within the ecliptic plane at all distances from the sun investigated by spacecraft. The heliosphere is probably dominated by plasma of solar origin within the ecliptic plane, although

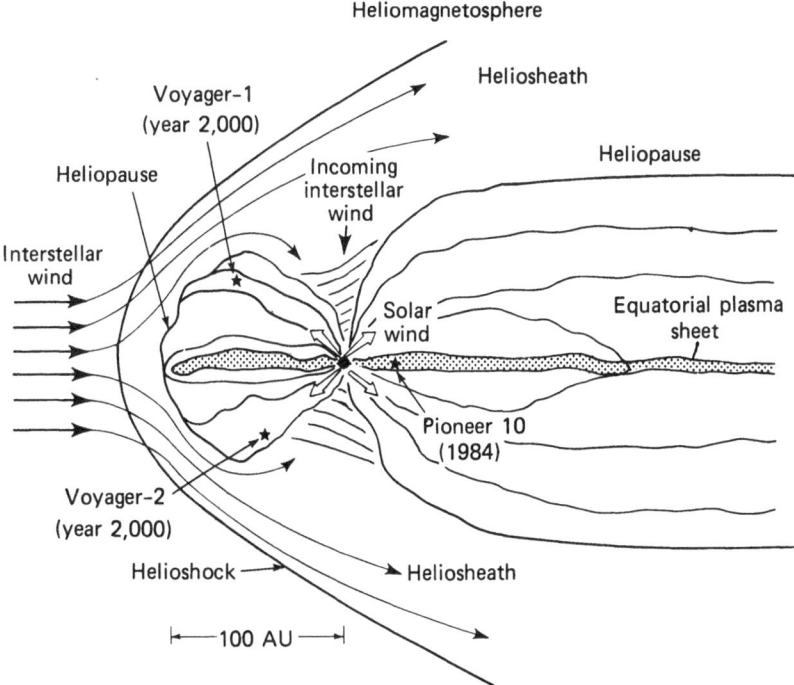

FIG. 33 - A sketch of a hypothetical heliomagnetosphere; many of the indicated features have either been observed or inferred (see text for explanation).

interstellar ions could possibly dominate at large heliocentric distances and over the solar poles. It is reasonable the speculate that a cusp or "neutral point" exist between the interstellar and interplanetary magnetic fields, where interstellar plasma enters the heliosphere and is recirculated within the system.

I have noted here the location of the Pioneer 10 spacecraft in the year 1984 ($\sim$ 30 AU), and of the Voyager 1 and 2 spacecraft in the year 2000 at distances of $\sim$ 75 AU and $\sim$ 60 AU, respectively. It is evident that if current estimates on the location of the heliopause are correct then it may be quite a while before we sample the interstellar medium.

## 7. CONCLUDING REMARKS

It is both interesting and instructive to watch the evolution of magnetospheric physics over the past 20 years. The initial exploration of the Earth's radiation belts was carried out by cosmic ray physicists who investigated the spatial and dynamical structure of the high energy particles, with the view that this was the most important aspect of the physics to be learned in the radiation belts. Gradually, it became evident that the high energy protons and electrons were strongly affected by the low energy plasma, where the bulk of the energy density resided, and by the waves. With the advent of improved instrumentation, the emphasis in the physics of magnetospheres shifted to the low energies with strong input from laboratory plasma physicists and new students in the field with a strong background in plasma physics. The new vocabulary was enriched with terms such as distribution functions, plasma waves, wave-particle interactions, plasma instabilities, anomalous resistivity, Alfven waves, Mach numbers, etc. The study of other planetary magnetospheres has only intensified the move in this direction, and the emphasis on high energy particles has consequently diminished.

A similar state of affairs exists in the study of solar and interplanetary particles, where single particle models have given way to collective interactions between the low energy plasma, waves, and suprathermal distributions. The vocabulary is not the same any more in that the energy spectra have been supplanted by distribution functions with characteristic temperatures, pressures, densities, and bulk flows, as opposed to power law exponents and particle anisotropies. It is a changing and challenging set of new physics, which promises to provide us with a much more comprehensive understanding of physical phenomena occurring in the world around us.

## Acknowledgements

This work has been supported by NASA under Task I of contract N-00024-78-C-5384 between The Johns Hopkins University and the Department of the Navy, and by NASA Grant NAGW-154 to The Johns Hopkins University.

# References

Akasofu, S.-I., 1979, Solar Phys., 64, 333.
Alexander, J. K. et al., 1981, J. Geophys. Res., 86, November.
Alfven, H. and C. G. Falthammar, 1963, Cosmical Electrodynamics, 2nd
  edition, Oxford, Great Britain.
Alfven, H., 1969, Comment to the author, Leningrad Conference.
Alfven, H., 1981, Cosmic Plasma, D. Reidel, Holland.
Armstrong, T. P. et al., 1981, J. Geophys. Res., 86, November.
Atreya, S. K. et al., 1974, Science, 184, 154.
Behannon, K. et al., 1981, J. Geophys. Res., 86, November.
Birkeland, K, 1908, The Norwegian Aurora Polaris Expedition, 1902-03, 1,
  Aschhoug, Christiania, Norway.
Bridge, H. S. et al., 1979a, Science, 204, 987.
Bridge, H. S. et al., 1979b, Science, 206, 972.
Bridge, H. S. et al., 1981, Science, 212, 217.
Broadfoot, A. L. et al., 1979, Science, 204, 979.
Broadfoot, A. L. et al., 1981, Science, 204, 206.
Carbary, J. F. et al., 1981, J. Geophys. Res., 86, November.
Cheng, A. F., 1980, Ap. J., 242, 812.
Colgate, S. A. et al., 1978, Space Plasma Physics, National Academy of
  Sciences, Washington, D. C., 70.
Desch, M. D. and M. L. Kaiser, 1981, Nature, special issue on Saturn.
Fillius, R. W. and C. E. McIlwain, 1980, J. Geophys. Res., 85, 5803.
Fisk, L. A., 1976, J. Geophys. Res., 81, 4633.
Frank, L. A. et al., 1980, J. Geophys. Res., 85, 5695.
Goertz, C. K., 1978, J. Geophys. Res., 82, 3145.
Goertz, C. K., 1980, Geophys. Res. Lett., 7, 365.
Gurnett, D. A. et al., 1981, Science, 212, 235.
Gurnett, D. A. and F. L. Scarf, 1981, in Physics of the Jovian
  Magnetosphere, Dessler (ed.), Cambridge (in press).
Hamilton, D. C. et al., 1980, Geophys. Res. Lett., 7, 813.
Hamilton, D. C. et al., 1981, J. Geophys. Res., 86, November.
Heikkila, W. J., 1972, in Critical Problems in Magnetospheric Physics,
  E. R. Dyer (ed.), U.S. National Academy of Science, Wash., D. C., 67.
Hundhausen, A. J., 1977, in Coronal Holes and High Speed Wind Streams,
  Zirker (ed.), Colorado University Press, 225.
Ipavich, F. M., 1974, Geophys. Res. Lett., 1, 149.
Kaiser, M. L. et al., 1981, Nature, special issue on Saturn.
Kennel, C. F., 1973, Space. Sci. Rev., 14, 511.
Kirsch, E. et al., 1981a, Geophys. Res. Lett., 8, 169.
Kirsch, E. et al., 1981b, Nature, special issue on Saturn.
Krimigis, S. M. and E. T. Sarris, 1979, Dynamics of the Magnetosphere,
  Akasofu (ed.), Reidel, 599.
Krimigis, S. M. and E. C. Roelof, 1981, in Physics of the Jovian
  Magnetosphere, Dessler (ed.), Cambridge (in press).
Krimigis, S. M. et al., 1975, Geophys. Res. Lett., 2, 561.
Krimigis, S. M. et al., 1979, Science, 206, 977.
Krimigis, S. M. et al., 1980, Geophys. Res. Lett., 7, 13.
Krimigis, S. M. et al., 1981a, J. Geophys. Res., 86, November.
Krimigis, S. M. et al., 1981b, Science, 212, 115.

Kulsrud, R. M., 1979, in Particle Acceleration Mechanisms in Astro-physics, Arons, McKee, Max (eds.), American Inst. of Physics, New York, 1979.
Maclennan, C. G. et al., 1981, J. Geophys. Res., 86, (submitted).
McDonald, F. B. et al., 1980, J. Geophys. Res., 85, 5813.
McNutt, R. L., Jr. et al., 1979, Nature, 280, 803.
Ness, N. F., 1979, Solar System Plasma Physics, Vol. II, Kennel, Lanzerotti, Parker (eds.), North Holland, 183.
Ness, N. F. et al., 1979a, Nature, 280, 799.
Ness, N. F. et al., 1979b, Science, 206, 966.
Ness, N. F. et al., 1979c, Science, 204, 982
Nishida, A., 1976, J. Geophys. Res., 81, 1771.
Obayashi, T., 1975, Solar Physics, 40, 217.
Olbert, S., 1968, in Physics of the Magnetosphere, Carovillano, McClay, Radoski (eds.), Springer-Verlag, New York, 641.
Pneuman, G. W. and R. A. Kopp, 1971, Solar Physics, 18, 258.
Potemra, T. A., 1978, Astrophys. Space Sci., 58. 207.
Pyle, K. R. and J. A. Simpson, 1977, Ap. J., 215, L89.
Roelof, E. C. et al., 1976, J. Geophys. Res., 81, 2304.
Rosenberg, R. L. and P. J. Coleman, 1969, J. Geophys. Res., 74, 5611.
Russell, C. T., 1979, in Solar System Plasma Physics, Vol. II, Kennel, Lanzerotti, Parker (eds.), North Holland, 207.
Russell, C. T. et al., 1980, J. Geophys. Res., 85, 8319.
Russell, C. T., 1981, J. Geophys. Res., 86, submitted.
Rygg, T. A. et al., 1974, J. Geophys. Res., 79, 4127.
Sarris, E. T. et al., 1981, Geophys. Res. Lett., 8, 349.
Scarf, F. L. et al., 1981a, Vistas in Astronomy, special issue on Jupiter.
Scarf, F. L. et al., 1981b, Nature, special issue on Saturn.
Scarf, F. L. et al., 1981c, Nature, special issue on Saturn.
Scholer, M. et al., 1979, Geophys. Res. Lett., 6, 707.
Simpson, J. A. et al., 1980, J. Geophys. Res., 85, 5731.
Siscoe, G. L., 1975, Icarus, 24, 311.
Siscoe, G. L., 1979, in Solar System Plasma Physics, Kennel, Lanzerotti, Parker (eds.), North Holland, 319.
Slavin, J. A. et al., 1980, J. Geophys. Res., 85, 7625.
Slavin, J. A. and R. E. Holzer, 1981, J. Geophys. Res., 86, (submitted).
Smith, E. J. et al., 1978, J. Geophys. Res., 83, 717.
Spreiter, J. R. and S. S. Stahara, 1980, J. Geophys. Res., 85, 7715.
Strom, R. G. et al., 1975, Sci. Am., 233, 26.
Thompsen, M. F., 1979, Rev. Geophys. and Space Phys., 17, 369.
Thorne, R. M., 1981, in Physics of the Jovian Magnetosphere, Dessler (ed.), Cambridge (in press).
Waite, J. H., Jr. et al., 1979, Geophys. Res. Lett., 6, 723.
Warwick, J. W. et al., 1979, Science, 204, 995.
Vaisberg, O. L. et al., 1976, in Physics of Solar Planetary Environments, D. J. Williams (ed.), American Geophysical Union, Wash., D. C., 904.
Van Allen, J. A. et al., 1980, J. Geophys. Res., 85, 5679.
Zwickl, R. D. et al., 1980, Geophys. Res. Lett., 7, 453.
Zwickl, R. D. et al., 1981, J. Geophys. Res., 86, November.

## STATUS OF THE FIREBALL CONCEPT IN
## THEORY AND EXPERIMENT

### E.L. Feinberg

P.N.Lebedev Physics Institute,
Moscow, USSR, 117924-GSP

The fireball concept introduced,
adopted, and widely used in cosmic
ray physics for more than two decades had been decidedly rejected by
accelerator physicists.    It is only
during the last 5-8 years that accellerator experiments in TeV range
have led to the cluster idea thus
marking convergence of two kinds of
particle physics.    Analysis shows
that the convergence is even more
essential than one might think. Heavy
nonresonant unstable objects have
natural place in any quantum field
theory while any multiperipheral
type model when applied to accelerator experiments needs such objects.
In general fireball concept seems to
have a far reaching support both in
theory and experiment.

## 1.   Introduction

Fireballs appeared in particle physics in 1958: Mie-
sowicz Cracow group [1], Niu [2], and Cocconi [3] came to a
conclusion that, sometimes at least, multiple production
of hadrons at laboratory energy $E_L \sim 10^3 - 10^4$ GeV proceeds
via two lumps of hadronic matter having the mass $\mathcal{M} \sim$
$\sim 3 - 5$ GeV/c$^2$, slowly moving in CMS and isotropically
decaying statistically into $\langle n \rangle \sim$ 6-10 particles. They were
called fireballs [3]. Very soon after that fireballs were
reported also at some 10 times smaller energy with only
one fireball per event [4].
The overwelming majority of cosmic ray physicists
accepted the idea at once. In the pattern adopted colliding particles after producing fireballs pass by possibly
somewhat excited, so as to decay into     few particles
conventionally called"isobaric"  or, later, "fragmentation",
as distinct to the more numerous products of fireballs,
the "pionization" ones. Accumulation of data at still higher energies has been steadily giving more evidence, in

particular in favour of larger, sometime gigantic fireballs
($\mathcal{M} \sim 100$ GeV/c$^2$, $<n> \sim 100-200$)[5] .

However exactly at the same years as fireball concept,
new accelerators appeared (Dubna, 10 GeV; CERN and
Brookhaven, $\sim 30$ GeV). Accelerator experimentalists
enthusiastically plunged into investigation of accessible
events with multiplicity much lower than in cosmic rays
and interpretable in terms of one meson exchange Feynman
graphs, one-reggeon exchange, dispersion relations, etc.

Ever since for almost two decades there have existed,
we may say, two different kinds of particle physics. For
cosmic ray people fireballs have been indisputable objects
invariably used in emulsion data treatment, in calculation
of extensive air shower development, etc. On the contrary
accelerator people and majority of theoreticians have neglected them entirely. It is only for the last 5-8 years
that situation has begun to change. Earlier various multiperipheral "comb like" models with pions, or resonances, or
partons as teeth of the comb, had been dominating (and they
hold to a considerable extent even now). However when the
accelerators approached the energy at which fireballs had
been seen in cosmic rays, contradictions with the simple
comb picture made it imperative to introduce various complications: branching of combs, rather heavy resonance exchange, etc. By and by those people who at one time would not
hear about fireballs started to consider some rather heavy
intermediate decaying objects. The extreme degree of this
trend was expressed in the Annual Report for 1973 of the
CERN Director W.Jentschke: as one of the main achievements
he mentioned the gained understanding of multiple production as following the pattern of Fig.1 [6]. Of course, this
is a typical one-fireball scheme which had been formulated
in cosmic rays some 15 years earlier.

However this extreme interpretation of the accelerator data is by no means dominating. More often people assume that hadrons are produced in groups of 3-4 particles with
a common mass $\mathcal{M} \sim 1.5 - 2.0$ GeV/c$^2$ called clusters. Thus,
after some two decades of acute counterposition we observe
convergence. I shall try to show that the convergence is
even more close than usually believed.

This convergence is due not only to the increase of the
machine energy. No less important is the fact that after many years of rejection of the quantum field theory (QFT) in
favour of axiomatic S-matrix theory, the Regge pole method,
etc., QFT is fully rehabilitated. We may accordingly again
use its quasiclassical approximation, we may study space-
time evolution of processes and particles, etc.

One of the most popular arguments against fireballs
has been that they are "exotic" objects. However now it is
clear that this is erroneous. There exists a natural place
for unstable heavy nonresonant objects in any QFT. But
their production rate, mass spectrum and decay law depend on

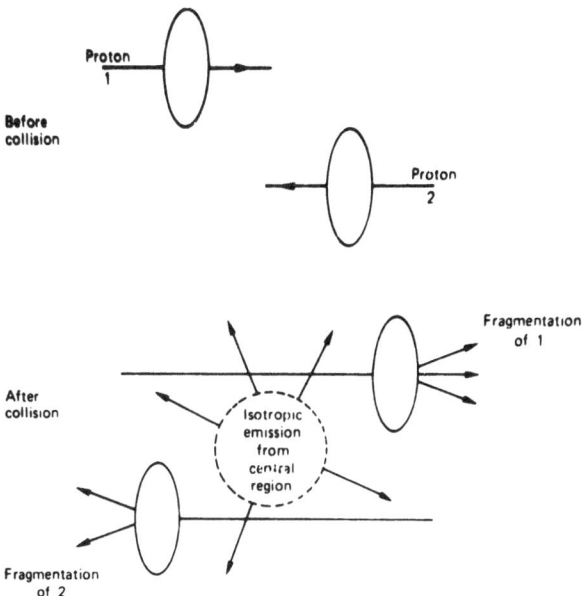

Proton 1

Proton 2

Before collision

After collision

Fragmentation of 1

Isotropic emission from central region

Fragmentation of 2

**Fig.1.** Pattern of the multihadron production
according to Ref. 6 interpretation of accelerator data at $E_L \sim 1000$ GeV.

the assumed dynamics, and here opinions diverge.
Very soon accelerators for $E_L \sim 10^2 - 10^3$ TeV will be
put into operation. In these incomparably better experimental conditions the cosmic ray fireball problem will find
its final solution.
Part II of this review covers the theoretical reasons
for fireballs; Part III - quite briefly - the cosmic ray
experimental situation; Part IV - the analysis of accellerator data on the basis of correlation studies. Part V
contains conclusions.

## II. Theoretical reasons for fireballs

### 1. Thermodynamic peripheral model

No doubt, cosmic ray people were prepared for the
fireball concept by theorists who had been since 1936
constructing thermodynamic models, - those by Heisenberg [7]
(another version in Refs. [8,9]), Fermi [10] and Pomeranchuck [11]
with the crowning hydrodynamical model by Landau [12] , - all
based on a common idea: since interaction is strong, two
colliding high energy nucleons stop releasing their energy
within the Lorentz contracted overlap volume having the CMS

thickness $\ell \sim m_\pi^{-1} 2 m_N / \sqrt{s}$ ($\sqrt{s} = \sqrt{2 m_N E_L}$ is the common CMS energy, $m_\pi^{-1} \sim r_o$ - the nucleon radius, $m_N$, $m_\pi$ - nucleon and pion masses). The vacuum boils up to a tremendous temperature producing new hadrons which interact with each other, their expansion and cooling as well as the final number and energy spectrum being determined by thermodynamics.

Such a quasiclassical approach is justified by large production multiplicity $n$ , i.e. by large number of excited degrees of freedom, by high quantum numbers. Accordingly production probability is governed mainly by the multidimensional phase space volume with equal probability of all final states at a given total energy. This is equivalent to microcanonical distribution. Herefrom direct way leads to thermodynamics. Nowadays this approach may be additionally supported by two arguments.

First, we know that e.g. at ISR energies ($\sqrt{s} \sim 60$ GeV) all produced particles, mostly pions, some $\langle n_\pi \rangle \sim 15 - 20$ in number, cover a rather small rapidity interval [x]$Dy \sim \sim 3-4$. Therefore relative rapidity of two "neighboring" particles is very small, $\Delta y \sim Dy / \langle n_\pi \rangle \sim 0.2-0.3$, and their longitudinal relative motion is nonrelativistic. It is highly improbable for these particles not to interact again and again.

Secondly, now we know that each produced pion consists of two valence quarks, of pairs of sea quarks and of gluons. Therefore the number of excited degrees of freedom is extremely large, many times larger than simply $3 \langle n_\pi \rangle$ . This clarifies considerably the miraculous success of hydrodynamical Landau model (originally it had been supposed to hold only for $E_L \gtrsim 1000$ GeV) even for laboratory energies $E_L \sim 10-100$ GeV where $\langle n \rangle$ is moderate. The theory of quark-gluon plasma has put thermodynamic approach on a much firmer basis [13]. A fireball becomes e.g. a superdense MIT bag of quark-gluon plasma expanding and finally decaying into many usual bags of normal hadronic density.

However all these models treated head-on collisions while actually collisions are peripheral. Fermi and Heisenberg tried to take this into account but with unsatisfactory results. E.g. Heisenberg [14] assumed the following scheme: only overlapped parts of colliding nucleons spend their energy $\Delta E$ to produce new particles as long as $\Delta E$ is sufficient for producing at least two pions. Since $\Delta E$ decreases presumably exponentially with impact

---

[x] Rapidity is $y = \frac{1}{2} \ln \frac{E + P_{||}}{E - P_{||}}$ , $E$, $P_{||}$ being energy and longitudinal momentum of a produced particle. It is approximately equal to pseudorapidity $\eta = \ln tg(\theta/2) \approx y$, $\theta$ being the production angle.

parameter $d$ increasing, $\Delta E \sim \exp(-m_\pi d)$, this gave for largest effective inelastic collision $d_{max}$, $d_{max} \sim m_\pi^{-1} \ln(E/m_N)$ and therefore an absurdly large cross section $\sigma_{in} \approx \frac{\pi}{4} m_\pi^{-2}(\ln(E_L/m_N))^2$ (e.g. at $E_L \sim 10^6$ GeV it exceeded the actual one already known in those years some 50-100 times). The source of error is quite significant and important for us: the classical treatment used here was not permissible [15]. In fact, for a collision with duration $\Delta t \sim \ell \sim m_\pi^{-1} \cdot 2 m_N/\sqrt{s}$ such a treatment does not violate the energy-time indeterminacy relation only if $\Delta E \gg 1/\Delta t \sim m_\pi \sqrt{s}/2m_N$ , or, essentially, if $\Delta E \sim \sqrt{s}$ , i.e. if nearly whole energy goes over into new particles. This may happen only for head-on collisions. Thus we come to the conclusion: the very process of peripheral ($d \gtrsim m_\pi^{-1}$) interaction must be treated quantum theoretically; the quasiclassical objects may arise only as subsystems of such a process. Accordingly, the models of Fig.2 were proposed (the Fig.2b model without the above motivation, was proposed also in Refs. [16] and became widely known).

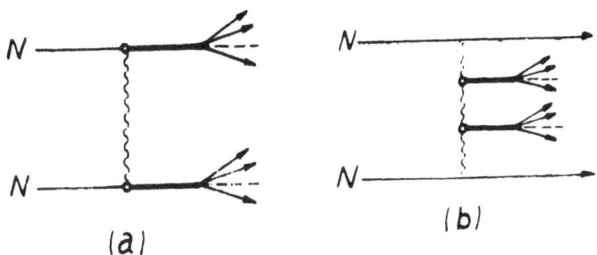

(a)   (b)

Fig.2. "Thermodynamical peripheral model" [15,16]: quasiclassical subsystems arise in the course of a quantum theoretically treated peripheral collision.

With this modification still another shortcoming of head-on collision thermodynamical models is done away with: the classical description of initial stage in these models is on the verge of contradicting indeterminacy relation $\Delta x \cdot \Delta p \gtrsim 1$ when we subdivide the overlap volume into elements for hydrodynamic treatment [17,18]. On the contrary, in the thermodynamic peripheral model of Fig.2 no serious restriction on the initial volume of the thermodynamical subsystem is imposed.

Moreover, violation of asymptotic freedom characteristic for head-on collision thermodynamic models is avoided.

Thus peripherality plus thermodynamical model lead to the fireball type schemes.

## 2. Multiperipheral model with fireballs

However, as has been said above, in accellerator physics a different way was tried. (Quasi)two- and even (quasi) three -body processes were successefully described by the one meson exchanges and - later - by Regge pole exchanges with appropriate phenomenological parameters: reggeon trajectory intercepts $\alpha_R(0)$ and slopes $\alpha'_R(0)$, formfactors etc. For larger multiplicities this naturally led, at first, to the multiperipheral Amati-Fubini-Stanghellini (AFS) model where interaction was effected by pion exchange and $\rho$ -mesons were produced in vertices (Fig.3a) 19 (see also the review paper 20 ), - a famous "comb" which leads

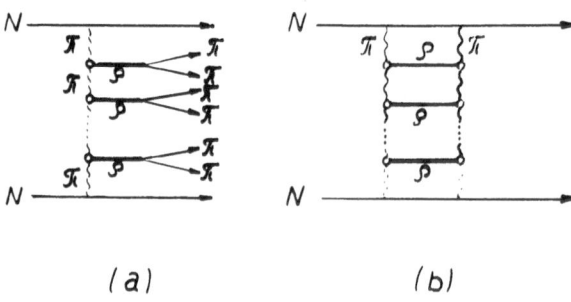

$(a)$ $(b)$

Fig.3. Multiperipheral AFS model [19]: a) inelastic "comb like" amplitude; b) the resulting "ladder amplitude" of the elastic shadow scattering.

to shadow elastic scattering "ladder" amplitude (Fig.3b). The AFS model has three remarkable properties.

1) It gives a reasonable multiplicity-energy dependence, $\langle n_\pi \rangle = a\, \ell n\,(E_L/m_N)$, however with $a$ too small (some 2-3 times) to fit experimental data; $a$ is determined by the number of final pions per vertex, $K$ (in the AFS model, $K = 2$).

2) The ladder gives Regge type behaviour of total cross section, $\sigma_{tot} \sim s^{\alpha_\rho(0)-1}$, however, with a too small Pomeron intercept $\alpha_\rho(0) \approx 0.3$ [21,22], thus predicting fast energy decrease, $\sigma_{tot,AFS} \sim E_L^{-0.7}$ in drastic contradiction to experiment.

3) Finally, it gives the Regge kind small angle elastic scattering, $d\sigma_{el}/d|q^2| \sim exp\{-\alpha'_\rho(0)|q^2|\ell n \frac{E_L}{m_N}\}$ being the transferred four momentum, however it cannot explain why the Pomeron slope fitting experiment, $\alpha'_\rho(0) \approx 0.2$ GeV$^{-2}$, differs so enormously from the slopes of all other trajectories ($\alpha'_R(0) \approx 1$ GeV$^{-2}$).

The analysis shows 21 that all three advantages of the AFS model can be preserved while all three mentioned shortcomings done away with if we modify it by making ladder steps much heavier, i.e. if instead of $\rho$ -mesons in vertices of Fig.3a we substitute heavy (mass $\mathcal{M} \sim$ 2-4 GeV/c²) objects, clusters or fireballs decaying into some 5-10 pions each. x) All this is true for asymptotics, when the number of vertices $\mathcal{N}$ is large, $\mathcal{N} \gg 1$. On purely kinematic grounds $\mathcal{N}$ , for heavy vertices, may attain the value $\mathcal{N}$ = 3-4 only at $E_L \gtrsim 10^4 - 10^5$ GeV (at $E_L = 10^3$GeV, $\mathcal{N} \approx 2$) 21. Let us add that the heavy vertex chain model was proposed (without the motivation of the kind described above) also by Hasegawa 23 who called them "H-quanta".

However this model is oversimplified. There is no reason for complete elimination of light resonance vertices, of low multiplicity diffraction dissociation, etc.

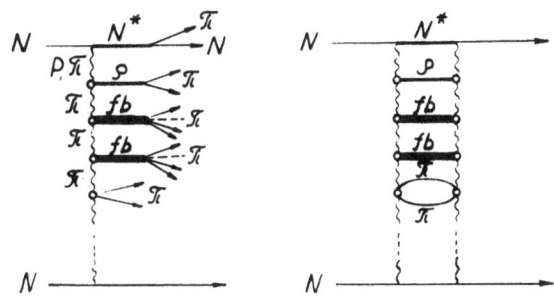

Fig.4. Hybrid Lebedev Institute multiperipheral ("multicluster")model (LIM) allowing for both low multiplicity vertices( $\rho$ -meson production, diffraction dissociation,two virtual hadron "elastic" scattering into real particle state,etc.) and high multiplicity fireball generation 24-27.

Accordingly a hybrid model of Fig.4 was proposed by the Lebedev Institute group 24,25 , allowing for all such processes. After careful analytic study, all parameters which could not be directly taken from experiment were fixed 26 . This model xx) was successfully used for

---

x) When this recipe was proposed the thus predicted slope $\alpha'_\rho(0) \approx 0.2$ GeV⁻² was considered unsuitable since in those times experimental estimate was erroneous, $\alpha'_P(0) \approx 1$ GeV⁻².

xx) In previous publications(24-27 etc.) this model was called "multicluster". However there exist also other multicluster models (e.g. the Independent Cluster Emission Model, also discussed below) and to avoid confusing we prefer from now on to refer to the model 24-27 as to the Lebedev Institute Model or LIM.

describing experimentally found inclusive and semiinclusive distributions, as well as correlations, of pions and nucleons produced in $\pi p$ and $pp$ collisions at $E_L$ = 28 - 200 GeV (partially also for 400 and 2000 GeV). The bank of some 70 000 exclusive computer simulated events exists and is being exploited for some 8 years each time when new experimental data appear. The authors are satisfied with the obtained 10-15% accuracy [20,27]. We may believe that the world of this model is rather close to the real one in which we live. It deserves mentioning that with parameters fixed in Ref. [26] for energies up to $E_L$ = 400 GeV in this model events with only two and three vertices dominate (four vertices chains contribute only 3 mb cross section), which therefore must be rather heavy. The fireball decay is treated thermodynamically [11].

There were also many attempts of describing experiments within multiperipheral models without fireballs. E.g. within the AFS model production of other rather light resonances ($\omega$, $f$, $A_2$) was assumed [28] and the nearly constant $\sigma_{tot}$ up to $E_L \sim 10^3$ GeV was obtained. Besides, all these resonances together with $P'$ trajectory and reggeized $\pi$ were allowed to substitute $\pi$-exchange [29], and this secured some additional success. However, in order to explain e.g. two-particle rapidity correlations (see below, Part IV) it turned out necessary to introduce also the so called enhanced branchings of various trajectories [30]. As a result the original graph of Fig.3 transforms into a combination of few elements each being a heap of many various entangled trajectories with mutual rescatterings, branchings, etc. Such a heavy element may equally well be considered a fireball interpreted in terms of mutual exchanges and interactions of multitude of trajectories of all kinds ($P, P', \rho, \omega, A_2, f, \pi$) taken into account.

Thus we see that heavy clusters are necessary to secure energy nearly independent cross section and correct values of multiplicity and of slope $\alpha'_P(0)$ of the pomeron trajectory within any multiperipheral model.

### 3. What are fireballs in the quantum field theory?

Are fireballs, i.e. nonresonant heavy lumps of matter, really,"exotic"? I wish to stress that this is entirely erroneous. Such objects decaying into stable particles are inevitable elements of any QFT [31]. Their decay process (e.g. validity of thermodynamic treatment) is,however, quite an independent problem.

Let us take quantum electrodynamics (QED), - the best studied QFT, - leaving hadrons aside for a moment. We may observe electron-positron, $e^+e^-$, ahhinilation (at CMS energy $\sqrt{s}$ ) into any lepton pair, $\ell^+\ell^-$, e.g. into $\mu^+\mu^-$, $\tau^+\tau^-$ or into $e^+e^-$ again (Fig.5a) .Here the heavy

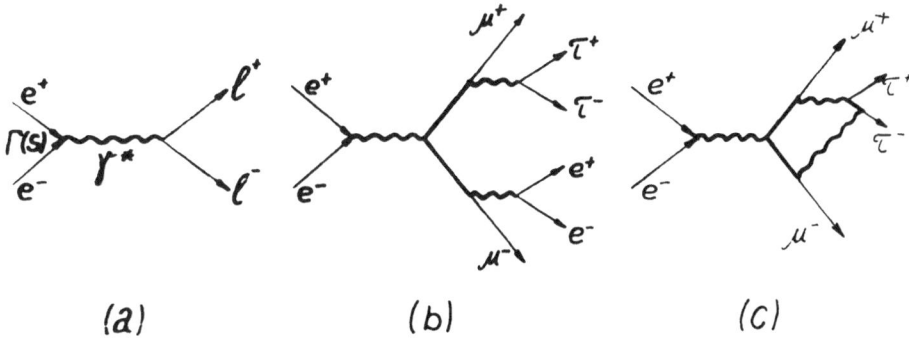

(a)    (b)    (c)

Fig.5. Electron-positron annihilation into leptons; a) lowest order in coupling constant $e^2$; b), c) - some of the higher order amplitudes.

$\gamma$ -quantum, $\gamma^*$ (mass $\mathcal{M}_{\gamma^*} = \sqrt{s}$ ), is exactly the intermediate nonresonant object. It decays into only two particles because we restricted ourselves to the lower order approximation in coupling constant $e^2 \ll 1$. E.g. in the third order we may have (very rare) processes of Figs. 5b, 5c. Here $\gamma^*$ behaves like a regular fireball. The mass spectrum of produced $\gamma^*$, i.e. the $\mathcal{M}_{\gamma^*} \equiv \sqrt{s}$ dependence of the total cross section for annihilation into leptons is $\sigma_{tot}^{annih} \sim |\Gamma(s)|^2 \mathcal{S}_{\gamma}(s)$ , $\Gamma(s)$ being the vertex function (a formfactor), $\mathcal{S}_{\gamma}(s)$ - the so called spectral function of the photon propagator (not to be confused with the designation of $\varrho$ meson!). It can be straightforwardly calculated in QED and is schematically shown in Fig.6 (full line). If now we take besides into account the possibility of annihilation into some resonance state, e.g. into a $\varrho^{\circ}$-meson, $e^+e^- \to \varrho^{\circ} \to \pi^+\pi^-$ , then a dashed resonance curve is to be superimposed on this rather smooth background.

All this can be qualitatively transferred onto any QFT. E.g. chromodynamics adds as a final state quark-anquark pair, $e^+e^- \to \gamma^* \to q\bar{q}$ .

If instead of $e^+e^-$ annihilation we consider a hadron collision then for large $\sqrt{s}$ the final states with various numbers of hadrons have matrix elements of similar order (in distinction to QED where the amplitude of Fig.5a greatly exceed those of Figs. 5b,5c). The example possible in QCD is shown for a $\pi N$ collision in Fig.7. Here $\pi^*$ is a "pion like" fireball.

Thus a nonresonant background (with superimposed resonant peaks which in hadron physics may be quite numerous) and therefore nonresonant heavy objects decaying into many particles are natural elements in any QFT. Such an object is simply a particle (with appropriate quantum numbers) displaced far from its mass shell in the time like direction. People assuming that heavy hadronic clusters are only resonances (maybe still unknown) introduce in fact a very specific hypothesis, namely that there is no smooth background in $\varrho(s)$. This does not hold in QED (see Figs. 5,6) and there is no reasons for such a hypothesis in hadron physics. Of course the rate of nonresonant cluster production depends critically on formfactors $\Gamma(s)$. They may a priori be very small for large $m = \sqrt{s}$. This is however equally possible for production of resonant objects.

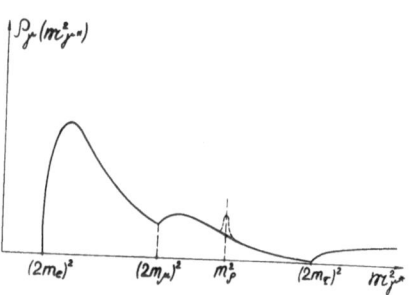

**Fig.6.** Schematic representation of the photon propagator spectral function $\varrho_\gamma$ influencing the $\gamma^*$ mass spectrum in QED (full line). Dashed peak shows modification due to possibility of a $\varrho$-meson production (abscissa is greatly distorted in order to show all lepton tresholds in a single graph; multiple pair tresholds at $(4m_e)^2$, $(6m_e)^2$, ...., etc. are omitted).

Summarizing the theoretical reasons for fireballs we see that: 1) Heavy clusters are necessary for any multiperipheral model compatible with known basic properties of multiple production. 2) These clusters may be both still unknown resonances and nonresonant fireballs. 3) Nonresonant clusters like fireballs are quite usual objects in any QFT. 4) Possibility of thermodynamic evolution and decay of fireballs (which in that case must be produced peripherally by some quantum field mechanism) seems to be rather well founded theoretically although not undoubtedly proved. A bag of dense quark gluon plasma is a good candidate for it.

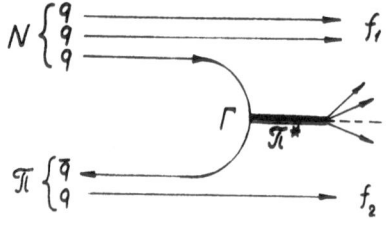

**Fig.7.** Possible graph for a $\pi N$ inelastic collision leading to formation of a $\pi^*$ fireball in QCD. $f_1$, $f_2$ are fragmentation jets.

## III. Fireballs in cosmic rays

Detection of fireballs means first of all selection of groups of particles nearly isotropically distributed in some Lorentz frame. Here advantage is taken of the remarkable property of any isotropic decay distribution: plotted versus rapidity y or pseudorapidity $\eta = -\ln tg(\theta/2) \approx y$ it can be rather well approximated by a gaussian curve, e.g.:

$$\frac{dN}{d\eta} = \frac{K}{\sqrt{2\pi}\,\delta}\, e^{-(\eta-\eta_c)^2/2\delta^2} \equiv G(\eta-\eta_c,\delta), \quad \delta \approx 0.8, \quad (1)$$

K being total number of particles in this "cluster" (K is substituted by $K^{ch}$, or $K^-$, etc. if only charged, or negative, etc. particle distribution , $N^{ch}$, $N^-$, etc. is considered); $\eta_c$ - its pseudorapidity as a whole. Thus products of any fireball cover an interval $(Dy)_{fb} \sim 1.5 - 2$ on the rapidity axis. Meanwhile according to already mentioned experiments at $E_L \sim 1000$ GeV, all produced particles cover only the $Dy \sim 3-4$ interval and $Dy$ is known to very slowly (maybe logarithmically) increase with $E_L$. Herefrom originates the first main difficulty for resolving fireballs: their products strongly overlap mutually and with products of decay of any light resonance on the rapidity axis.

Moreover there exists a second complication. As has long been known (in cosmic rays since $\sim 25$ years back), at a given $E_L$ features of inelastic collisions fluctuate tremendously. E.g. the inelasticity coefficient k - a fraction of energy spent for production of new particles - has equal probability from $k \approx 0.20$ up to $k \approx 0.90$. Multiplicity n also strongly fluctuates, its dispersion being $D \sim \langle n \rangle / 2$, i.e. much larger than for a Poisson distribution, $D \sim \sqrt{\langle n \rangle}$ . All this is closely connected with superposition of various mechanisms of production having comparable probabilities. E.g. diffraction dissociation of one or both impinging particles gives two small multiplicity groups (or a particle and a group) with very large rapidity interval between them. A pionization fireball is expected to give a high multiplicity group within the interval. The multiperipheral AFS comb must give a plateau, etc. Accordingly in individual experimental events particles are scattered over the (pseudo )rapidity axis in a variety of different,often completely irregular, patterns. Only superposition of many such events, i.e. an inclusive distribution has some unique regular shape. But it is hard to draw herefrom any conclusion on fireballs. Two approaches to the problem are possible. One is carefull analysis and sorting of individual events, e.g. use of the so called Duller-Walker plot, etc., well known to all cosmic ray physicists. It is hardly necessary to describe them here. It is sufficient to repeat that cosmic

ray people find enough evidence for reality of fireballs
with thermodynamic distribution of their decay products.
Extensive world wide observations have been summarized recently in Ref. 5   as testifying in favour of three kinds
of fireballs: the small ones (nominated by these authors
as "Mirim" or "heavy quanta, H") with average mass $\mathcal{M} \sim$
$\sim$ 2-3 GeV/c$^2$ and decay temperature $T_k \approx$ 0.13 GeV; medium
ones ("Açu", or "superheavy quanta, SH") with $\mathcal{M} \sim$
$\sim$ 15-30 GeV/c$^2$ and $T_k \approx$ 1 GeV, and gigantic ones ("Guaçu"
or "ultraheavy quanta", UH) with $\mathcal{M} \sim$100-300 GeV/c$^2$ and
still higher (2-4 times?) $T_k$ . It cannot be said that such
a sharp division into three distinct classes is accepted
by all cosmic ray physicists. However a somewhat weaker
statement seems to be accepted, namely: 1) mass spectrum
of fireballs at $E_L \sim$ $10^3$-$10^7$ GeV extends up to $\mathcal{M} \sim$100
GeV/c$^2$; 2) average transverse momentum of decay products
tends to increase somewhat with $E_L$ ; 3) at larger $E_L$ it
is more probable to meet larger $\mathcal{M}$ .

The second way is the study of various correlations
within inclusive and semiinclusive distributions. This is
possible if statistics is extensive. Thus it suits accelerator experiments and has been used in variety of approaches. Let us turn to it.

## IV. Clusters and fireballs in accelerator studies

The general principle of correlation studies is clear:
if we choose two particles with rapidities $y_1$ and $y_2$  wi-
thin rapidity interval $|y_1 - y_2| \ll 2\delta \approx$ 1-2, then there is a
considerable chance that they originate from the same
isotropically decaying cluster (although there must be
some overlapping with products of other clusters). On the
contrary, for $|y_1 - y_2| > 2\delta$  particles almost definitely
belong to different clusters. In the first case, if we find
any "short range ordering", SRO, then this reflects correlation within a cluster. In the second one, "Long range ordering", LRO, is due to mutual correlation of clusters.They
may be of kinematic (energy, momentum, charge etc.compensa-
tion) or dynamic (additional interaction between particles
or clusters )origin.

It is impossible to adequately describe all theoretical
and experimental researches on various types of correlations, as well as to avoid missing some essential papers.
Apologizing for it in advance I wish to refer first of
all to some review (or semireview) papers   31b,32-37,60
and to consider two impressive examples.

1) The simplest kind of correlations - the two particle rapidity correlation - is measured by the correlation
function

$$C(y_1, y_2) = \frac{1}{\sigma_{in}} \frac{d^2\sigma(y_1, y_2)}{dy_1 \, dy_2} - \frac{1}{\sigma_{in}^2} \frac{d\sigma(y_1)}{dy_1} \frac{d\sigma(y_2)}{dy_2} \qquad (2)$$

or by the correlation coefficient

$$R(y_1, y_2) = \frac{C(y_1, y_2)}{\frac{1}{\sigma_{in}^2} \frac{d\sigma(y_1)}{dy_1} \frac{d\sigma(y_2)}{dy_2}} = \frac{\sigma_{in} \, d^2\sigma(y_1, y_2)/dy_1 \, dy_2}{\frac{d\sigma}{dy_1} \frac{d\sigma}{dy_2}} - 1, \qquad (3)$$

which both are zero for independent particle emission, $d^2\sigma/dy_1 dy_2 = d\sigma/dy_1 \cdot d\sigma/dy_2$ . These quantities were measured on accel-erators and revealed strong SRO, both between particles of the same ($R^{++}$ or $R^{--}$) and of any ($R^{CC}$) charge, e.g. Fig.8: at $y_1 = y_2 = 0$, $R^{CC}(0,0) \approx 0.6$. This shows directly that there are many correlated particles within a decaying clus-

ter. Simple AFS multi-peripheral model cannot explain it (in this model correlation between produced $\varrho$-mesons has even negative sign and at $y_1 = y_2 = 0$ a mini-mum is predicted instead of a maximum 30 ).Special efforts to get a large R were undertaken, such as allowing for all light resonances ($\varrho, f, \omega, A_2$), both exchanged reggeized and produced real.But success, as has been said in Sect.II.3, could be achived only if moreover there were included branchings of Pomeron and nonpomernn ladders "enhanced" by arbitrarily many addi-tional exchanges around the vertices 30.

We have already re-marked that such a complicated pattern may be understood as intro-duction of clusters with

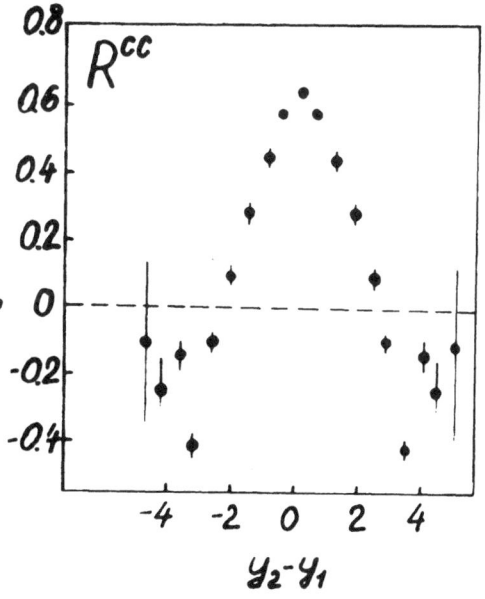

Fig.8. Two charged particle rapidity correlation coeffi-cient $R^{CC}(y_1, y_2)$ at $E_T$=200 GeV pp collision (experiment).

multifarious interactions within them disentangled in

terms of multitude of exchanged Regge trajectories.

At the same time multicluster models with sufficiently heavy clusters give R = 0.6 quite naturally. For instance, in accordance with Fig.8 showing that LRO is small,people often consider Independent Cluster Emission Model (ICEM) assuming that there is no correlation between clusters as a whole and moreover that they are,on the average at least identical. This last assumption is very essential as we shall see. Therefore this model would be more properly cal- led IICEM - Independent Identical Cluster Emission Model. It can be found analytically within it that [38]

$$C(y_1, y_2) = A_0 \cdot \left( \frac{1}{\sigma_{in}} \frac{d\sigma}{dy} \right)_{y = \frac{y_1 + y_2}{2}} \cdot G\left(y_1 - y_2; \delta\sqrt{2}\right) \quad (4)$$

with $G$ from eq.(1), and

$$A_0 = \langle K(K-1) \rangle / \langle K \rangle \quad (5)$$

being the "clustering parameter" of particles having spe- cific charge, or any charge, etc.

Experiment indeed does follow eq.(4) with $A_0 \approx$ $\approx 2.4 - 2.7$ if charged particles are studied .According to the ICEM eq.(5), this means that the number of charged particles within the cluster is $\langle K^{ch} \rangle \approx 3-4$ and the total multiplicity is estimated as $\langle K \rangle \approx 5-6$ [61].

But the same procedure may be used for semi-inclusive distribution, namely for each overall multiplicity n separately. Here ICEM predicts [39] :

$$C_n(y_1, y_2) = A_0(n) \frac{1}{\sigma_n} \frac{d\sigma_n}{dy} G\left(y_1 - y_2; \delta\sqrt{2}\right) - \frac{1 + A_0(n)}{n} \frac{1}{\sigma_n^2} \frac{d\sigma_n}{dy_1} \frac{d\sigma_n}{dy_2} \quad (6)$$

$(d\sigma_n/dy$ is taken for $y = (y_1 + y_2)/2$ ). The experiment gives for $A_0(n)$ the data of Fig.9 [40] (for charged particles). We see that $A_0(n)$ increases with n for $n/\langle n \rangle > 1.5$. If we again use the ICEM eq.(5) then, for the main region of n, $A_0(n) \approx 1.2$, we get $\langle K^{ch} \rangle \approx$ 2 or $\langle K \rangle \approx 3$, in essential disagreement with the previous estimate. The new one enab- les to reduce clusters to light three particle resonances. Although this leaves without consideration $A_c(n)$ for $n/\langle n \rangle > 1.5$ where much heavier clusters are needed (and besides unconceivably excludes $\varrho$ meson production, etc.), accelerator people are often satisfied with such a conc- lusion.

I wish however to stress that both these figures most probably still underestimate the size of essentially con- tributing clusters. In fact, assumption of identical clusters is a non-realistic simplification. Many authors conjectured that it is important to consider at least a double component model, one being low multiplicity dif-

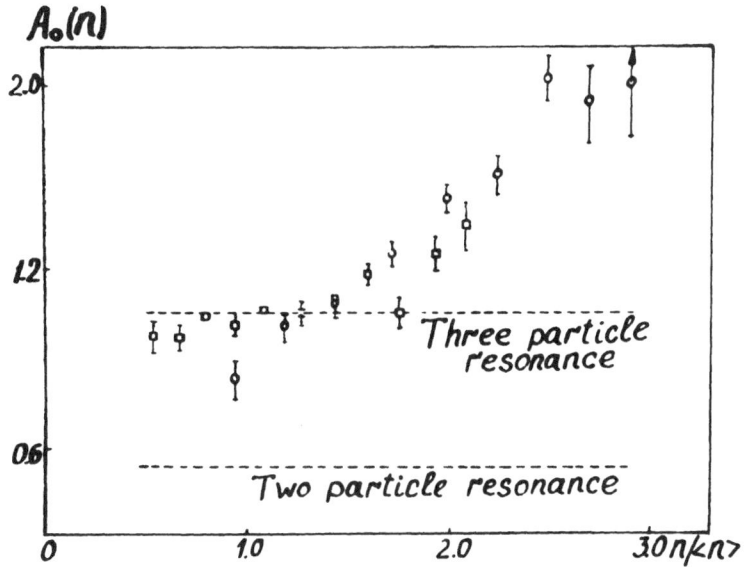

**Fig.9.** The clustering parameter $A_o(n)$ (in ICEM equal to $\langle K(K-1)\rangle/\langle K\rangle$) at 200 GeV pp collision versus $n/\langle n\rangle$ , $n$ being multiplicity (everything for charged particles) [40].

fraction dissociation (D), the other pionization (ND), – multiperipheral or containing heavier clusters (Refs. 38, 41-43 etc.). It was shown e.g. in Ref.[43] that by fitting seven arbitrary parameters and neglecting interference of the two components one can explain within such a model a variety of exclusive and semiinclusive distributions (see also [44] ). It was also shown that although the D-component has smaller production cross section than ND it essentially influences the so called "Mueller moment" $f_2$ – the integral over rapidities of $C_n(y_1-y_2)$ [37].

Such a double component model can be considered a simplified version of the Lebedev Institute Model [24,25] (LIM) shown in Fig.4. As has been already said, the bank of events computer simulated according to this model [26,27] enabled the authors to satisfactorily describe not only available inclusive distributions but correlations as well. In particular it was used to find the average cluster size. It is important that it could be done in two

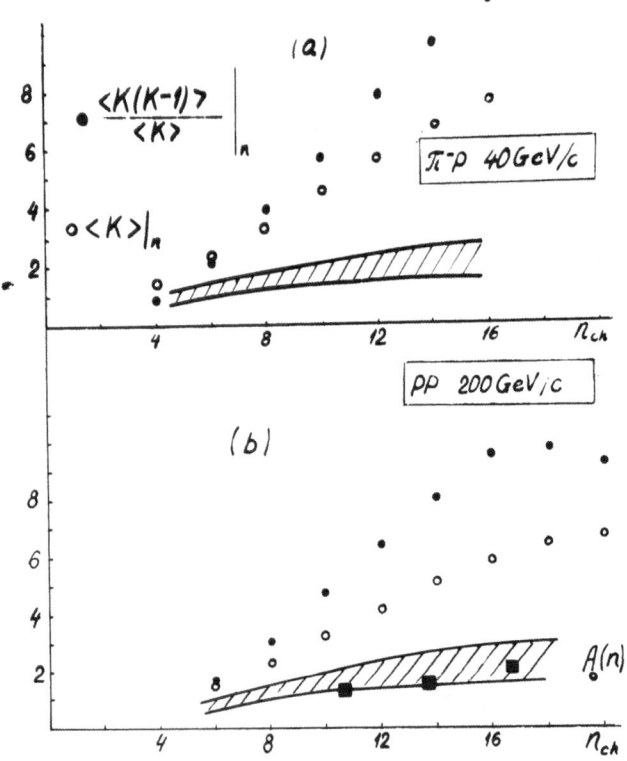

Fig.10. The clustering parameter $A_o(n)$ for $E_T = 200$ GeV and 40 GeV calculated according to eq.(6) from experimental data (shaded strip indicating the experimental uncertainty) and from the Lebedev Institute Model bank of computer simulated data (black squares) together with true values of $\langle K(K-1)\rangle/\langle K\rangle$ and $\langle K\rangle$ obtained directly from the same bank of data. According to ICEM, $A_o(n)$ should be equal to $\langle K(K-1)\rangle/\langle K\rangle$ (everything is given for charged particles)

different ways. First of all the authors, acting as experimenters usually do when they know only a) $d\sigma_n/dy$ and b) $C_n(y_1,y_2)$ [40], used for these quantities the bank data and found $A_o(n)$ values from eq.(6). They are shown in Fig.10 by squares which fall within the shaded strip showing $A_o(n)$ found from real experimental data (with

uncertainties of the experiment taken into account). The
agreement is obvious and this fact once again supports LIM
as a realistic one. According to the ICEM eq.(5) these
should be understood as $\{<K(K-1)>/<K>\}_{sm}$ for charged
particles. However, the LIM bank data gives full information
concerning each simulated event. In particular the number of
particles $K^{ch}$ for each vertex of the multicluster chain
in each event is known from the procedure of computer simulation. Therefore, both, $<K(K-1)>/<K>$ and separately $<K>$ for
each given n could be found directly without using eq.(5)
based on ICEM. This second way gave $<K(K-1)>/<K>$ shown
by black points in the same Figure. They drastically differ
from the ones drawn according to ICEM. This clearly shows
complete inadequacy of ICEM for extracting cluster size from
experiment. The directly found $<K>$ values are shown on the
same Figure by open circles. They demonstrate that usual
procedure based on ICEM greatly underestimates the true
size of clusters. At $n^{ch}$ = 16-20 (thus for $n^{ch}/<n^{ch}> \gtrsim 2$)
an average cluster can contain quite many - some six -
charged particles.

   b) As another example we take the azimuthal two particle correlations.
   Rapidity $y$ is first of all a characteristics of longitudinal motion. But fireball products must also mutually
compensate their transverse momenta. Thus their azimuthal
angles $\varphi_i$ in the transverse plane must be correlated. If
we take a particle at some $y_1$ as a reference point and
measure the angles $\varphi_i$ of all particles at some other
$y = y_2 = y_1 + \Delta y$ (Fig.11) then, if they all belong to the same
cluster, the number of particles with $|\varphi_i| > \pi/2$,
$N(\Delta y, |\varphi_i| > \frac{\pi}{2}) \equiv N_>$ , must be on the average somewhat
larger than $N(\Delta y, |\varphi_i| < \frac{\pi}{2}) \equiv N_<$ , say, to
exceed it by one particle. Therefore, the ratio:

$$B(\Delta y) = (N_> - N_<)/(N_> + N_<) \qquad (7)$$

is a measure of such a correlation. For $\Delta y = 0$ and large
$<K>$ we have, very roughly, $B(0) \approx 1/<K>$ (due to
other cluster contribution, of course, $B(0) \gtrsim 1/<K>$ ).
   Experiments indeed show that $B(\Delta y)$ exceeds zero and
this is particularly impressive if we select events with
$n_{ch} \geq 6$ (Fig.12), i.e. apply the same criterium that
was used by cosmic ray people in 1958 [1]. According to
the estimate, $B(0) \approx 1/<K>$ , the effective clusters here
are large: $<K^{ch}> \sim$ 5-10. In Ref. [46] they are called
"superclusters". LIM satisfactorily describes this effect.
It is important that effective size of clusters for events
$n_{ch} \geq 6, \Delta p_\perp > 0.2$ GeV/c increases with $E_L$ , - $B(0)$
decreases from 0.22 at 40 GeV down to 0.093 at
$\sqrt{s}$ = 52.5 GeV. No doubt, there is a need of additional
studies of this phenomenon.

I shall not describe
other correlation resear-
ches. The effects studied
extensively include rapi-
dity gap distributions
between neighboring [47]
and not neighbouring [48,49]
(on the rapidity axis)
particles; selection of
maximum length gaps [50];
fluctuation analysis [51];

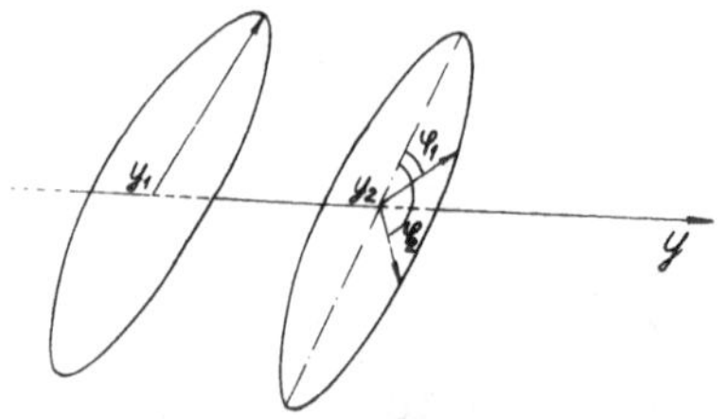

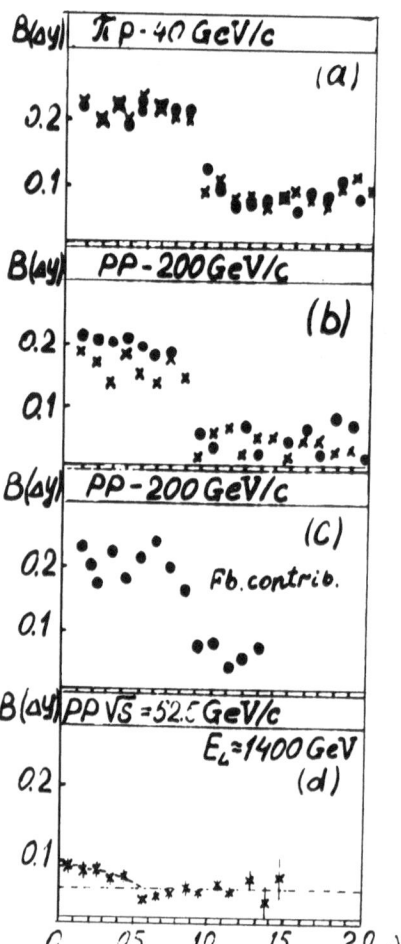

Fig.11. Designations for
azimuthal correlations

Fig.12. Azimuthal correlation parameter $B(\Delta y)$
(eq.(7) for events with
$n \geq 6$, $\Delta P_\perp > 0.2$ GeV/c
(charged particles) accord-
ing to experiment (cros-
ses, x ) and to the Lebedev
Institute Model (points, ● )
[27] : a) $\pi^- p$, 40 GeV [45];
b) pp, 205 GeV [60] ; c) pp,
200 GeV according to the
model when only fireball
contribution is taken into
account; d) pp, $\sqrt{s}$ = 52.5
GeV ($E_L \approx 1400$ GeV) [49]

study of rapidity disper-
sion [52]; charge transfer
between backward and for-
ward hemispheres [53] and
across any point on the
rapidity axis [54]; two particle correlation versus
the two particle combined
mass [55] ; correlation of
lengths of neighbouring
gaps (short after short
ones or long after short
ones) [56,57] , etc. Detailed
statistical characteristics
have been invented and theoretically studied [58,59].

In many papers the basic pattern is that of Independent (Identical) Cluster Emission Model. The dominating conclusion is: production goes via clusters which are too large to be reduced only to light meson resonances ($\varrho$, $\omega$, $f$, $\eta$, $A_2$), since the number of particles produced per cluster is $\langle K \rangle = 3-4$, or $\langle K^{ch} \rangle = 2$, but too small to be a cosmic ray fireballs. However some authors mention necessity of "including a sizeable component of 4 - and 5-body resonances" (Ref. [32], p.180).

Gap distribution analysis also suggests larger clusters, e.g. $\langle K \rangle \sim 6$ [57]. We already mentioned controversy between estimates of the clustering parameter from inclusive an semi-inclusive distributions [61], etc.

As it was stressed above the reduction to clusters of a single type, e.g. as in ICEM, seems to be unsatisfactory (another essential unsatisfactory element of ICEM is neglection of mutual "repulsion" in rapidity space of neighbouring clusters characteristic for multiperipheral model; as has been shown by special investigation this also leads to underestimation of cluster size [62] ). At the same time mixing of two different components leads to complicated effects. The Lebedev Institute Model which seems to be more realistic, means symbollically,

" average cluster"= $\langle$ low multiplicity clusters (resonances + diffr.diss. + ...) + heavy fireballs $\rangle$.

Entering this "formula" fireballs, according to the proper accel-erator data analyses (see Figs.10, 12), are rather heavy and even at $E_L \sim 200 - 1400$ GeV can effectively have $\langle K \rangle = 6-10$ in events with $n_{ch} \geq 6$ or (what is for these $E_L$ practically the same) with $n/\langle n \rangle \gtrsim 1.5-2$ . It deserves stressing that the effective total number of vertices $\eta$ in the chain, including heavy clusters in LIM (Fig.4) is rather small; $\eta \approx 2-3$ at $E_L \lesssim 200$ GeV, $\eta \approx 3-4$ at $E_L \sim 2000$ GeV [27]. For instance, an "average" cluster $\langle K \rangle = 4$ may result from one initial proton kept unchanged, one $\varrho$ — meson (K = 2) and one fireball, K = 9. However it must be stressed that such an "averaging" over various contributing mechanisms must take them with different weights for different kinds of effects (e.g. correlations) studied. This may be an explanation of divergencies between various authors estimating K .

Therefore accelerator data at $E_L \sim 200-2000$ GeV does not contradict the cosmic ray fireball concept with masses $\mathfrak{M} \sim 2-5$ GeV/c$^2$ and even gives sensible evidences for it if properly interpreted.

## V. Conclusions

With the machine energy increasing one can observe steady convergence of cosmic ray concept of fireball production mechanism with the multiperipheral model

accepted in accelerator physics. At present, available
machine data ($E_L < 2000$ GeV) reveals clusterization of produced hadrons which does not contradict the "usual fire-
ball" ($\mathfrak{M} \approx 2-5$ GeV/c$^2$) essential contribution. This data
even supports it if properly analyzed.

Any multiperipheral model badly needs heavy clusters
in order to secure a) absolute value and energy dependence
of multiplicity; b) energy behaviour of cross section, and
c) Pomeron trajectory slope, which all would not drastically contradict experiment.

These heavy clusters may in principle be only still
unknown resonances. However such a special hypothesis would
contradict our knowledge concerning properties of propaga-
tors in the QFT theoretically well studied e.g. in quantum
electrodynamics and most plausible in any quantum field
theory.

In any QFT there is a natural place for nonresonant
heavy objects. They may be simply particles displaced
from the mass shell in the time-like direction, in particular bare or "semibare" particles $31c$ , or partons.

Decay of these nonresonant objects is a special problem. Thermodynamical decay is a rather well founded theoretically possibility. It suits hadronic experiment remar-
kably well. But a "quark"- or "gluon-fireball" may also
decay in a different way, like the one so popular now for
large $P_\perp$ jets.

Thermodynamically (and hydrodynamically) evolving
objects of older models (Heisenberg – Fermi- Pomeranchuck-
Landau) are satisfactorily grounded for actual peripheral
collisions only if they are understood as subsystems ariz-
ing due to quantum mechanical interaction of colliding
hadrons. This naturally leads, in the case of a multiperipheral model, to heavy (fireball) vertices. It is not
sufficient to work with a multiperipheral model with identical vertices (clusters). Various mechanisms (and various
vertices) essentially contributing at high energies (production of resonances and of nonresonant heavy objects,
diffraction dissociation "elastic" scattering of two virtual
particles into two final real ones, etc.) are to be taken
into account in general case. In different observable
effects they can tell with quite different relative weights.
The LIM experience leads to a satisfactory description of
such a complicated pattern.

Summing up theoretical arguments and cosmic ray and
acce lerator data properly analyzed we should admit that
the fireball concept is in a rather good shape.

References

1  Ciok, P. et al., 1958, Nuovo Cim. 8, 66.
2  Niu, K., 1958, Nuovo Cim. 10, 994.

3  Cocconi, G., 1958, Phys. Rev. 111, 1699.
4  Dobrotin, N.A., and Slavatinsky, S.A., 1960, Proc.
   Rochester Conference; 1962, Nucl. Phys. 35, 152.
5  Lattes, C.M.G., Fujimoto, Y., and Hasegawa, S., 1980,
   Phys. Reports 65, 151.
6  Jentschke, W., 1973, CERN Annual Report, p.11.
7  Heisenberg, W., 1936, Zs. Phys. 101, 533.
8  Heisenberg, W., 1939, Zs. Phys. 113, 61.
9  Heisenberg, W., 1949, Zs. Phys. 126, 569.
10 Fermi, E., 1950, Progr. Theor. Phys. 5, 570.
11 Pomeranchuk, I.Ya., 1951, Doklady (Comtes Rendus) Akad.
   Nauk SSSR 78, 889 (in Russian).
12 Landau, L., 1953. See Collected Papers, D. Ter Haar,
   ed., Gordon and Breach, New York, 1965.
13 Shuryak, E.V., 1980, Phys. Reports 61, 71.
14 Heisenberg, W., 1952, Zs. Phys. 133, 65.
15 Feinberg, E.L., and Chernavsky, D.S., 1951, Doklady
   (Comptes Rendus) Akad. Nauk SSSR, 81, 795; 1953, ibid.,
   91, 511 (in Russian).
16 Takagi, S., 1952, Progr. Theor. Phys. 7, 123;
   Kraushaar, W.L., and Mark, L.J., 1954, Phys. Rev. 93,
   326.
17 Blokhintzev, D., 1957, Sov. Phys. - JETP, 32, 350.
18 Novakowski, J., and Cooper, F., 1974, Phys. Rev. D9,771.
19 Amati, D., Stanghellini, A., and Fubini, S.,1962,
   Nuovo Cim. 26, 846.
20 Dremin, I.M., and Dunajevsky, A.M., 1975, Phys. Reports
   18, 159.
21 Feinberg, E.L., and Chernavsky, D.S., 1964, Sov. Phys.
   - Uspekhi, p.1.
22 Tow, Don M., 1970, Phys. Rev. D2, 154.
23 Hasegawa, S., 1963, Progr. Theor. Phys. 29, 128.
24 Dremin, I.M., Royzen, I.I., White, R.B., and Chernavsky,
   D.S., 1964, Intern. Conf. on High Energy Physics,
   Dubna, 1964; 1965, Sov. Phys. - JETP 21, 633.
25 Akimov, V.N., Chernavsky, D.S., Dremin, I.M., and
   Royzen, I.I., 1969, Nucl. Phys. B14, 285.
26 Volkov, E.I. et al., 1973, Sov. Journ. Nucl. Phys.
   17 (N 2) and 18 (N 1); 1974, ibid., 20(N1).
27 Dremin, I.M., Orlov, A.M., and Volkov, E.I., 1978,
   Review of the results of the multiperipheral cluster
   model, P.N.Lebedev Institute preprint N 247.
28 Boreskov, K.G., Kaidalov, A.B., and Ponomarev, L.A.,
   1971, ITEP Preprint N 950.
29 Levin, E.M., and Ryskin, M.G., 1973, Sov. Journ.
   Nucl. Phys. 17 (N 2).
30 Levin, E.M., and Ryskin, M.G., 1975, Sov. Journ.
   Nucl.Phys. 21 (N 2).
31 Dremin, I.M., and Feinberg, E.L., a) Proceed. of the
   IX Intern. Symposium on Multipart. Dynamics, Tabor,
   1978; b) 1979, Sov. Journ. of Particles and Nuclei,
   10 (N 5), 394; c) Feinberg, E.L., 1980, Uspekhi Fiz.

Nauk $\underline{132}$, 255; Sov. Phys. - Uspekhi, $\underline{23}$(N 10).

32 Slansky, R., 1974, Phys. Reports $\underline{11}$, 99.

33 Foa, L., 1975, Phys. Reports $\underline{22}$, 1.

34 De Tar, C., Proceed. 18-th Intern. Conf. on High Energy Physics, Tbilisi, 1976, $\underline{1}$, Paper A3-4.

35 Bopp, F.W., 1978, Rivista Nuovo Cim. $\underline{1}$, 1.

36 Giacomelli, G., and Jacob, M., 1979, Phys. Reports $\underline{55}$, 1.

37 Bialas, A., 1973, Invited Talk at the 6-th Symposium on multiparticle Hadrodynamics, Pavia.

38 Pirila, F., and Pokorski, S., 1973, Phys. Lett. $\underline{43B}$, 502; Lett. Nuovo Cim. $\underline{8}$, 141.

39 Berger, E.L., 1975, Nucl. Phys. $\underline{B85}$, 61; 1974, Phys. Lett. $\underline{49B}$, 369.

40 Amendolia, S.R., et al., 1975, Nuovo Cim. $\underline{31A}$, 17.

41 Fialkowski, K., 1972, Phys. Lett. $\underline{41B}$, 379; Bialas, A., Fialkowski, K., and Zalewski, K., 1972, Nucl. Phys. $\underline{B48}$, 237.

42 Fialkowski, K. and Miettinen, H.I., 1973, Phys. Lett. $\underline{43B}$, 64.

43 Harari, H., and Rabinovici, E., 1973, Phys. Lett. $\underline{43B}$, 49.

44 Ranft, G., and Ranft, J., Phys. Lett. $\underline{49B}$, 286.

45. Angelov, N., et al., 1976, JINR-Dubna preprint.

46 Basile, M., et al., 1977, Nuovo Cim. $\underline{39A}$, 441.

47 Quigg, C., Pirila, P., and Thomas, G.H., 1975, Phys. Rev. Lett., $\underline{34}$, 290.

48 Adamovich, M.I., et al., 1976, Nuovo Cim. $\underline{33A}$,183.

49 Gershkovich, A.M., and Dremin, I.M., 1976, Short Comm. on Phys. (Lebedev Inst.), $\underline{1}$, 6.

50 Chew, G.P., 1973, Phys. Rev. $\underline{D7}$,934.

51 Ludlam, T., and Slansky, S., 1973, Phys. Rev. $\underline{D8}$, 1408.

52 Berger, E.L., Fox, G.C., and Krzywicki, A., 1973, Phys. Lett. $\underline{43B}$, 132.

53 Chao, A.W., and Quigg, C., 1974, Phys. Rev. $\underline{D9}$, 2016.

54 Krzywicki, A., and Weingartner, D., 1974, Phys. Lett. $\underline{50B}$, 265.

55 Thomas, G.H., Mini-Rapporteur talk, 18-th Intern. Conf. on High Energy Phys., Tbilisi, 1976, $\underline{1}$, Paper A3-19.

56 Snider, D.R., 1975, Phys. Rev. $\underline{D11}$,140.

57 Sengupta, P.K., et al., 1979, Phys. Rev. $\underline{D20}$, 601; Roy, S., et al., 1980, Phys. Rev. $\underline{D21}$, 2497.

58 Mueller, A.H., 1971, Phys. Rev. $\underline{D4}$, 150.

59 Krzywicki, A., 1977, Proc. Internat. Sympos. on Multihadron Dynamics, Tutzing.

60 Whitmore, J., 1974, Phys. Reports $\underline{10}$, N 5.

61 Ganguli, S.N., and Roy, D.P., 1980, Phys. Reports $\underline{67}$, 201.

62 Orlov, A.M., 1980, Sov. Journ. Nucl. Phys. $\underline{32}$ (N2).

# RECENT ADVANCES IN
# ELEMENTARY PARTICLE PHYSICS

L. VAN HOVE

CERN, Geneva

(Paper not received)

Achevé d'imprimer par le Service de Documentation du CEN Saclay
nᵒ 81 - 0034 - janvier 1982
DÉPOT LÉGAL : 1er trimestre 1982

Edité par le Commissariat à l'Energie Atomique
ISBN 2 - 7272 - 0056 - 0 (Edition complète)
ISBN 2 - 7272 - 0070 - 6 (Volume XII)